Transport and Thermal Properties of f-Electron Systems

Transport and Thermal Properties of f-Electron Systems

Edited by

G. Oomi

Kumamoto University
Kumamoto, Japan

H. Fujii and T. Fujita

Hiroshima University
Hiroshima, Japan

Springer Science+Business Media, LLC

Library of Congress Cataloging-in-Publication Data

Transport and thermal properties of f-electron systems / edited by G.
Oomi, H. Fujii and T. Fujita.
        p.   cm.
    Proceedings of the Hiroshima Workshop on Transport and Thermal
Properties of f-Electron Systems, T²-PfS, held August 30-September
2, 1992, in Greenpia Yasuura, Hiroshima, Japan.
    Includes bibliographical references and index.
    ISBN 978-1-4613-6243-2   ISBN 978-1-4615-2868-5 (eBook)
    DOI 10.1007/978-1-4615-2868-5
    1. Electron transport--Congresses. 2. Thermal electrons-
-Congresses.   I. Oomi, G. II. Fujii, H. III. Fujita, T.
IV. Hiroshima Workshop on Transport and Thermal Properties of f
-Electron Systems (1992 : Hiroshima-shi, Japan)
QC176.8.E4T73 1993
530.4'1--dc20   93-32228
                                                             CIP

Proceedings of the Hiroshima Workshop on Transport and Thermal Properties of f-Electron Systems,
T^2-PfS, held August 30–September 2, 1992, in Greenpia Yasuura, Hiroshima, Japan

ISBN 978-1-4613-6243-2

© 1993 by Springer Science+Business Media New York
Originally published by Plenum Press New York in 1993
Softcover reprint of the hardcover 1st edition 1993

PREFACE

The Hiroshima Workshop on Transport and Thermal Properties of f-Electron Systems, T^2PfS, was held in the hotel Greenpia Yasuura on the shores of the Seto Inland Sea near Hiroshima, Japan from August 30, to September 2, 1992, as a satellite meeting of the International Conference of Strongly Correlated Electron Systems in Sendai. The purpose of this workshop was to bring together those scientists who are actively involved in the research of 4f- and 5f-electron systems; particularly the transport and thermal properties such as electrical resistivity, Hall effect, thermoelectric power, thermal conductivity, thermal expansion and specific heat. Hence, the organizing committee limited the number of participants to 60; 25 from abroad and 35 from Japan. In the workshop, all the sessions consisted of oral presentations; 25 invited talks and 5 contributed talks, including at least 10 minutes of discussion for each presentation.

The program was divided into the following five topics: [1] Kondo-lattice semiconductors, [2] superconductivity of f-electron systems, [3] anomalous transport and thermal properties of 4f- and 5f-compounds, [4] low-carrier heavy-electron systems and [5] theoretical investigation of heavy-electron and mixed-valence states. This division of topics has been retained in the organization of papers in this volume. Almost all of the invited and contributed papers are included. These papers include excellent reviews of both the recent advances and historical background of each topic. We believe this book would be a tutorial text for researchers working in the field of solid state physics.

We thank all the authors for their efforts to submit the papers in a timely manner. The concluding remarks given by Professor T. Kasuya, was a major contribution to both the meeting and the book. We would like to thank him for accepting this difficult task. We are indebted to the members of International Advisory Committee for their suggestions concerning the invited speakers and assisting with the scope of this workshop. We would like to thank the members of the organizing committee, who established the program and conducted the workshop successfully. We especially thank T. Takabatake, secretary, T. Hihara, program, G. Oomi, publications, K. Kojima, general affairs and J. Sakurai, coordinator for their hard work. Special thanks are also due to the session chairmen for running the sessions and to the referees for their advice on the papers.

H. Fujii and T. Fujita
Chairmen of the $T^2PfS'92$

Hiroshima, September 1992

CONTENTS

Kondo semiconductors

$4f$ and $5f$ compounds

Superconductivity of f-electron systems

Theory

KONDO SEMICONDUCTOR CeNiSn

T. Takabatake[1], G. Nakamoto[1], H. Tanaka[1], H. Fujii[1], S. Nishigori[2],
T. Suzuki[2], T. Fujita,[2] M. Ishikawa[3], I. Oguro[3], M. Kurisu[4] and
A.A. Menovsky[5]

[1]Faculty of Integrated Arts and Sciences, Hiroshima University, Hiroshima
  730, Japan
[2]Faculty of Science, Hiroshima University, Higashi-Hiroshima 724, Japan
[3]Institute for Solid State Physics, University of Tokyo, Tokyo 106, Japan
[4]Faculty of Engineering, Iwate University, Morioka 020, Japan
[5]Van der Waals-Zeeman Laboratorium, Universiteit van Amsterdam, 1018 XE
  Amsterdam, The Netherlands

1. INTRODUCTION

Since the discovery of semiconducting behavior of SmB_6 in 1969,[1] the formation of a small energy gap in 4f-electron systems has been the subject of intensive studies. The presence of energy gap of several 10 K was found in valence-fluctuating (VF) compounds gold SmS, TmSe and YbB_{12} with cubic structures.[2-4] A simple picture of these systems is that the hybridization of the 4f electron states and the conduction band leads to the small gap at the Fermi level.[5] However, the detailed mechanism of the gap formation remains unsettled. Recently, CeNiSn has been found to be the first example of a cerium compound showing the behavior of a small-gapped semiconductor.[6] From the activation-type resistivity, the gap energy E_g was estimated to be 6 K. This compound crystallizes in an orthorhombic structure $(Pn2_1a)$,[7] which is closely related to the ε-TiNiSi type structure. Subsequently, similar gap formation has been found in $Ce_3Pt_3Bi_4$ and CeRhSb,[8,9] where the values of E_g are 70 and 8 K, respectively. The latter crystallizes in the same type of structure as CeNiSn, whereas the former in the cubic $Y_3Au_3Sb_4$-type structure. These findings have renewed the interest in the problem of the insulating ground state of the Kondo lattice.[10,11]

In this article, we review the experimental studies of the transport, magnetic and thermal properties of CeNiSn.[12-17] The results obtained on single crystalline samples demonstrate that an anisotropic gap opens in the heavy-fermion bands as temperature is reduced below 6 K. Because of the small gap energy of several Kelvins, we expect strong effects of magnetic field and pressure on the gapped state. The strength of hybridization in this compound can be controlled by substituting Co, Cu and Pt for Ni in the nonmagnetic sublattice. The substituted samples serve as systems where one can examine the coherence effect on the gap formation. Furthermore, we compare the physical properties of CeNiSn with those of the

Transport and Thermal Properties of f-Electron Systems
Edited by G. Oomi *et al.*, Plenum Press, New York, 1993

isostructural compound CePtSn, the latter of which is an antiferromagnetic Kondo compound with $T_N = 7.5$ K.[18,19]

2. SAMPLE PREPARATION AND CHARACTERIZATION

Polycrystalline samples were prepared from stoichiometric starting materials by arc-melting in a purified argon atmosphere. The samples were homogenized by annealing in quartz ampoules for 10 days at 1000°C. Single crystals were grown from Ames Laboratory Ce by a Czochralski technique in a triarc furnace or by a floating-zone method in an infrared mirror furnace. From metallographic examination and electron-probe microanalysis (EPMA) of the as-grown crystal, impurity phases of $CeNi_2Sn_2$ and cerium oxides were detected at the tail end and on the surface. As shown in Fig. 1, a line-shaped impurity phase of $Ce_3Ni_4Sn_3$ of 1~10 μm width was detected even in the central part of the crystal. The volume fraction of this phase is less than about 1% of the sample, which is much smaller than that assumed by Kasuya[20] to explain the anomalous properties of CeNiSn based on a two-phase model. The impurity phase of $Ce_3Ni_4Sn_3$ could not be eliminated by starting with off-stoichiometric compositions of $Ce_{1.05}NiSn_{1.03}$ and $Ce_{1.07}NiSn_{1.05}$. However, any deviation from the 1-1-1 stoichiomerty larger than the resolution of about 0.3 at.%. was not detected by EPMA for the host phase in spite of the off-stoichiometric starting compositions.

The unit cell of CeNiSn consists of four formula units and hence contains an even number of valence electrons irrespective of the valence states of the Ce ions. We note here that in CeRhSb, which exhibits similar semiconducting behavior, the number of valence electrons is supposed to be same as in CeNiSn. In the orthorhombic structures of the two compounds, the Ce atoms form a zigzag chain along the a axis. Recently, band structure calculations on CeNiSn have been performed by Yanase and Harima[21] using a self-consistent LAPW method. According to their calculations, CeNiSn is a semimetal with a hole Fermi surface on the Δ axis and two electron Fermi surface centered at the X points. The overlap of the valence band with the conduction band is about 600 K. Since the 4f components are about 80 % both at the top of the valence band and the bottom of the conduction band, we expect large effect of strong correlation among 4f electrons on the transport and magnetic properties of this compound.

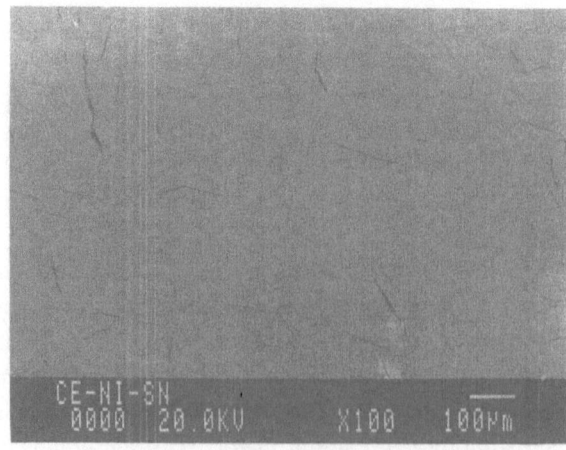

Fig. 1. Scanning electron micrograph of a CeNiSn crystal showing the impurity phase of $Ce_3Ni_4Sn_3$.

3. BASIC PROPERTIES OF CeNiSn

3.1 Magnetic Susceptibility

In Fig. 2, magnetic susceptibility $\chi(T)$ of a single crystalline sample of CeNiSn along the three principal axes is compared with that of CePtSn along the a axis labelled as $\chi_a(T)$.[13,19] We note here that the relationship $\chi_a > \chi_b > \chi_c$ at low temperatures is the same for the two compounds. Hence, we may attribute the magnetic anisotropy in CeNiSn to the effect of the hybridization similar to that in CePtSn where Ce ions are almost trivalent. However, the temperature dependence is weaker in CeNiSn at low temperatures below 100 K, which is an indication of valence fluctuation in this compound. The VF character has been confirmed by inelastic neutron scattering experiments,[22] by which no well-defined crystal-field (CF) excitations are observed. The maximum in $\chi_a(T)$ at 12 K for CeNiSn is not associated with a long-range magnetic order as will be discussed later. Instead, it can be attributed to the development of antiferromagnetic correlations among quasiparticles as observed in neutron scattering experiments.[23]

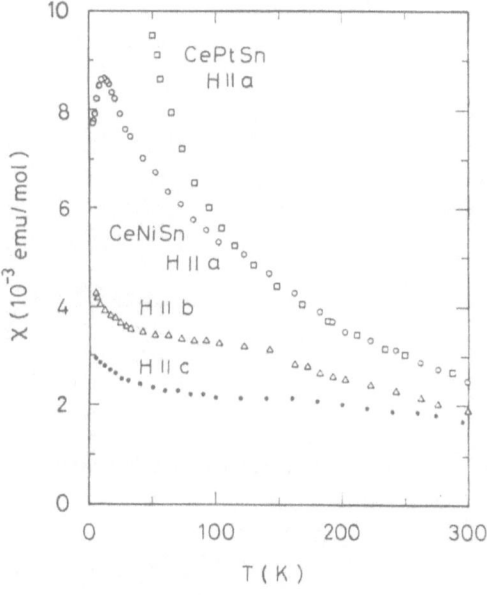

Fig. 2. Magnetic susceptibility vs temperature for single-crystalline CeNiSn (H//a, H//b, H//c) and CePtSn (H//a).

3.2 Electrical Resistivity

Figure 3 represents the resistivity $\rho(T)$ of CeNiSn and CePtSn as a function of temperature.[13,19] As temperature is decreased from 300 K, the $\rho(T)$ curves of CeNiSn initially show a quasi-logarithmic increase in all directions, whereas those of CePtSn are metallic. From a local maximum around 100 K for $\rho_a(T)$ and 60 K for $\rho_b(T)$, the Kondo temperature T_K of CeNiSn is estimated to be about 100 K for the degenerate $J = 5/2$ manifold. Another lnT dependence for $\rho_a(T)$ appears between 15 and 40 K, which is similar to that

3

found in the magnetic part of the resistivity in CePtSn.[19] The close similarity suggests the presence of another Kondo scale of about 30 K for the CF ground state in CeNiSn although the CF level scheme is not well defined.

A local maximum in $\rho_a(T)$ appears at the same temperature of 12 K at which $\chi_a(T)$ shows the peak. The decrease in both $\rho_a(T)$ and $\rho_b(T)$ below 12 K indicates that the system goes into the coherent scattering regime. With further decrease of temperature below 6 K, the resistivities turn to increase. An activation-type variation appears only in a small temperature range between 2.5 and 4.6 K. Nevertheless, the gap energy $E_g/k_B = T_g$ in the formula $\rho(T) = \rho_0 \exp(E_g/2k_B T)$ was deduced as 1.0, 4.8 and 8.0 K for the a, b and c axes, respectively. Such a large difference in T_g implies an anisotropic gapping of the density of states on the Fermi surface. The value of T_g along the c axis is still one order of magnitude smaller than that reported for $Ce_3Pt_3Bi_4$. As shown in the inset of Fig. 3, $\rho_c(T)$ and $\rho_b(T)$ pass through a weak maximum at 0.4 and 0.6 K, respectively, and decrease with further decreasing temperature. A plausible explanation for this temperature dependence is that the gapping of the quasiparticle band is incomplete or there exist impurity bands in the gap.

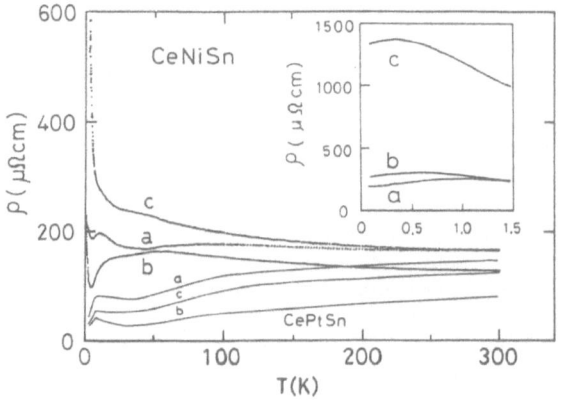

Fig. 3. Electrical resistivity vs temperature for single-crystalline CeNiSn and CePtSn.

3.3 Hall Effect

The temperature dependence of the Hall coefficient R_H for three configurations H//a (I//b), H//b (I//c) and H//c (I//b) is represented in Fig. 4 from Ref. 15. The three curves rise rapidly with decreasing temperature below 100 K and exhibit a positive peak around 9 K, followed by a precipitous drop to a negative value. The positive R_H peak of the size of $5 \times 10^{-3} cm^3/C$ is in common with those found in heavy-fermion compounds like $CeAl_3$ and $CeRu_2Si_2$.[24] Acording to the theory of Fert and Levy,[25] the strong temperature dependence of R_H with a positive peak arises from the contribution of the intrinsic skew scattering. Further, the extraordinary part of R_H is proportional to the product of $\chi(T)$ and magnetic resistivity $\rho_m(T)$. Between 100 and 10 K, R_H(H//a) approximately follows the product. Since the peak temperature of R_H in heavy-fermion systems is regarded as the

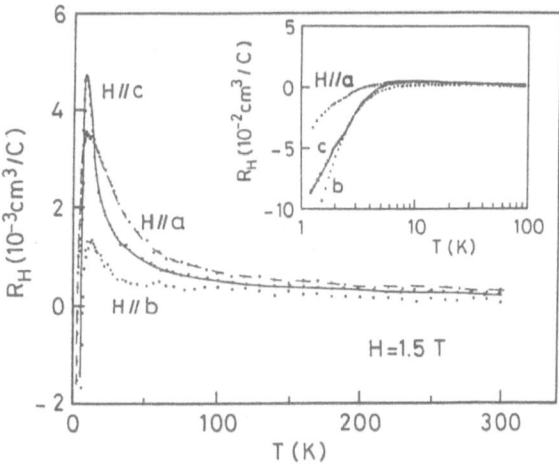

Fig. 4. Hall coefficient of CeNiSn as a function of temperature taken for three configuration; H//a (I//b), H//b (I//c) and H//c (I//b) (after ref. 15).

onset of coherence,[24] the above result indicates that the electronic state in CeNiSn gradually goes into a coherent scattering regime below 9 K. This temperature is somewhat lower than the maximal temperature in $\chi_a(T)$ and $\rho_a(T)$ at 12 K. The strong decrease in R_H below 5 K can be ascribed to the reduction of carrier density caused by the opening of the energy gap in the density of states. If we assume a single type of carriers, the concentration at 1.3 K is estimated to be 4.6×10^{-3} per formula unit from the data for H//c.

3.4 Specific Heat

The specific heat C of a single crystalline sample of CeNiSn is shown in Fig. 5 in a plot of C/T vs T^2 for T < 16 K. No appreciable anomaly exists near 12 K, and hence the peaking in both $\chi_a(T)$ and $\rho_a(T)$ at 12 K does not originate from a long-range magnetic order. The value of C/T decreases almost linearly with T^2 down to nearly 6 K and then suddenly diminishes. This temperature dependence is consistent with the opening of a gap below 6 K as inferred from the transport properties. Furthermore, the large value of C/T of 0.2 J/K²mol near 6 K indicates the development of a narrow band of heavy quasiparticles antecedent to the gap opening. Using the relation between γ and T_K derived for a single Kondo impurity, $T_K = 0.68R/\gamma$ (R is the gas constant),[26] T_K is estimated as 28 K. This temperature is within the lower temperature range where $\rho_a(T)$ increases as ln T, and thus this T_K can be regarded as the Kondo temperature for the CF ground state.

The magnetic contribution to the specific heat, C_m, was estimated by subtracting the data of LaNiSn from that of CeNiSn. As shown in the inset of Fig. 5, C_m/T reveals a pronounced maximum near 6.7 K. The magnetic entropy up to 20 K amounts to only half of Rln2. Between 1.5 and 5 K, C_m/T shows a linear variation, $C_m/T = \gamma + AT$, as opposed to the T^2 dependence of C_m/T in usual metallic systems. The observed temperature dependence is consistent with the renormalized density of states with a V-shaped gap near the Fermi level E_F, which was first proposed based on the results of NMR experiments.[27] The nuclear-spin lattice relaxation rate $1/T_1$ of ^{119}Sn in CeNiSn was found to be proportional to T^3 between 0.4 and 1.3 K. The proposed density of states is proportional to $|E-E_F|/E_g$, which yields a T^3 dependence for $1/T_1$ and a T^2 dependence for C_m at low temperatures below $T_g = E_g/k_B$.

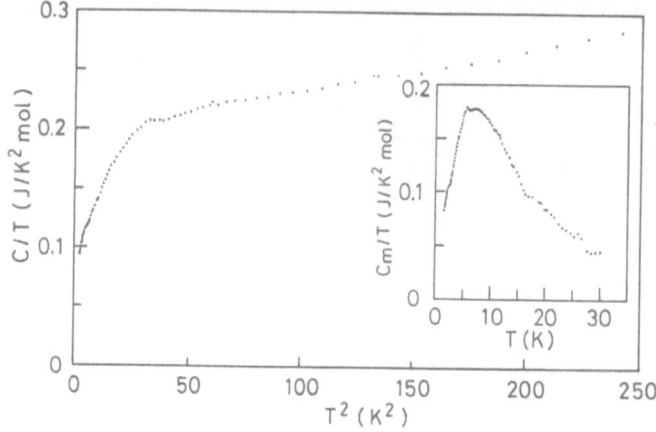

Fig. 5. Specific heat divided by temperature C/T vs T^2 for CeNiSn. The inset shows the magnetic contribution to the specific heat divided by temperature C$_m$/T vs T.

3.5 Magnetic Correlations Sudied by NMR and μSR

Both NMR and μSR techniques have been employed to study the magnetic correlations in CeNiSn at low temperatures. Kyogaku et al.[28] extended the temperature range of NMR experiments down to 80 mK. They observed a strong deviation of $1/T_1$ of ^{119}Sn from the T^3 behavior below 0.4 K. With further decrease of temperature below 0.13 K, $1/T_1$ decreases exponentially together with an increase in the line width. The results suggest that the pseudogap state becomes unstable below 0.4 K and a spin excitation gap is induced by the development of quasistatic magnetic correlations.

Krazer et al.[29] revealed from μSR experiments that CeNiSn exhibits properties typical of a paramagnet moving towards magnetic order below 1 K. However, no transition into long-range order was observed down to 33 mK. The formation of extended spin correlation up to short-range order is deduced from the unusual dependence of muon spin relaxation rate and muon spin precession frequency on external field.

4. HIGH-MAGNETIC FIELD STUDIES

4.1 Magnetization

The field dependence of magnetization M(H) of single crystal CeNiSn at 1.3 K is represented in Fig. 6 from Ref. 16. The M(H) curve only along the *a* axis exhibits a weak metamagnetic-like transition near 13 T. It is more clearly seen in the derivative susceptibility dM/dH vs H. However, the increase in M associated with the transition is much smaller than that found in nonmagnetic heavy-fermion compounds like CeRu$_2$Si$_2$.[30] This fact suggests that the weak transition in CeNiSn is not due to the suppression of antiferromagnetic intersite interactions as found in CeRu$_2$Si$_2$ but due to the collapse of the pseudogap. Above 20 T, M$_a$(H) increases linearly with increasing field and attains 0.3 μ$_B$/Ce at 36 T. The size of the magnetization is only one fourth of that found in the isostructural, antiferromagnetic compound CePtSn, 1.2 μ$_B$/Ce.[19] The small and linearly increasing moment in CeNiSn can be interpreted as a result of strong Kondo-type interaction persisting even after the pseudogap has collapsed. Assuming the effective moment of 0.3 μ$_B$/Ce, the magnetic energy at the transition field of 13 T corresponds to the thermal energy of 2.6 K, which is comparable to the gap energy estimated from the *a*-axis resistivity.

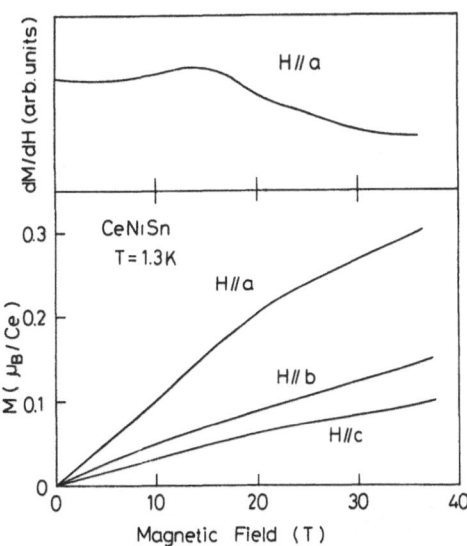

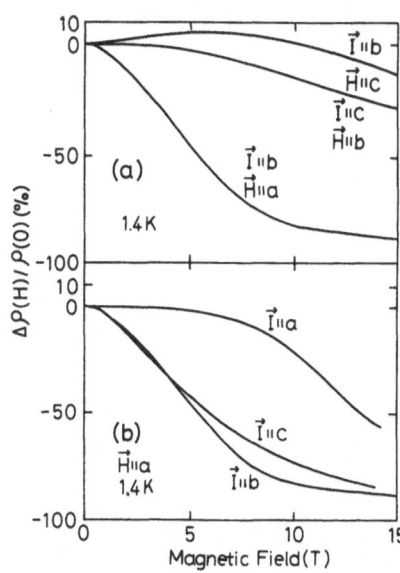

Fig. 6. Magnetization of CeNiSn along the three principal axes at 1.3 K (after ref. 16).

Fig. 7. Magnetoresistance of CeNiSn at 1.4 K (after ref. 14).

4.2 Magnetoresistance

Strong suppression of the energy gap in CeNiSn by application of magnetic field was first demonstrated by the magnetoresistance measurements on a polycrystalline sample.[12] The resistivity at 1.7 K was found to decrease from 730 to 200 $\mu\Omega$cm as the field is raised to 24 T. The results obtained on single crystalline samples show very strong anisotropy.[14] The normalized magnetoresistance $\Delta\rho(H)/\rho(0)$, where $\Delta\rho(H) = \rho(H)-\rho(0)$, at 1.4 K are presented in Figs. 7(a) and 7(b). At H = 15 T, the negative magnetoresistance for H//a attains -88%, which is much larger than for H//b and H//c. Thus, the energy gap is most sensitive to the magnetic field applied along the easy a axis. For H//a, the dependence of magnetoresistance on the current direction was further examined. In Fig. 7(b), the field dependence of $\Delta\rho(H)/\rho(0)$ for I//c is similar to that for I//b, whereas $\Delta\rho(H)/\rho(0)$ for I//a is almost constant up to 4 T and then gradually decreases with increasing field. These results suggest a strong anisotropic scattering mechanism under magnetic fields.

Shown in Fig. 8 is the temperature dependence of the resistivity along the three principal axes in fields of 0, 12 and 14 T. At H = 0 T, $\rho_a(T)$ exhibits a maximum near 11.4 K, which has been ascribed to the development of antiferromagnetic correlations. In a field of 14 T parallel to the a axis, this maximum is almost smeared out. The strong suppression of the upturn below 6 K is a result of the gap suppression by magnetic fields. A drastic effect occurs in the resistivity for the configuration I//b and H//a, which indicates metallic behavior and is in contrast to the semiconductor-like behavior for H//c. Furthermore, at temperatures below 4 K, it obeys a T^2 dependence with a coefficient of 1.0 $\mu\Omega$cm/K^2 and a residual resistivity of 39 $\mu\Omega$cm. The size of this coefficient is typical for moderately heavy fermion systems.

7

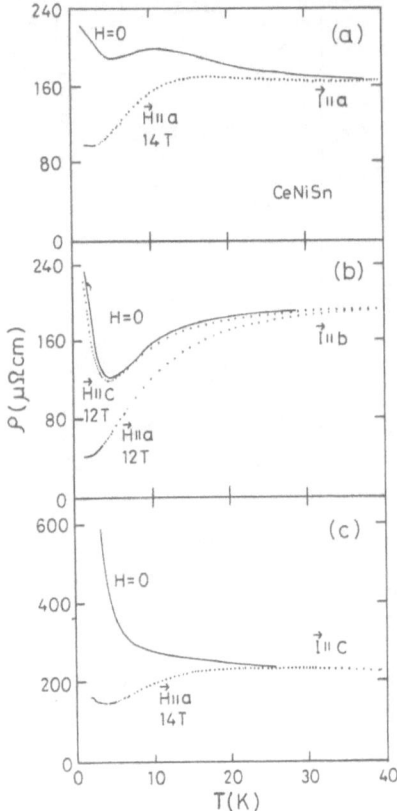

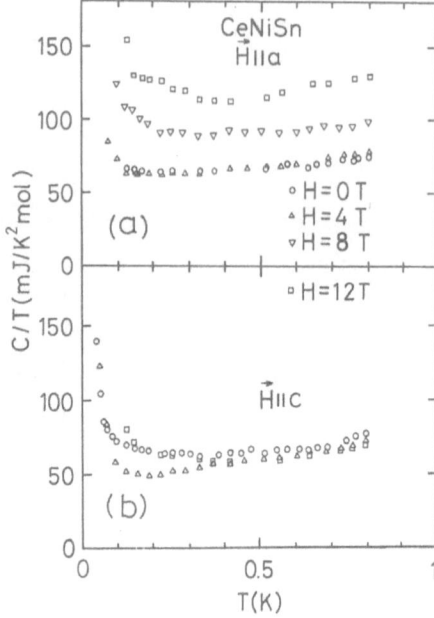

Fig. 8. Electrical resistivity vs temperature for CeNiSn in external fields 0, 12 and 14 T for electrical currents along the three principal axes (after ref. 14).

Fig. 9. Specific heat of CeNiSn plotted as C/T vs T in magnetic fields for (a) H//*a* and (b) H//*c* (after ref. 14).

4.3 Specific Heat in Magnetic Fields

The suppression of the energy gap by magnetic field was further studied by specific-heat measurements.[14] Temperature variations of the specific heat of CeNiSn in magnetic fields parallel to the *a* and *c* axes are shown in Figs. 9(a) and 9(b), respectively. At H = 0 T, C/T is almost proportional to T between 0.3 and 0.8 K and the linear extrapolation to T= 0 K yields a γ value of 57 mJ/K^{2}mol. This size of γ value seems to be too large to be ascribed to the contribution from impurity phases. Rather, it may be the contribution from the residual density of states at E_F in the pseudogap, as inferred from the saturation of the resistivity below 1 K. The origin of the upturn in C/T below 0.2 K is not clear yet.

When magnetic field is raised to 12 T, the value of C/T is strongly enhanced for H//*a*, whereas it is almost unchanged for H//*c*. For H//*a*, the field dependence of C/T was measured at 0.15, 0.42 and 0.76 K.[14] The values of C/T at these temperatures stay constant for H < 4 T and then increase monotonically from about 70 to 125 mJ/K^{2}mol. This large enhancement is consistent with the heavy-fermion behavior in $\rho_b(T)$ at 12 T in Fig. 8(b). These results support the idea that the density of states in the minimum of the V-shaped pseudogap is increased by application of magnetic field along the easy axis of magnetization. In this anisotropic suppression of the pseudogap, a strong spin polarization of the renormalized band should play an important role.

5. HIGH-PRESSURE STUDIES

The effect of application of pressure on the gapped state in CeNiSn has been studied by Kurisu et al.[31,32] from resistivity measurements. As a typical result, the a-axis resistivity is shown in Fig. 10. With increasing pressure, the upturn in $\rho_a(T)$ below 6 K is strongly suppressed, and the local maximum at 12 K shifts to higher temperatures. Above 12 kbar, the resistivity shows metallic behavior with a single maximum around 100 K. This maximum, which originates in the interplay of the Kondo effect and the CF effect, shifts also to higher temperatures. Generally, application of pressure on cerium-based Kondo-lattice compounds increases the degree of the 4f-ligand hybridization so that the renormalized band broadens.[33] As a result, the value of T_K is increased and hence the maximal temperature of the resistivity due to the Kondo effect is elevated. In fact, we find large shift of both the peaks in $\rho_a(T)$, which are originally at 12 and 65 K, respectively, to 30 and 180 K. The broadening of the renormalized band should also be responsible for the suppression of the gap at low temperatures. In other words, the gap formation in CeNiSn is

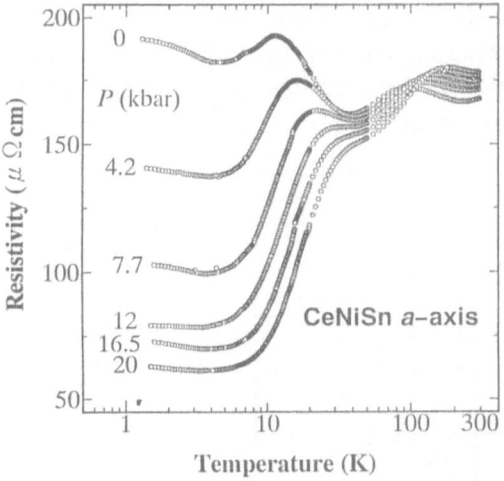

Fig. 10. Electrical resistivity vs lnT for single-crystalline CeNiSn along the a axis under various applied hydrostatic pressures.

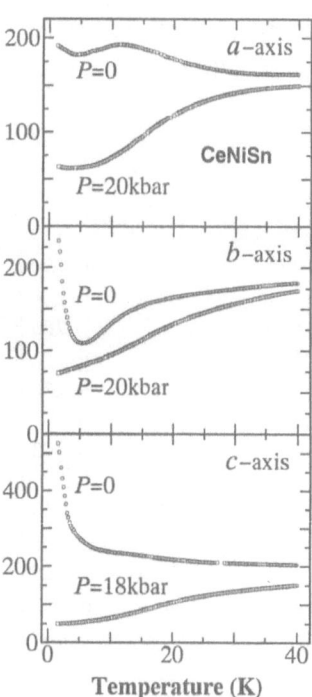

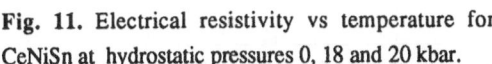

Fig. 11. Electrical resistivity vs temperature for CeNiSn at hydrostatic pressures 0, 18 and 20 kbar.

strongly suppressed by the increase of the degree of hybridization. We will see in the next section that the reduced hybridization by negative chemical pressure also suppresses the gap.

It is interesting to compare the effect of pressure on the resistivity with that of magnetic field. Comparing the resistivity curves in Figs. 8 and 11, we notice that the effect of application of 20 kbar is comparable with that of a magnetic field of 14 T applied along the a axis. However, a simple scaling is rather difficult because the negative magnetoresistance is significant only below 30 K, whereas the overall temperature dependence is changed by application of pressure.

6. EFFECT OF ALLOYING

Low-temperature properties of heavy-fermion compounds are generally very sensitive to a small amount of substitution.[34] We expect that the substitution of 3d element Co or Cu for the Ni atom in CeNiSn would change the number of conduction electrons. If the Ni atom is replaced by the 4d or 5d atom in the same column in the periodic table, i.e., Pd or Pt, then the unit cell volume would increase without changing the number of conduction electrons so much. Keeping this in mind, we have performed systematic studies of the substituted systems $CeNi_{1-x}T_xSn$ with T= Co, Cu and Pt.

Figure 12 shows variations of the lattice parameters with x. The substitution of Co for Ni hardly changes the lattice parameters, whereas that of Cu leads to a linear increase in the three parameters. By the substitution of Pt, the c parameter increases significantly but the a parameter slightly decreases. For $CeNi_{1-x}Cu_xSn$, we previously reported that the energy gap disappears near x= 0.1 and then a long-range antiferromagnetic order develops for x $\geq$ 0.13.[12] Hereafter, we compare the magnetic and thermal properties of the Co and Pt substituted systems. Temperature dependences of the magnetic susceptibility $\chi(T)$ of

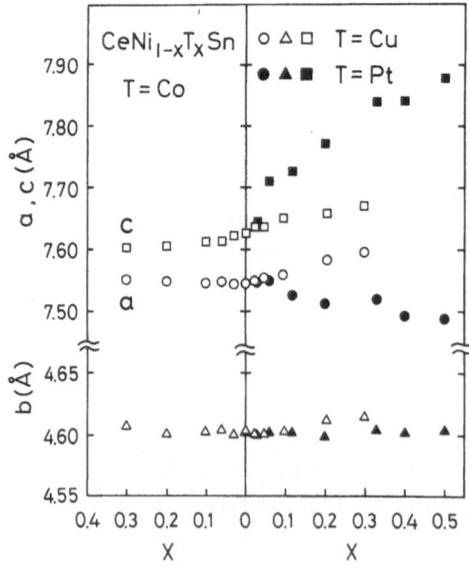

Fig. 12. Variations of lattice parameters of $CeNi_{1-x}T_xSn$ (T = Co, Cu and Pt).

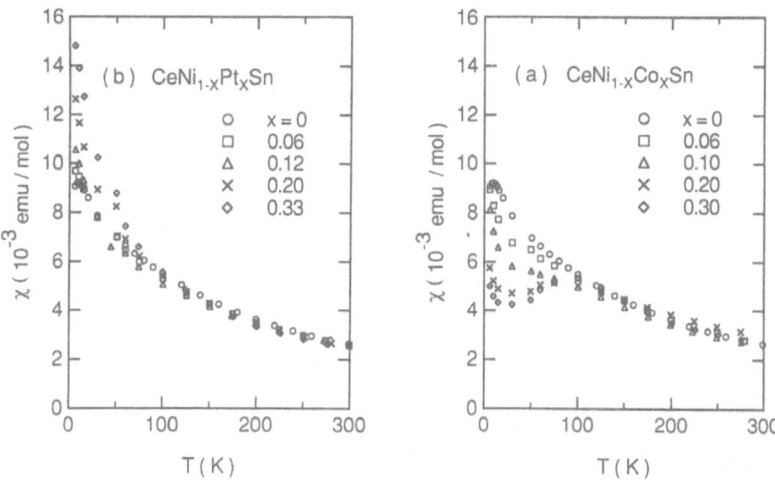

Fig. 13. Magnetic susceptibility vs temperature for field-aligned polycrystalline samples of (a) CeNi1-xCoxSn and (b) CeNi1-xPtxSn.

polycrystalline samples of $CeNi_{1-x}T_xSn$ for T= Co and Pt are shown in Figs. 13 (a) and 13(b), respectively. The measurements were performed on powdered samples which were pre-aligned by application of a magnetic field of 14 T at 4.2 K.

The data of CeNiSn with a maximum around 12 K agree with those of $\chi(T)$ along the a axis of the single crystalline sample as was shown in Fig. 2. This agreement guarantees that the powders are aligned along the easy a axis of the magnetization. In Fig. 13(a), the values of $\chi(T)$ for T<100 K decrease as x is increased in $CeNi_{1-x}Co_xSn$. At higher Co concentration $x \geq 0.2$, a broad maximum appears around 100 K and the value of the paramagnetic Curie temperature estimated from the Curie-Weiss fitting exceeds -130 K. These features are the signals for the transition to the VF regime. The Pt substitution has an opposite effect on the system and leads to the Kondo regime. In Fig. 13(b), the low-temperature value of $\chi(T)$ increases with increasing x in $CeNi_{1-x}Pt_xSn$, resulting in the recovery of the Curie-Weiss behavior down to low temperatures. The paramagnetic Curie temperature remains constant at about -70 K.

The trend toward either the VF or the Kondo regime is confirmed by the result of magnetization measurements at 4.2 K. As shown in Fig. 14(a), the slope of the magnetization curve decreases with increasing Co concentration. A downward curvature near 4 T for $x \geq 0.1$ may indicate the saturation of impurity contribution. All the magnetization curves of the Pt substituted samples are almost linear with field. The value of M(H) at H = 14.6 T increases linearly with increasing Pt concentration, whereas it decreases with Co concentration at almost the same rate of $dM/dx = 0.42\ \mu_B$.

As mentioned above, Co substitution for Ni in CeNiSn may decrease the number of conduction electrons. The Fermi level accordingly lowers toward the level of the unrenormalized 4f states, E_f. For a Cerium impurity in metal, the Kondo temperature T_K is proportional to $\exp[-1/|J|\ N(E_F)]$, where J is the effective exchange integral given by

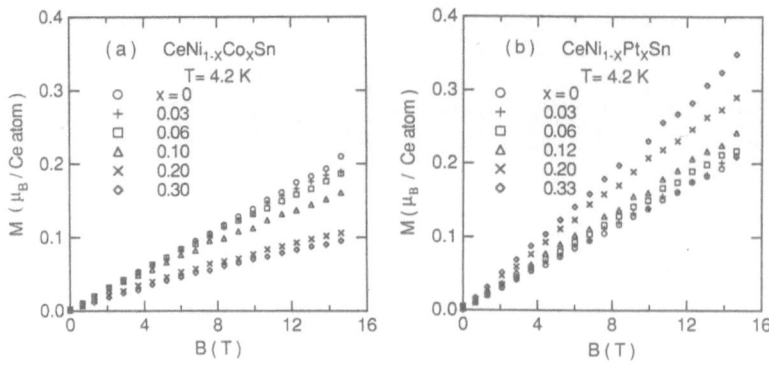

Fig. 14. Magnetization vs magnetic field for powdered polycrystalline samples of (a) $CeNi_{1-x}Co_xSn$ and (b) $CeNi_{1-x}Pt_xSn$.

$J = |V_{k,f}|^2/(E_f - E_F)$ with the hybridization matrix element $V_{k,f}$ between the 4f state and the conduction electrons.[35] Then, T_K is expected to be strongly enhanced by Co substitution and the system is converted to the VF regime, as is experimentally observed. On the contrary, the negative chemical pressure induced by the Pt substitution may weakens the hybridization of the 4f states with the Sn-5p and Ni-3d states. In fact, recent photoemission study by Nohara et al.[36] have revealed that the hybridization weakens significantly on going from CeNiSn to CePtSn. The decrease of $V_{k,f}$ in the above expression for T_K results in the lowering of the value of T_K. This is consistent with the transition into the Kondo regime observed for the Pt substituted system .

The effect of the substitution on the gapped state in CeNiSn was studied by the measurements of resistivity and specific heat. However, the temperature dependence of resistivity of polycrystalline samples of $CeNi_{1-x}T_xSn$ (T= Co and Pt) was found to be very sample-dependent even for a fixed value of x. This is presumably caused by the preferred orientation of the sample, as is expected from the highly anisotropic behavior in the the resistivity of the single crystalline sample (see Fig. 3). This situation did not allow us to study how the gap energy changes as a function of x.

The low-temperature specific heat of $CeNi_{1-x}T_xSn$ (T= Co and Pt) changes smoothly with x as shown in Fig. 15.[37,38] For pure CeNiSn, the sudden decrease in C/T below 6 K originates from the gap formation in the density of states. With increasing Pt concentration, the value of C/T increases and an upturn in C/T appears for $x \geq 0.12$. This change implies that the density of quasiparticle states grows within the gap and eventually a heavy fermion band is formed. The Kondo temperature for the x = 0.2 sample is estimated to be 25 K from an analysis of C/T using the isolated Kondo impurity model.[38] The pronounced peak at 2.1 K for x = 0.33 indicates the development of a long-range magnetic order for higher Pt concentration. This trend is consistent with the increased magnetic moment with Pt concentration, as was shown in Fig.14. We recall here that the transition into an antiferromagnetically ordered state is induced by the Cu substitution at a smaller concent-

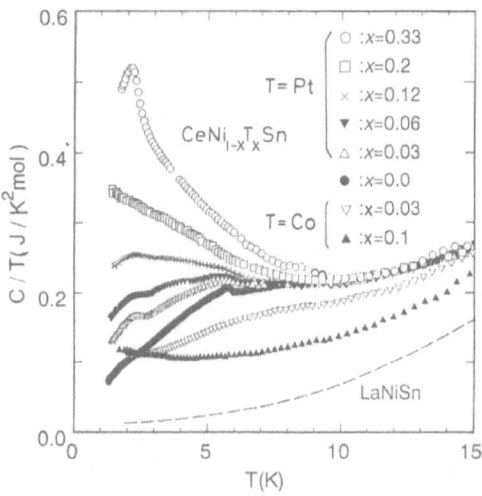

Fig. 15. Specific heat divided by temperature C/T vs T for $CeNi_{1-x}T_xSn$ (T= Co and Pt).

ration x = 0.13.[12] In this case, the increase in both the number of 3d electrons and the unit-cell volume may promote the transition.

On the other hand, the substitution of only 3 at.% Co is enough to reduce the value of C/T substantially throughout the temperature range below 30 K. Thereby, the anomaly due to the gap formation below 6 K is also smeared out. This in turn suggests that a moderate mass-enhancement is a necessary condition for the gap formation. Comparing the results of the Co, Cu and Pt substituted systems, we find that any replacement of about 10% of the Ni sublattice in CeNiSn closes the energy gap. This fact implies that the loss of the periodicity in the Kondo-lattice is very destructive for the gap formation.

7. CONCLUDING REMARKS

In this article, we reviewed our experimental attempts to understand the semiconductor-like behavior in CeNiSn at low temperatures. The results of transport, thermal and magnetic measurements characterize this compound as a rare example located near the crossover between the Kondo and VF regimes. The highly anisotropic behavior reflects the strong hybridization of the 4f states with the Sn-5p and Ni-3d states in the orthorhombic structure. The Kondo temperature of this system is about 28 K. At low temperatures below 12 K, coherence of local spin fluctuations develops, as indicated by the peaking in both $\rho_a(T)$ and $\chi_a(T)$ at 12 K and in $R_H(T)$ at 9 K. An anisotropic gap opens in the heavy-quasiparticle bands as temperature is reduced below 6 K. This semiconducting state with a charge excitation gap of several Kelvins is destructible by applying either hydrostatic pressure of 20 kbar or magnetic field of 13 T along the easy a axis. Substitutions of Co and Pt for a few percent of Ni in CeNiSn destroy the energy gap and shift the system toward the VF and the heavy-fermion regimes, respectively. The combined results indicate that the gap formation in CeNiSn is very sensitive to the degree of hybridization of the 4f and the conduction electron states. It is also emphasized that the development of coherence in the Kondo lattice is crucial for the gap formation.

ACKNOWLEDGMENT

The authors acknowledge fruitful collaborations with J. Sakurai, G. Kido, K. Sugiyama, M. Date, T. Hiraoka, M. Kyogaku, Y. Kitaoka, K. Asayama, D. Jaccard, U. Ahlheim, C. Geibel, F. Steglich, A. Krazer and G.M. Kalvius. We are benefited from valuable discussion with A. Yanase, S. Namatame, A. Fujimori, M. Kasaya, M. Kohgi, Y. Kuramoto and T. Kasuya. This work was supported in part by a Grant-in-Aid for Scientific Research and for International-Joint Research Program from Ministry of Education, Science and Culture of Japan.

REFERENCES

1. A. Menth, E. Buehler and T.H. Geballe, Phys. Rev. Lett. 22:295(1969).
2. A. Jayaraman, V. Narayanamurti, E. Bucher and R.G. Maines, Phys. Rev. Lett. 25:1430(1970).
3. P. Haen, F. Holzberg, F. Lapierre, T. Penny and R. Tournier, in *Valence Instabilities and Related Narrow Band Phenomena*, ed. by R.D. Parks (Plenum, New York, 1977) p. 1495.
4. M. Kasaya, F. Iga, M. Takigawa and T. Kasuya, J. Magn. Magn. Mater. 47&48:429(1985).
5. N.F. Mott, Phil. Mag. 30:403(1973).
6 T. Takabatake, Y. Nakazawa and M. Ishikawa, Jpn. J. Appl. Phys. Suppl. 26-3:547(1987).
7. I. Higashi, K. Kobayashi, T. Takabatake and M. Kasaya, to be published in J. Alloys Compds. (1992).
8. M.F. Hundley, P.C. Canfield, J.D. Thompson, Z. Fisk and J.M. Lawrence, Phys. Rev. B42:6842 (1990).
9. S.K. Malik and D.T. Adroja: Phys. Rev. B43:6277(1991).
10. P.S. Riseborough, Phys. Rev. B45:13984(1992).
11. P. Schlottmann, Phys. Rev. B46:998(1992).
12. T. Takabatake, Y. Nakazawa, M. Ishikawa, T. Sakakibara, K. Koga and I. Oguro, J. Magn. Magn. Mater. 76&77:87(1988).
13. T. Takabatake, F. Teshima, H. Fujii, S. Nishigori, T. Suzuki, T. Fujita, Y. Yamaguchi, J. Sakurai and D. Jaccard, Phys. Rev. B41:9607(1990).
14. T. Takabatake, M. Nagasawa, H. Fujii, G. Kido, M. Nohara, S. Nishigori, T. Suzuki, T. Fujita, R. Helfrich, U. Ahlheim, K. Fraas, C. Geibel and F. Steglich, Phys. Rev. B45:5740(1992).
15. T. Takabatake, M. Nagasawa, H. Fujii, M. Nohara, T. Suzuki, T. Fujita, G. Kido and T. Hiraoka, J. Magn. Magn. Mater. 108:155(1992).
16. T. Takabatake, M. Nagasawa, H. Fujii, G. Kido, K. Sugiyama, K. Senda, K. Kido and M. Date, Physica B172:177(1992).
17. T. Takabatake and H. Fujii, to be published in Jpn. J. Appl. Phys. Ser.8 (1992).
18. J. Sakurai, Y. Yamaguchi, S. Nishigori, T. Suzuki and T. Fujita, J. Magn. Magn. Mater. 90&91: 442(1990).
19. T. Takabatake, H. Iwasaki, G. Nakamoto, H. Fujii, H. Nakotte, F.R. de Boer and V. Sechovsky, to be published in Physica B182(1992).
20 T. Kasuya, J. Phys. Soc. Jpn. 61:1863(1992).
21. A. Yanase and H. Harima, Progr. Theor. Phys. Jpn. Suppl. 108:19(1992).
22. M. Kohgi, K. Ohoyama, T. Osakabe and M. Kasaya, J. Magn. Magn. Mater. 108:187(1992).
23. T.E. Mason, G. Aeppli, A.P. Ramirez, K.N. Clausen, C. Broholm, N. Stücheli, E. Bucher and T.T.M. Palstra, Phys. Rev. Lett. 69:490(1992).
24. F. Lapierre, P. Haen, R. Brigs, A. Hamezić, A. Fert and J.P. Kappler, J. Magn. Magn. Mater. 63&64: 338(1987).
25. A. Fert and P.M. Levy, Phys. Rev. B36:1907(1987).

26. N. Andrei, K. Furuya and J.K. Loewenstein, Rev. Mod. Phys. 55:331(1983).

27. M. Kyogaku, Y. Kitaoka, H. Nakamura, K. Asayama, T. Takabatake, F. Teshima and H. Fujii, J. Phys. Soc. Jpn. 59:1728(1990).

28. M. Kyogaku, Y. Kitaoka, K. Asayama, T. Takabatake and H. Fujii, J. Phys. Soc. Jpn. 61:43(1992).

29. A. Kratzer, G.M. Kalvius, T. Takabatake, G. Nakamoto, H. Fujii and S.R. Kreitzman, Europhys. Lett. 19:649(1992).

30. J.M. Mignot, J. Flouquet, P. Haen, F. Lapierre, L. Puech and J. Voiron, J. Magn. Magn. Mater. 76&77:97(1988).

31. M. Kurisu, T. Takabatake and H. Fujiwara, Solid State Commun. 68:595(1988).

32. M. Kurisu, T. Takabatake and H. Fujii, in this issue.

33. J.D. Thompson, J. Magn. Magn. Mater. 63&64:358(1987).

34. N. Grewe and F. Steglich, in *Handbook on the Physics and Chemistry of Rare Earths*, ed. by K.A. Gschneidner, Jr. and L. Eyring (Elsevier, Amsterdam, 1991) Vol.14, p. 343.

35. B. Cornut and B. Coqblin, Phys. Rev. B5:4541(1972).

36. S. Nohara, H. Namatame, A. Fujimori and T. Takabatake, to be published in Phys. Rev. B (1992).

37. T. Fujita, T. Suzuki, S. Nishigori, T. Takabatake, H. Fujii and J. Sakurai, J. Magn. Magn. Mater. 104-107:1415(1992).

38. S. Nishigori, H. Goshima, T. Suzuki, T. Fujita, G. Nakamoto, T. Takabatake, H. Fujii and J. Sakurai, to be published in Physica B (1992).

NON FERMI-LIQUID GROUND STATE IN THE HEAVY FERMION COMPOUNDS

Farkhad G. Aliev

Physics Department, Moscow State
University, 119899, Moscow, Russia
and
Dpto Fisica de la Materia Condensada,C-III
Universidad Autonoma de Madrid, 28049
Madrid, Spain

INTRODUCTION

Heavy fermions, one of the most astonishing fields of condensed matter physics, beginning from the work of Andres et.al.[1], is possibly feeling at present a third wave of interest. The first stage (see reviews[2,3]) characterized mainly by "quantitative progress" due to a very rapid growth of the number of the new heavy fermion compounds: "normal" as well as superconducting ones. In this period usually two alternative types of the HF ground state were treated: Fermi-liquid (FL) or gapped[2] as in SmB_6. At the same time the simplified scheme[3] according to which heavy fermions were considered as concentrated Kondo-systems was used. The second stage came with work[4] as a possible repel point associated with the understanding of the fact, that the FL ground state of the HFS may be characterized by the presence of weak magnetic coherent correlations[4,5] .

At present it is almost obvious that the "classical" FL is not a certain attribute of the ground state of intermetallic heavy fermion compounds. Recently two new HFS ($CeNiSn$ [6,7] and $Ce_3Bi_4Pt_3$ [8]) were unexpectedly found to show a gapped ground state. While for the CeNiSn compound a small gap at the Fermi level ($E_g \sim$ 6-10K) seems to exist[6,7] within a many-body resonance (of width about 50K), for $Ce_3Bi_4Pt_3$ only an observation of the gap of about 35K was reported[8]. On the other hand, another type of non Fermi-liquid ground state: "Marginal Fermi-Liquid" (MFL) was recently[9-11] shown to determine the low temperature properties in the dilute U limit of cubic $U_{0.2}Y_{0.8}Pd_3$ and the electron heat capacities of doped UPt_3 and UBe_{13}.

In the coherent regime HF could obey Fermi-statistics and have a well defined Fermi-surface[12]. The anisotropic character of the parameter J of spin-flip exchange between f- and d- states near the Fermi level could result in the strongly anisotropic reconstruction of the Fermi surface[13]. In the specific case characterized by a number of electrons on the magnetic center n equal to unit (n=1) the renormalized Fermi-surface will have zero volume[14]. Lacroix[15] showed that for n near 1 and for a rather high J the ground state should be nonmagnetic, while for a moderate J, the transition into a magnetic ground state for n near unit will proceed (Fig.1a). Possible disposition of SmB_6, $Ce_3Bi_4Pt_3$ and CeNiSn HFS also shown on this diagram.

GAPPING OF THE ELECTRON SPECTRUM IN CeNiSn

Let us now try to go inside the anomalous ground state of the HFS CeNiSn. In this chapter analyzing low temperature transport[6,7,17,18], thermal[19-21] and NMR[22-24] properties of polycrystalline and single crystalline samples we will "draw" possible excitation spectra near the Fermi-level. In the next part the transformation of the CeNiSn ground state through the effect of external influences (pressure and alloying) will be examined.

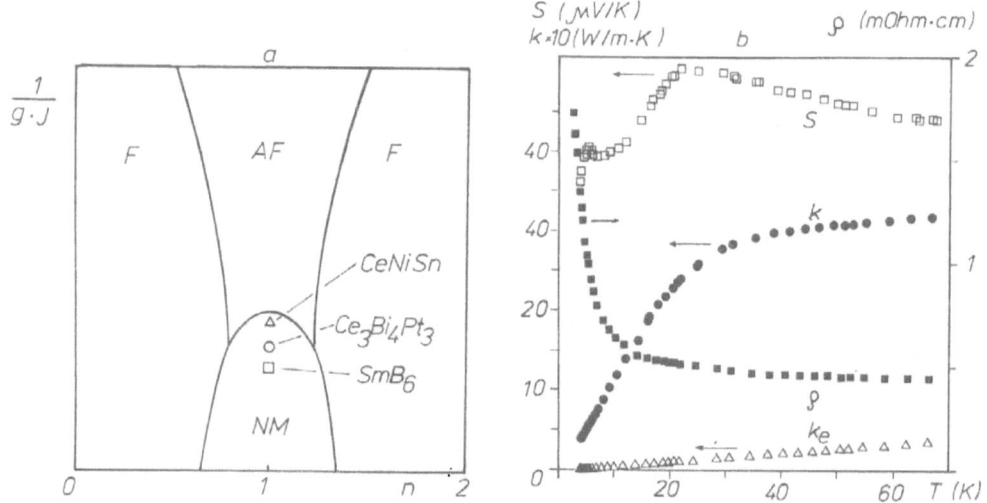

Figure 1 a) Phase diagram of Kondo-lattice[16] .
b) Transport characteristics[21] of polycrystalline CeNiSn.

Figure 1b shows representative temperature dependences of resistivity ρ , Seebeck coefficient S and thermal conductivity k (electron part k_e) for polycrystalline CeNiSn samples[21]. From thermoelectric properties[25], besides the reported in early work the maximum near 70K (possibly originated[26] from a crystal field splitting effect $\triangle_{CF}$ of about 7meV) we see a clear evidence of two other maxima in the temperature range corresponding to the gapping of the spectrum: at below 20K and 3-5K. A qualitatively analogous behavior is also shown by the Seebeck coefficient of the single crystals[18]. Moreover near 4K a noticeable bend on the ln ρ vs 1/T dependence exists[17,27]). Curiously, that the existence of the second low temperature maximum on the electron heat capacity in the gapped state (Fig.2a) was explained as being due to the presence of impurity phases[20].

In our opinion, a clearer proof and demonstration of the intrinsic character of the second characteristic temperature scale in the gapped ground state of CeNiSn comes from the temperature dependence of the thermal expansion[19,21] α (T) (Fig.2b). For a qualitative analysis of the α(T) behavior below 1K we will use the theory of the negative thermal expansion in Kondo lattices[28]. This model proposes a strong correlation between the existence of a pseudogap in the electron spectrum near the Fermi level and negative value at T→0. Modelling of the NMR data by different gap structures showed[22] the best correspondence of experimental data with a linear dependence of g(E) on energy E (g(E)~E). Here we propose that this simple g(E) dependence should be modified in the vicinity of E_F due to at least two facts. The first reason is that neither a linear, nor a parabolic dependence of g(E) near the Fermi energy can explain large negative

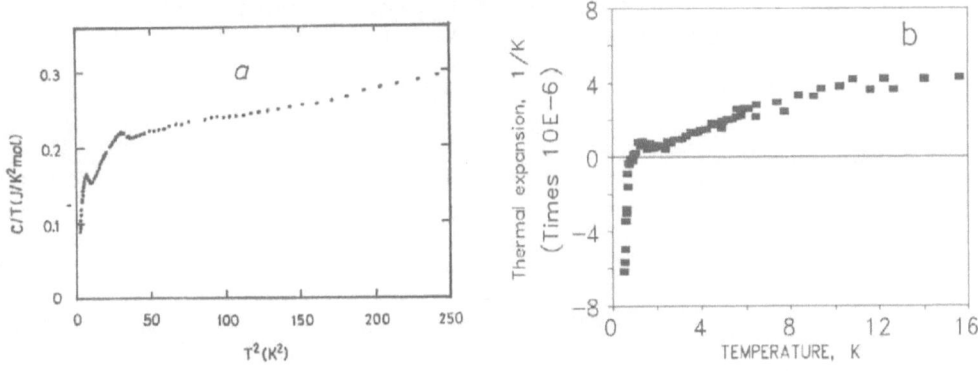

Figure 2 a) Linear term of the heat capacity[20] and b) thermal expansion[19] vs temperature of CeNiSn.

values in the frame of the theory[28]. Only a bend in the g(E) dependence near E_F could produce a rather high negative α values[28]. The other reason to modify the early proposed[22] density of electron states inside the gap comes from the more recent NMR experiments which revealed some deviation of the $1/T^1$ vs T dependence from the g(E) fitting curve below 1.5K followed by quasi saturation[23] of the $1/T^1$ signal at T<0.5K.

The strongly anisotropic CeNiSn crystal structure[29,30] with Ce-atoms forming a system of (bc) planes, separated by planes containing Ni and Sn atoms results in the anisotropic parameter J, mirrored by the anisotropy of the lattice[30], magnetic and transport[18] properties (Fig.3a,b). Along "more magnetic" a-direction the transport gap is about 2-3 times lower than in the b-c plane. The critical magnetic field of about 13T suppressing the gapping along a-axis[32] is consistent with effective magnetic moment $\mathcal{M}_{eff} \sim 0.3\ \mathcal{M}_B$ and the gap of 2K.

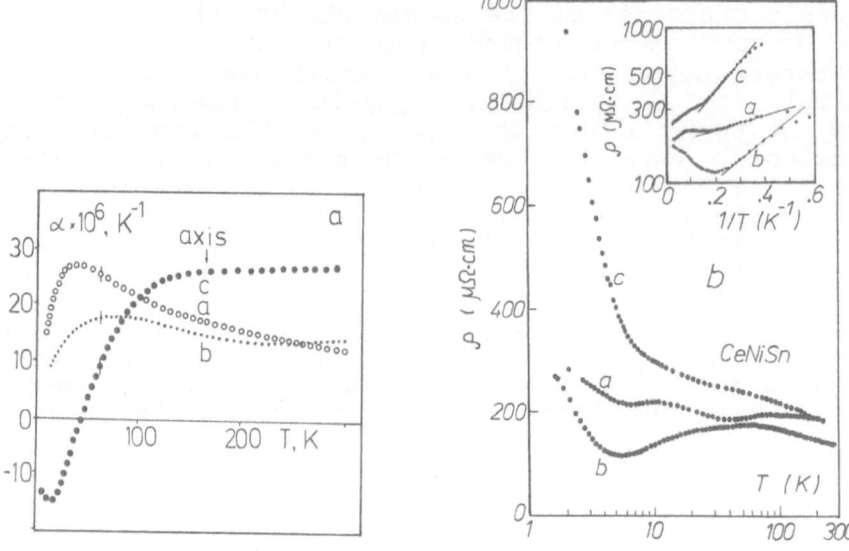

Figure 3. a) Anisotropy of thermal expansion[30] and
b) Anisotropy of resistivity[18] for CeNiSn

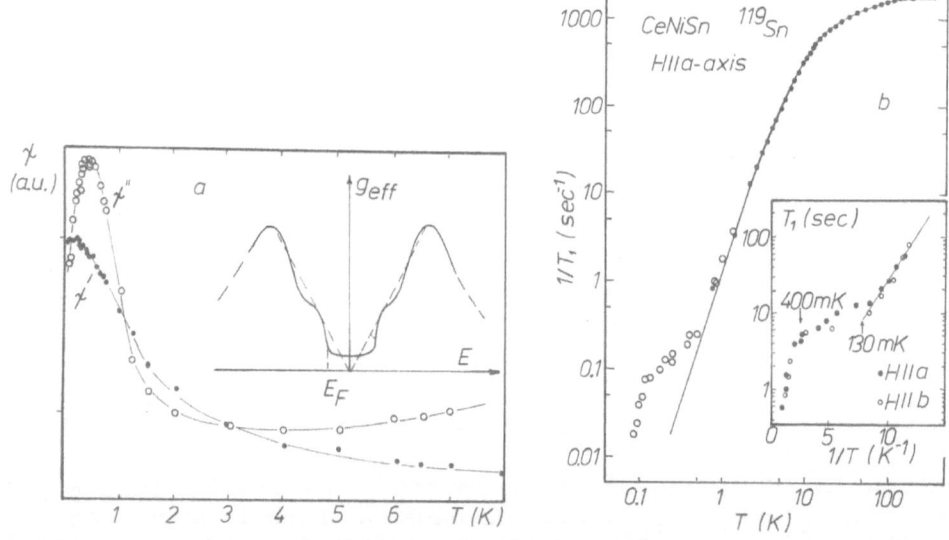

Fig.4a) Magnetic susceptibility[21] b) NMR[24] vs T for CeNiSn.

Now we will try to draw schematically the character of
the density of states discussed before, neglecting the
possible anisitropy between b and c - directions. The
effective density of states (see insert on Fig.4a) being sum
$g_{eff}(E) = g_a(E) + g_{bc}(E)$ seems to be rather close to the one
proposed to fit the NMR- data[22] and at the same time will
describe the negative thermal expansion coefficient at $T \to 0$
as well as the two-fold anomalies in the gapped region
observed in the temperature dependences of the specific heat
and Seebeck coefficient.

The long range coherent interaction of the Kondo-reduced magnetic Ce moments in CeNiSn possibly becomes important[19,21] below 1K. The maximum[21] on the imaginary part $\chi''(T)$ near 0.5K (Fig.4a) may be considered as an indication of the existence of this transition. More likely, the process of formation of the ground state having as a consequence the appearance of the gap in magnetic excitation spectra near E_F is complete below T=0.5K. In fact, recently Kuogaki et.al.[24] also reported magnetic instability in CeNiSn at very low temperatures: the development of static magnetic correlations inducing a "spin gap" of about 0.25K was seen[24] (Fig.4b).

EFFECT OF THE SUBSTITUTION AND PRESSURE ON THE GAPPED GROUND STATE OF CeNiSn

Let us now discuss the influence of external factors, for example, the change of composition or the effect of hydrostatic pressure on the ground state of CeNiSn. Of course, a correct analysis should be rather complicated, because the proposed spectrum (Fig.4a) is itself a function of the temperature. Nevertheless it is reasonable to suppose that the substitution of Ce by La in $Ce_{1-x}La_xNiSn$, increasing[21] the effective volume V of the lattice according to the derivative $dV/dx \approx 7Å^3$, for small x could induce a "negative pressure effect" and also will transform the gap mainly along b-c plane. In fact, a very small (x<0.05) La substitution, having only a negligible effect on the upper gap value[33], shifts the CeNiSn system towards a magnetic instability below 5K. This may be seen also in the temperature dependence of the thermal expansion[21] of polycrystalline $Ce_{0.97}La_{0.03}NiSn$, presented in Fig.5. It is interesting to note that the same type of anomaly near T ∼ 5-6K, possibly indicating the proximity of the ground state of CeNiSn to "weak" antiferromagnetism, was also recently deduced from the temperature dependences of volume thermal expansion[31] and thermal conductivity[34] of CeNiSn single crystals along the b-direction. On the other hand, studies of thermal expansion under pressure[31] revealed that a hydrostatic pressure of about 8 kbars completely suppresses the AF instability at T ∼ 6K, reducing the absolute values more than twice down to those corresponding to data on polycrystals[19,21]. Therefore we can conclude that, in comparison with single crystals the polycristalline CeNiSn samples seem to be "pressed".

To our knowledge, no studies of the thermal expansion of single crystalline CeNiSn samples have been performed below 4K. Based on the analysis presented here, we propose that an anomalously strong effect will show up in the thermal expansion of CeNiSn at T<2K under a hydrostatic pressure of only a few kbars. Also we can suppose that the AF instability seen in the single crystals near 6K may be the reason of the partial suppression of the gap value along the a-direction. In this case, studies of the anisotropic transport of CeNiSn under pressure could even reveal an initial increase of the $E_g{}^a$ value under pressure. The studies of the electron transport in CeNiSn under pressure carried out previously were done using polycrystalline samples[17,27]. In fact, in these experiments only the pressure dependence of the higher gap value was analyzed. A reanalysis of these data plotting

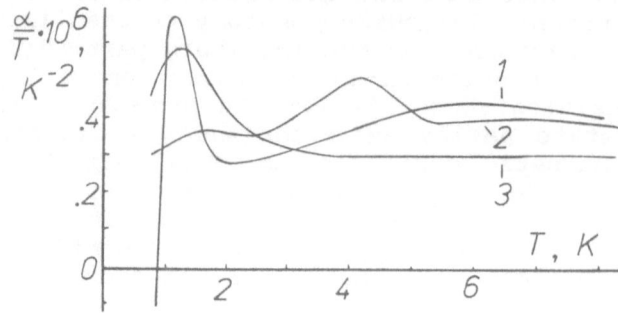

Fig.5 Linear term in thermal expansion[21]: CeNiSn (curve 1), $Ce_{0.97}La_{0.03}NiSn$ (2) and $Ce_{0.9}La_{0.1}NiSn$ (3).

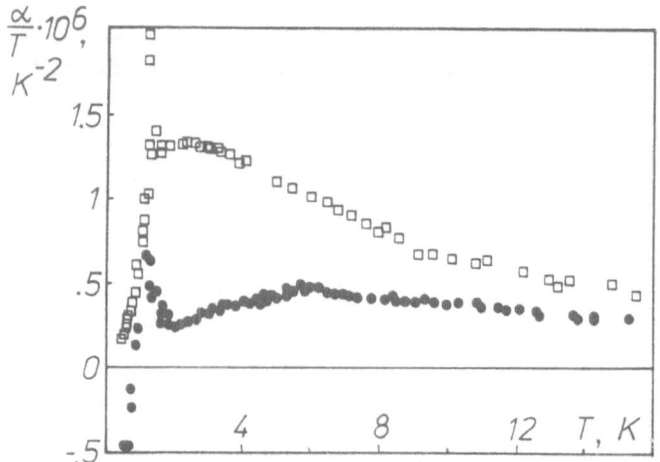

Figure 6 Thermal expansion[21] on T of CeNiSn (●) and $CeNi_{0.9}Cu_{0.1}Sn$ (□).

two gaps (E_g^{bc} as E_{g1}) for (7<T<20)K and E_g^a as E_{g2} for (2<T<6)K) shows that even in "pressed" polycrystalline samples a small initial increase of E_{g2} at p< 6kbars followed by a more rapid (in comparison with E_{g1}) gap suppression at p>10kbars possibly exists.

Coming back to the effect of alloying, let us consider shortly[21,35] the other stoichiometric substitution: $CeNi_{1-x}Cu_xSn$. A change of the free electron density of states in the Ni-Cu sheets could mainly affect the s-f interaction along the a-direction and therefore a smaller gap value. The experimentally observed substantial increase of the low temperature α/T values[21] (as well as of the linear term in the heat capacity[35]) of $CeNi_{0.9}Cu_{0.1}Sn$ (Fig.6) very likely reflects the disappearance of the gap with the development of the magnetic ground state. The physical reason for the strengthening of magnetism in $CeNi_{0.9}Cu_{0.1}Sn$, in comparison with CeNiSn, may be the decrease of the Kondo temperature originated from the weakening of the s-f exchange due to substitution of the narrow Ni conduction electron zone by the wide Cu one.

$Ce_3Bi_4Pt_3$: GAPPING OF THE ELECTRON AND MAGNETIC EXCITATION SPECTRA

We will shortly discuss below the main characteristics of the other gapped intermetallic Ce-compound: $Ce_3Bi_4Pt_3$. A cubic crystal structure and a rather "high" value of the gap in the transport characteristics[36] of about 35K results in the more stable to external influences (in comparison with CeNiSn) nonmagnetic ground state. In fact, the effect of the substitution of Ce by La in $(Ce_{1-x}La_x)_3Bi_4Pt_3$ leads to a rather rapid (but without transition into a magnetic state) decrease of the energy gap, as well as to a corresponding increase of the linear C/T term[36]. So, for $Ce_3Bi_4Pt_3$, no magnetic instability of the ground state was reported in agreement the position on the phase diagram proposed for this compound (Fig.1a). Nevertheless, the gap structure of $Ce_3Bi_4Pt_3$ also seems to be not very simple. From the $\ln \rho$ vs 1/T dependence it is clear that as with CeNiSn, here it is difficult to fit curve with only one E_g parameter[36]. As another evidence of the nontrivial character of the gapping process in $Ce_3Bi_4Pt_3$ the existence of the gap in the magnetic excitation spectra[37] may be considered where an absolute value at least twice higher than those found from transport data has been observed[36].

NON-FERMI-LIQUID BEHAVIOR IN THE METALLIC GROUND STATE

The absence of the normal FL features in the gapped state of some heavy fermion compounds, analyzed above, seems to be originating from the strong localization of the electrons near the Fermi-level and the closeness to an AF magnetic instability. In metallic compounds the formation of the FL ground state normally is considered in the frame of the formalism, were existence of short-range electron-electron interaction leads to the renormalization of the effective mass of the quasiparticles. As a consequence, low temperature properties of such interacting systems resemble those of a noninteracting Fermi-gas. Nevertheless, from a number of experimental characteristics in the dilute and concentrated limits, specially from the temperature dependences of the electron part of the specific heat, C_e, an important experimental evidence of non-Fermi-liquid behavior of HFS with a metallic ground state was also seen recently[9-11,38].

In the Kondo-alloy[9] $U_{0.2}Y_{0.8}Pd_3$ instead of a saturation of the added specific heat C_e/T at $T \to 0$, predicted for the normal Kondo-effect[3], a logarithmic divergence $C_e(T)/T \sim \ln T$ between 0.6 and 16K was observed. The authors[9] argued that their data provide evidence of two-channels quadrupolar

Kondo-effect (TCQKE) in the MFL ground state. In this approach, for a small (in comparison with crystal electric fields) spin-orbital coupling and for orbital channels (f) of more than double value of the scattered spin moment S (i.e. for f > 2S) a nontrivial infrared fixed point, which controls the low temperature properties, appears[39].

In the other explanation[38] a break of the FL behavior in $U_{0.2}Y_{0.8}Pd_3$ was described as being a consequence of the long-range correlations between quasiparticles. These interactions gradually appear with decreasing of the temperature and result in a second order phase transition at T=0. The evidence of this transition comes from experimentally observed proper scaling of the thermodynamic properties[38]. The later observations of the MFL behavior in the electronic

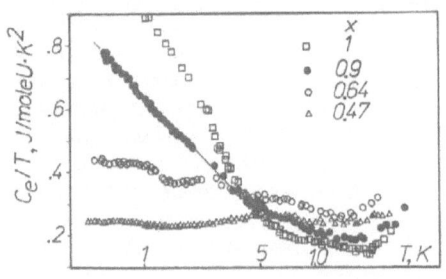

Figure 7 Term C_e/T vs lnT for $U_xTh_{1-x}Be_{13}$ compounds[11].

specific heat of the two doped concentrated heavy fermion systems[10,11]: $U(Pt_{0.94}Pd_{0.06})_3$ and $U_{0.9}Th_{0.1}Be_{13}$ support this hypotesis. In the former compound a weak antiferromagnetic transition near 6K exists. In $U_{0.9}Th_{0.1}Be_{13}$ an anomalous linear C_e/T vs ln(T) dependence is seen at least between 0.6K and 6K (see Fig.7). In this alloy the closeness to the antiferromagnetic instability may originate from the weak magnetic transition[40] in the superconducting state of $U_{1-x}Th_xBe_{13}$ (0.02<x<0.04). Let us also note that the observation of the C_e/T ln(T) behavior in $U_{0.2}Y_{0.8}Pd_3$ also proceeds in the U-concentration range close to the magnetic instability. First of all $U_{0.3}Y_{0.7}Pd_3$ already shows spin-glass behavior[9] and then the concentration dependence of this spin-glass transition temperature extrapolates to $T_{SG}=0$ value corresponding to $U_{0.2}Y_{0.8}Pd_3$ composition[41].

Finally, the lack[10,42] of the decrease of the resistivity predicted by TCQKE ($\rho \sim -AT$) in the ground state of $U(Pt_{0.94}Pd_{0.06})_3$ and $U_{0.9}Th_{0.1}Be_{13}$ compounds show that to describe MFL behavior in these heavy fermion systems new theories taking into account a possible magnetic phase transition at zero temperature should be developed. On this way a Kondo-hole approach[43] possibly could be fruitful.

CONCLUDING REMARKS

The analysis of the recent experimental data shows that in a number of heavy fermion compounds a simple Fermi-liquid approach to the normal ground state properties could break down. In Kondo-lattices with n=1 and a moderate J like in CeNiSn electron-electron correlations may result in an almost nonmagnetic, low conducting ground state, exceptionally unstable to external influences. An increase of the J value could stabilize the nonmagnetic character of the ground state, as it proceeds in $Ce_3Bi_4Pt_3$. For good conducting HF systems non-Fermi-liquid behavior in the ground state also exists. In fact this type of ground state is seen in the normal state of high-T_c superconductors[44] as well as in the dilute and concentrated limit of HFS[9-11,38] and is usually described within the frame of MFL terminology. A possible connection between two different types of non-Fermi-liquid ground state mentioned above could originate from the predicted[45] for MFL character of the excitation of both charge and spin polarizability, which gives a gapping at E_F of the spectral function for the quasiparticles having energy just below the Fermi-level.

ACKNOWLEDGMENTS

The author wishes to thank V.V.Pryadun, M.Yu. Kulikov, I.Grishchenko and for help in experiment and to N.B.Brandt, V.V.Moshchalkov, R.V.Scolozdra, R.Villar, S.Vieira, M.A.Lopez de la Torre, C.L.Seaman and P.A.Alekseev for discussions. The author gratefully acknowledges financial support from the Ministry of Science and Education of Spain. This work was also supported by grant MAT 92-0170 from Plan Nacional de Materiales and by Russian State Programm on Superconductivity.

REFERENCES

1. K.Andres, J.E.Graebner, H.R.Ott, Phys.Rev.Lett. 35: 1779 (1975).
2. G.R.Stewart, Rev. Mod. Phys. 56: 755 (1984).
3. N.B.Brandt and V.V.Moshchalkov,Adv.Phys.33:373 (1984).
4. S.Barth,H.R.Ott,F.N.Gygax,Phys.Rev.Lett.59:2991(1987)
5. E.Blount,C.M.Varma,G.Aeppli,Phys.Rev.Lett.66:512(1990).
6. T.Takabatake,Y.Nakazawa,M.Ishikawa,Jpn.J.Appl.Phys.26: 547 (1987).
7. F.G.Aliev, N.B.Brandt, V.V.Moshchalkov, M.K. Zalyalutdinov, G.I.Pak, R.V.Scolozdra, J. Magn.& Magn. Mat. 76-77: 295 (1988).
8. M.F.Hundley, P.C.Canfield, J.D.Thompson, Z.Fisk and J.M.Lawrence, Physica B171: 254 (1991).
9. C.L.Seaman,M.B.Maple,B.W.Lee,S.Ghamaty,M.S.Torikachvili J.S.Kang,J.W.Allen,D.L.Cox,Phys.Rev.Lett.67:2882(1991).
10. J.S.Kim,B.Andraka,G.R.Stewart,Phys.Rev.B45:12081(1992)
11. F.G.Aliev,A.V.Andreev,I.O.Grishchenko, R.Villar and S.Vieira, to. be publ. in proceedings of ICPTM-92.
12. P.Coleman, Phys. Rev. B28: 5255 (1983).
13. R.Martin, Phys.Rev.Lett. 48: 362 (1982).
14. R.Jullien,J.N.Fields,S.Doniach,Phys.Rev.B16:4889(1977).

15. C.Lacroix, Sol.St.Comm. 54:991(1985).
16. C.Lacroix,J.M.&M.M.63-64:239(1987).
17. F.G.Aliev,N.B.Brandt,V.V.Moshchalkov,M.K.Zalyalyutdiv,
    R.V.Scolozdra,and G.I.Pak Pis'ma ZETPh 48:536(1988).
18. T.Takabatake, F.Teshima, H.Fujii, S.Nishigory,
    T.Suzuki,T.Fujita, Y.Yamaguchi, J.Sakurai and D.Jaccard
    Phys.Rev. B41: 9607 (1990).
19. F.G.Aliev, A.I.Belogorochov, V.V.Moshchalkov,
    R.V.Scolozdra, M.A.Lopez de la Torre, S. Vieira and R.
    Villar, Physica B171: 381 (1991).
20. T.Takabatake, Y.Nakazawa, M. Ishikawa, I. Sakakibara,
    K.Kogoand I.Oguro, J.M.&M.M. 76-77: 87 (1988).
21. F.G.Aliev, R.Villar, S.Vieira, M.A.Lopez de la Torre,
    R.V.Scolozdra and M.B.Maple subm. in Phys. Rev B.
22. M.Kyogaku, Y.Kitaoka, H.Nakamura, K.Asayama,
    T.Takabatake, F.Teshima and H.Fujii, J.Phys.Soc.Jap.
    59:1728(1990).
23. M.Kyogaku,Y.Kitaoka,H.Nakamura,K.Asayama,T.Takabatake,
    T.Teshima and H.Fujii,Physica B171:235(1991)
24. M.Kuogaku,Y.Kitaoka,K.Asayama,T.Takabatake and
    H.Fujii,J.Phys.Soc.Jap.61:43(1992).
25. F.G.Aliev,N.B.Brandt,V.V.Moshchalkov,R.V.Scolozdra
    Preprint N.4 Physics Department, MSU (1988).
26. F.G.Aliev, V.V.Moshchalkov, M.K.Zalyalyutdinov,
    G.I.Pak,R.V.Scolozdra, P.A.Alekseev, V.N.Lazukov,
    I.P.Sadikov, Physica B163: 358 (1990).
27. M.Kurisu, T.Takabatake, H.Fujiwara, prepint
28. C.Bastide and C.Lacroix,Sol.St.Comm.59:121(1986).
29. R.V.Scolozdra, O.E.Koretscaya and Yu.K.Gorelenko,
    Inorganic Mater.20:604(1984).
30. F.G.Aliev,V.V.Moshchalkov,R.V.Scolozdra,M.A.Lopez de
    la Torre, S. Vieira, R. Villar, P.A.Alekseev,
    E.S.Klement'ev, V.N.Lazukov,I.P.Sadikov, G.A.Ivanov
    Smolenskii and I.D.Datt J.Moscow Phys.Soc.1:311(1991).
31. Y.Uwatoko, G.Oomi, T.Takabatake, H.Fujii, preprint.
32. T.Takabatake, M. Nagazawa, H.Fujii, G.Kido, M.Nohara,
    S.Nishigori, T. Suzuki, T. Fujita, R. Helfrich,
    U.Ahlheim, K.Fraas,C.Geibel and F.Steglich, Phys.Rev.
    B45:5740 (1992).
33. F.G.Aliev and M.K.Tambiev (to be publ.)
34. Y.Isikawa, K.Mori, Y. Ogiso, K. Oyabe and K. Sato
    J.Phys.Soc.Jap. 60: 2514 (1991).
35. T.Takabatake, Y.Nakazawa, M.Ishikawa, Preprint.
36. M.F.Hundley, P.C.Canfield, J.D.Thompson, Z.Fisk,
    J.M.Lawrence, Phys.Rev. B42: 6842 (1990).
37. A.Severing, J.D.Thompson, P.C.Canfield, Z.Fisk
    and P.S.Riseborough, Phys. Rev. B44: 6832 (1991).
38. B.Andraka and A.M.Tsvelik,Phys.Rev.Lett.67:2886(1991).
39. P.Nozieres and A Blandin J.Phys. 41: 193 (1980).
40. H.R.Ott,H.Rudigier,Z.Fisk,J.L.Smith,Phys.Rev.B31:1651
    (1985).
41. C.L.Seaman, private communication.
42. F.G.Aliev, A.V.Andreev, N.B.Brandt, V.V.Moshchalkov
    and V.Kovacik, Fiz.Tv.Tela. 29: 596 (1987).
43. R.Sollie and P.Schlottmann,J.Appl.Phys.69:5478(1991).
44. M.Gurvitch and A.T.Fiory,Phys.Rev.Lett.59:1337(1987).
45. C.M.Varma,P.B.Littlewood,S.Schmitt- Rink, E.Abrahams
    and A.E.Ruckenstein, Phys. Rev. Lett. 63: 1996 (1989).

MECHANISM OF THE APPEARANCE OF AN ENERGY GAP IN MIXED-VALENT RARE-EARTH COMPOUNDS

Mitsuo Kasaya
Department of Physics, Faculty of Science, Tohoku University
Sendai 980, Japan

INTRODUCTION

The origin of an energy gap at the Fermi level in mixed-valent SmB_6, gold-phase SmS, YbB_{12}, CeNiSn and $Ce_3Pt_3Bi_4$ has been a subject of controversy.[1-6] In case of SmB_6, for instance, valence of Sm ions is about +2.6, and then the number of conduction electrons per formula unit has been believed to be 0.6, because RB_6 compounds with $R=Ca^{2+}$ and Eu^{2+} are semiconductors and $La^{3+}B_6$ is a monovalent metal. How to construct an energy gap and how to obtain a semiconducting property in such a metallic state have been an attractive theme for theorist. Models of hybridization gap,[7] Wigner-crystal formation[8] and coherence pseudogap[9] have been proposed so far. From an experimental point of view, it has been required to distinguish the gap due to the effect of f-electrons from the one in the non-f electronic states. Furthermore, to find a requisite condition for the gap opening, it is also necessary to find other exemplifications, especially, in isomorphous compounds in which rare-earth ions exist as a well-defined valence state in one compound and as different valence state in another one, both of which show semiconducting properties. Recently, based on our experimental finding of the appearance of semiconducting properties in both $Ce_3Au_3Sb_4$ with well-defined Ce^{3+} ions and the mixed-valent compound $Ce_3Pt_3Sb_4$, we have pointed out that the valence of Ce ions in $Ce_3Pt_3X_4(X=Bi,Sb)$ is formally 4+ and the origin of the energy gap is not so much different from those of $Ce_3Au_3Sb_4$ and $La_3Au_3Sb_4$.[10] Furthermore, we pointed out that there are common features in the band structures of their reference systems without f electrons.[10,11] For Sm- and Yb-based mixed-valent compounds, for instance, the band calculations of CaB_6[(ref.12)] and YB_{12}[(ref.13)] predicted that non-f $R^{2+}B_6$ and non-f $R^{2+}B_{12}$ can be semiconductors, i.e., band structures of compounds with rare-earth sites replaced by non-f 2+ ions have a gap at the Fermi level. As for Ce-based compounds, the band structure of the reference system for CeNiSn has not yet been calculated, but band calculations of $La_3^{3+}Au_3Sb_4$[(ref 10)] and $Th_3^{4+}Ni_3Sb_4$[(ref.14)] suggest isoelectronic non-f $R_3^{4+}Pt_3X_4(X=Bi,Sb)$ to be semiconductors, i.e., the band structure of compounds with rare-earth site replaced by non-f 4+ ions has a band gap at the Fermi level. The valences 2+ and 4+ correspond to

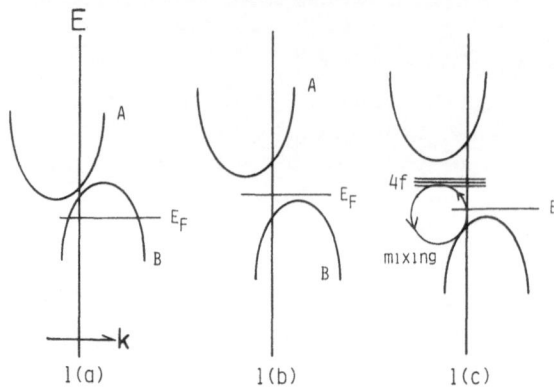

Fig.1. Schematic pictures of LaNiSn(a), non-f R^{4+}NiSn and $Th_3^{4+}T_3Sb_4$(T=Pt,Ni) (b), and CeNiSn and $Ce_3Pt_3Sb_4$(c).

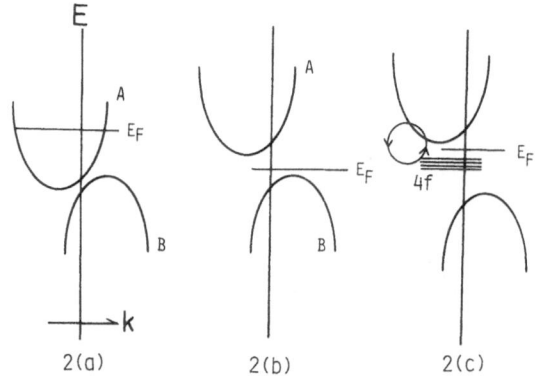

Fig.2. Schematic pictures of YB_{12}(a), CaB_6 and non-f $R^{2+}B_{12}$(b), and YbB_{12}(c).

ionic states of nonmagnetic ($J=0$) Yb(Sm) and Ce, respectively. Quite recently, band calculations of YbB_{12},[15] SmB_6,[15] CeNiSn,[15] LaNiSn[ref.16] and $Ce_3Pt_3Sb_4$[ref.17] were performed by Yanase, Harima and Takegahara. In the following, with the help of the results of these band calculations, we review a more detailed description of our model and the physical meaning of the origin of an energy gap in these exotic mixed-valent compounds. Main part of this manuscript will be published elsewhere.[18]

RESULTS AND DISCUSSION

First, we discuss calculated band structures of the reference systems for Ce-based mixed-valent compounds. As described above, the calculation of the band structure for $Th_3^{4+}Ni_3Sb_4$,[14] which is isoelectronic with $Th_3^{4+}Pt_3Sb_4$ and can be a reference system for mixed-valent $Ce_3Pt_3Sb_4$, revealed that $Th_3^{4+}Ni_3Sb_4$ is an insulator with a well-defined energy gap between the Sb $5p$ valence and Th $6d$ conduction bands. In the following, such a band structure is shown schematically as one in Fig.1(b). As for the reference system of CeNiSn, there is no band calculation for hypothetical non-f

R^{4+}NiSn. Therefore, as the next best policy, results of band calculation of metallic LaNiSn will be discussed. It is revealed that for any fixed wave vector $\mathbf{k}$, for instance at Γ, T, Z points and so on, the energy of branch A is higher than the energy of branch B, and there is no crossing between branch A and branch B.[16,18] This result is shown schematically in Fig.1(a). Branch A and branch B correspond to branch A and B in Fig.3 in ref.18, respectively. As the rare-earth site is varied from La to non-f R^{4+}, the number of electrons increases by four electrons per unit cell, because the unit cell of ϵ-TiNiSi-type structure contains four formula units. Therefore, if the bottom of branch A is higher than the top of branch B for any direction of $\mathbf{k}$ in non-f R^{4+}NiSn, electrons occupy branch B completely and the Fermi energy E_F is located in the band gap, leading to semiconducting properties in non-f R^{4+}NiSn. Therefore, electronic structures of non-f R^{4+}NiSn can be the same as those of non-f $Th_3^{4+}Ni_3Sb_4$ shown schematically in Fig.1(b). Under the existing circumstances, we mention that a necessary condition for the appearance of semiconducting property is satisfied in non-f R^{4+}NiSn. In other words, our prediction for semiconducting Ce-based mixed-valent compounds that their reference systems with rare-earth sites replaced by non-f R^{4+} ion are semiconductors or, to say the least, semi-metals, is realized in the reference system for $Ce_3Pt_3Sb_4$ and CeNiSn.

Next, band structures of reference systems for the Sm- and Yb-based compounds are discussed. For the reference system of SmB_6, results of band calculation of $Ca^{2+}B_6$ showed a well-defined energy gap[12] which is schematically shown in Fig.2(b). As for the reference system of YbB_{12}, there exist no non-f $R^{2+}B_{12}$ compounds. Therefore, no band calculation of, for instance, $Ca^{2+}B_{12}$, has been performed. Then, as an alternative, we quote results of band calculation of $Y^{3+}B_{12}$ performed by Harima et al.[13] New results of YB_{12} taking account of spin-orbit interaction are shown schematically in Fig.2(a).[15,16] For any fixed $\mathbf{k}$, energy of branch A is higher than the energy of branch B, similar to the case of LaNiSn. Therefore, as the rare-earth site is varied from Y to non-f R^{2+}, number of electrons decreases by one electron per primitive cell and band structure of non-f $R^{2+}B_{12}$ can be expected to that shown in Fig.2(b). It is worthwhile to note that a similar shift of the band structure has been observed also between metallic LaB_6 and semiconducting CaB_6.[12,19] Therefore, there is no exception which contradicts the common features we pointed out previously. The most important conclusion obtained from these discusions is that $Ce_3Pt_3Sb_4$(CeNiSn) and YbB_{12}(SmB_6) can be semiconductors, in principle, if valences of rare-earth ions are formally 4+ and 2+, respectively and $4f$ levels of interest are located in the band gap. Here we used terminology "formally 4+" in the sense that four electrons of rare-earth ion shift towards anions to fill the valence band.

The relation between the band gap observed in the non-f reference systems and the energy gap in the f-electron systems is as follows. In the case of Ce-based compounds, empty $4f$ levels are located above the Fermi energy E_F, as shown schematically in Fig.1(c). In the figure, we showed empty $4f$ levels corresponding to spin-orbit split $j=5/2$ levels. The spin-orbit split $j=7/2$ levels are located above those of $j=5/2$, but play no important role in the origin of an energy gap, therefore they are not shown in the figure. Nevertheless, the valence of Ce ions in Ce-based semiconducting compounds is formally 4+ and then the valence band is fully occupied by electrons. The mixed-valent originates from mixing between the occupied branch B and empty $4f$ levels. This mixing will modify the size of the energy gap, but does not necessarily cause the shift of the Fermi energy, i.e., semiconducting properties will be conserved. Such a picture is consistent with the results of band calculations of CeNiSn[ref 15] and $Ce_3Pt_3Sb_4$.[17] It is to be noted that this mixing introduces a considerable amount of $4f$ character in the

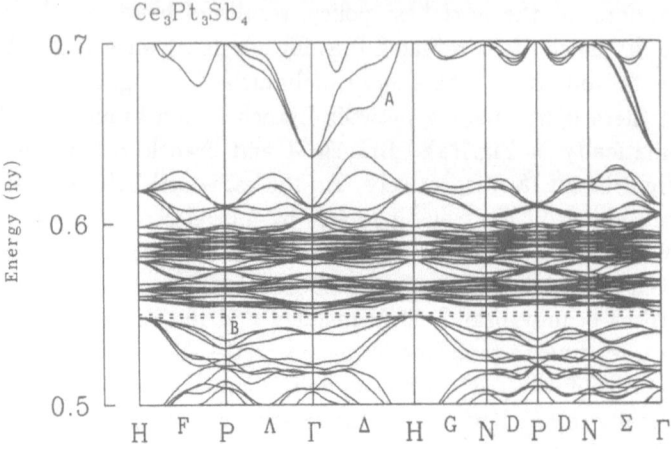

Fig.3. Self-consistent LAPW energy band structure for $Ce_3Pt_3Sb_4$.[17] Branch A and branch B correspond to branch A and branch B in Fig.1(b).

valence band of CeNiSn ($N_{4f} \sim 1$),[16] and then the valence band of CeNiSn does not correspond directly to branch B of Fig.1(b). Band calculation suggests that branch B and $4f$ levels mix strongly on the P-axis, resulting in hybridization gap between $4f$ level and the reconstructed valence band with considerable amount of $4f$ character.[15] Although a overlapping of branch B and $4f$ levels occurs at another direction of **k** in the band calculation, opening of an energy gap for any direction of **k** is possible in princile. On the other hand, in the case of $Ce_3Pt_3Sb_4$, results of band calculation shown in Fig.3 suggests that spin-orbit split j=5/2 and j=7/2 levels are located within a band gap of $Th_3Pt_3Sb_4$, i.e., branch A and branch B shown in Fig.3 correspond directly to the conduction band and valence band of $Th_3Pt_3Sb_4$.[14] Therefore, no hybridization gap in any **k** occurs in $Ce_3Pt_3Sb_4$. Minimum of $4f$ level and maximum of the valence band are located at Γ and H points, respectively. Such an energy gap can be termed as a band gap. The observed energy gaps, 455 K for $Ce_3Pt_3Sb_4$[ref.10], may correspond to this gap.

In the case of YbB_{12}, occupied $4f$ levels are located below E_F as shown in Fig.2(c). Therefore, the valence of Yb ion is formally 2+. In Fig.2(c), spin-orbit split j=5/2 levels are not shown. Needless to say, there exists mixing between the empty branch A and occupied $4f$ levels, resulting in mixed-valent in YbB_{12}. Overall features of the calculated band structure of YbB_{12} are quite similar to those shown in Fig.2(c). It is to be noticed that, in this case, branch A mixes strongly with $4f$ level at X-point, resulting in hybridization gap between the reconstructed conduction band and $4f$ level. Electron-hole symmetry is applicable to Figs.1(c) and 2(c). Here, it is worthwhile to mention about the definition of the hybridization gap more in detail. We show a conventional hybridization gap and an actual one which is realized in YbB_{12} in Fig.4(a) and Fig.4(b), respectively. It is to be noted that the hybridization gap is important only at around X-point. In other words, direct gap at another fixed **k** originates from a band gap. Minimum of energy separation between the $4f$ levels and the conduction band, which is important in the transport properties, may be dominated by this hybridization gap.

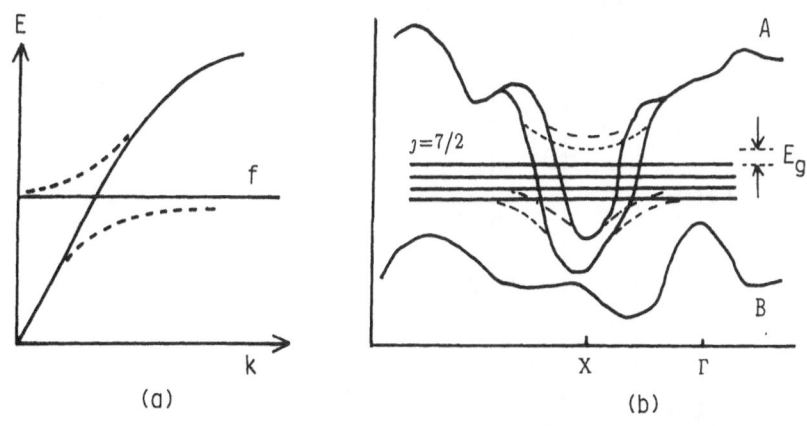

Fig.4. Conventional hybridization gap(a) and actual hybridization gap realized in YbB_{12}(b).

The observed energy gap of YbB_{12}, 90 K,[3] may correspond to this gap.

In the case of SmB_6, number of f-electron is five or six and then the band structure is rather complicated. However, the origin of the energy gap of SmB_6 may be the same as those of CeNiSn and YbB_{12}. As a representative example, we discuss mechanism of the appearance of the hybridization gap in SmB_6 more in detail. Calculated results of the band structure of SmB_6 by Yanase and Harima[15] are shown in Fig.5. In SmB_6, spin-orbit split $\jmath=5/2$ and 7/2 levels are located above the top of branch B. Moreover, the Fermi energy is located between spin-orbit split $\jmath=5/2$ and 7/2 levels. This situation is different from those of Ce- and Yb-based compounds. At the first glance, it looks like as if our prediction for Sm-based compounds that the band structure of reference system with rare-earth sites replaced by non-f R^{2+} ions has a gap at the Fermi level was not important in the case of SmB_6. The answer to this suspicion is shown in Fig.6. We consider two bands, P and Q, and only $\jmath=5/2$ levels for simplicity(Fig.6(a)). If band P and band Q have the same symmetry(Fig.6(b)), a crossing between band P and band Q does not occur, yielding branch a' and branch B with finite energy separation for any fixed **k**. Band calculation suggests that such a condition is realized in the electronic structure of RB_6. Then, the branch a' is mixed with one of the $4f$ levels, making a hybridization gap at around X-point. In this case, there exist six electrons above the bottom of branch a'. Then, E_F is located at the position shown in Fig.6(b), leading to semiconducting properties in SmB_6. On the other hand, in the case of different symmetry between band P and band Q, crossing between the two bands occurs, as shown in Fig.6(c). In such a case, band P mixes with one of the $4f$ level, while band Q does not. Then, a hybridization gap does not appear. Even if band Q can mix with another $4f$ level, a hybridization gap does not open. Therefore, for any fixed **k**, the condition that the energy of branch A exceeding the energy of branch B is necessary for non-f $R^{2+}B_6$ to be a semiconductor and also for the appearance of semiconducting properties in SmB_6. Here, it is to be noted that branch a' of Fig.6(b) corresponds to branch A' of Fig.5 which split from branch A by mixing with $4f$ level. We note in parentheses that there exists another hybridization gap in between branch A and spin-orbit split $\jmath=7/2$ level. The mechanism of the appearance of the latter hybridization gap is the same as that of the former. Furthermore, the latter hybridization gap is essentially the same as one in YbB_{12}.

SmB$_6$

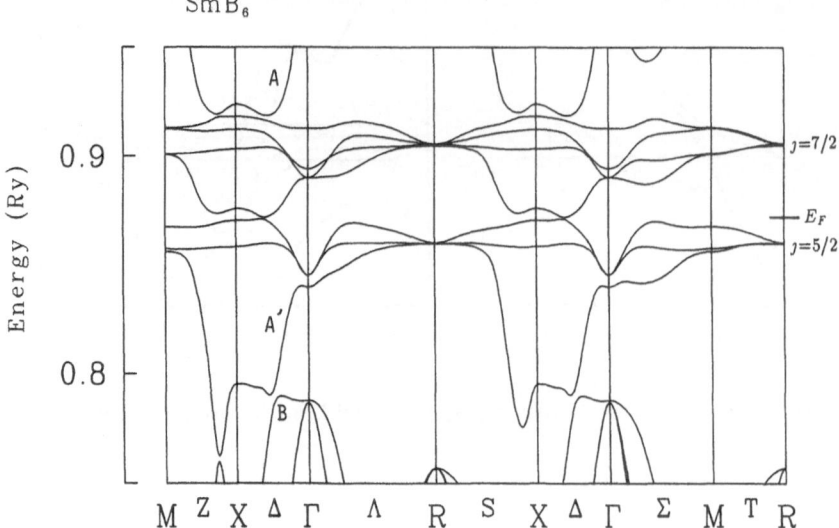

Fig.5. Calculated band structure of SmB$_6$(after ref.15). E_F means the Fermi energy.

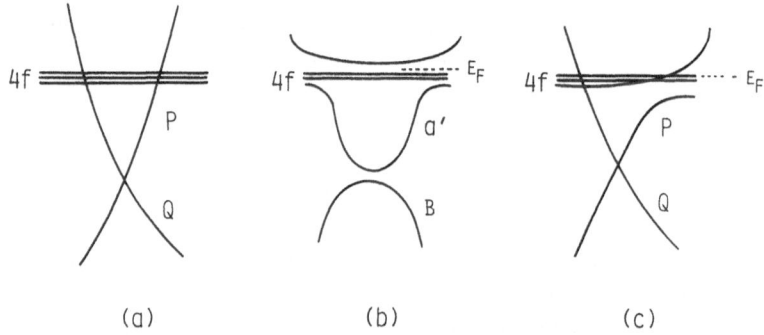

Fig.6. Schematic pictures for the explanation of the opening of hybridization gap in SmB$_6$(after ref.18). We consider two bands P, Q and $4f$ levels prior to mixing(a). If P and Q have the same symmetry, branch a' which is derived from mixing between P and Q is mixed with $4f$ level, resulting in the opening of hybridization gap(b). If P and Q have a different symmetry, P and Q can not mix with $4f$ level simultaneously. Then a hybridization gap does not appear(c).

Finally, it may be worthwhile to comment on the physical property of Ce$_3$Au$_3$Sb$_4$ and CePdSn. In the case of Ce$_3$Au$_3$Sb$_4$, band calculation of La$_3$Au$_3$Sb$_4$ showed that non-f R$_3^{3+}$Au$_3$Sb$_4$ can be semiconductors. Experimentally, Ce$_3$Au$_3$Sb$_4$ is also a semiconductor.[10] Therefore, Ce$_3$Au$_3$Sb$_4$ does not belong to the category we described above. Then band calculation of Ce$_3$Au$_3$Sb$_4$ can not probably explain the semiconducting property. In fact, valence of Ce ion in Ce$_3$Au$_3$Sb$_4$ is well-localized 3+ state and then Ce$_3$Au$_3$Sb$_4$ belongs to the Kondo regime. In such a case, framework of the electronic structure of Ce$_3$Au$_3$Sb$_4$ may be the same as that of La$_3$Au$_3$Sb$_4$, but a localized $4f^1$ level is located within a valence band in the former. Therefore, f electrons are not concerned with the origin of an energy gap. As for CePdSn, this compound is isoelectornic with

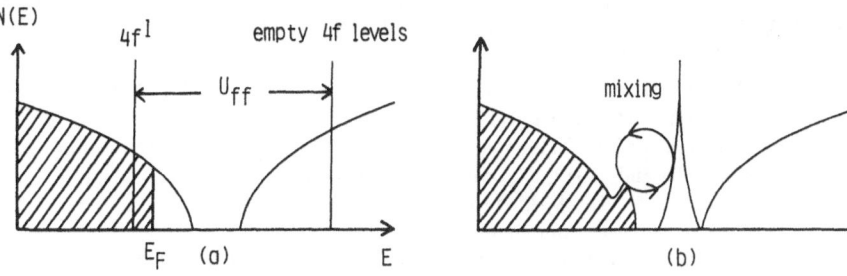

Fig.7. Schematic electronic structures of CePdSn(a) and CeNiSn(b).

CeNiSn. Therefore, CePdSn could be a semiconductor if Ce ion is in a mixed-valent regime and band calculation can create an energy gap at the Fermi level in principle. However, actual CePdSn shows a typical metallic behavior and valence of Ce ion is a well-localized 3+ state.[20] Then, CePdSn belongs to metallic Kondo regime. This may be due to weak mixing between Ce $4f$ and Pd $4d$ electrons. In such a case, correlation energy U_{ff}, which is not taken into account in the band calculation, becomes important. Schematic electronic structures of CePdSn and CeNiSn are shown in Fig.7.[11]

In conclusion, our previous prediction that the common features observed in non-f reference systems of mixed-valent semiconducting compounds play an important role in the origin of an energy gap is supported by recent band calculations. We suggest that the common features we predicted can be a powerful guide to look for new semiconducting mixed-valent rare-earth compounds. For example, mixed-valent Eu compound can be a semiconductor, if band structure of compounds with rare-earth site replaced by non-f 3+ ions has a gap at the Fermi energy. It is surprising that electronic structures of not only metallic mixed-valent compounds such as CeNi and CeSn$_3$[ref.21,22] but also of semiconducting compounds in mixed-valent regime can be explained by the band calculation.

ACKNOWLEDGMENTS

The author gratefully acknowledges Professors A. Yanase, K. Takegahara and Dr. H. Harima for useful comments and communicating their results prior to publication.

REFERENCES

1. J. C. Nickerson, R. M. White, K. N. Lee, R. Bachmann, T. H. Geballe, and G. W. Hull, Jr., Physical properties of SmB$_6$, *Phys. Rev.* B3:2030 (1971).
2. A. Jayaraman, P. Dernier, and L. D. Longinotti, Study of the valence transition in SmS induced by alloying, temperature, and pressure, *Phys. Rev.* B11:2783 (1975).
3. M. Kasaya, F. Iga, M. Takigawa, and T. Kasuya, Mixed valence properties of YbB$_{12}$, *J. Magn. Magn. Mat.* 47&48:429 (1985).
4. P. Wachter and G. Travaglini, Intermediate Valence and the hybridization model: a study on SmB$_6$,"gold" SmS and YbB$_{12}$, *J. Magn. Magn. Mat.* 47&48:423 (1985).
5. T. Takabatake, F. Teshima, H. Fujii, S. Nishigori, T. Suzuki, T. Fujita, Y. Yamaguchi, J. Sakurai and D. Jaccard, Formation of an anisotropic energy gap in the valence-fluctuating system CeNiSn, *Phys. Rev.* B41:9607 (1990).
6. M. F. Hundley, P. C. Canfield, J. D. Thompson, Z. Fisk, and J. M. Lawrence, Hybridization gap in Ce$_3$Bi$_4$Pt$_3$, *Phys. Rev.* B42:6842 (1990).
7. N. F. Mott, Rare-earth compounds with mixed valencies, *Phil. Mag.* 30:403 (1974).

8. T. Kasuya, K. Takegahara, T. Fujita, T. Tanaka, and E. Bannai, Valence fluctuation state in SmB_6, *J. Phys.* (Paris) 40:C5-308 (1979).

9. R. M. Martin, Fermi-surface sum rule and its consequences for periodic Kondo and mixed-valence systems, *Phys. Rev. Lett.* 48:362 (1982).

10. M. Kasaya, K. Katoh, and K. Takegahara, Semiconducting properties of the isomorphous compounds, $Ce_3Au_3Sb_4$ and $Ce_3Pt_3Sb_4$, *Solid State Commun.* 78:797 (1991).

11. M. Kasaya, H. Suzuki, K. Katoh, M. Inoue, and T. Yamaguchi, Structural and magnetic properties of ternary rare earth compounds RPt(Au)Sb and RPd(Ni)Sn, Special issue of *Jpn. J. Appl. Phys.* as "Physical Properties of Actinide and Rare Earth Compounds" (in press).

12. A. Hasegawa and A. Yanase, Electronic structure of CaB_6, *J. Phys.* C12:5431 (1979).

13. H. Harima, A. Yanase, and T. Kasuya, Energy band structure of YB_{12} and LuB_{12}, *J. Magn. Magn. Mat.* 47&48:567 (1985).

14. K. Takegahara, Y. Kaneta, and T. Kasuya, Electronic band structure of $Th_3Ni_3Sb_4$ and Th_3X_4(X =P,As,Sb), *J. Phys. Soc. Jpn.* 59:4394 (1990).

15. A. Yanase and H. Harima, Band calculations on YbB_{12}, SmB_6 and CeNiSn, *Prog. Theor. Phys. Suppl.* No.108:19 (1992).

16. H. Harima: private communication.

17. K. Takegahara, H. Harima, Y. Kaneta, and A. Yanase, Origin of gap formation in $Ce_3Pt_3Sb_4$ and $Ce_3Pt_3Bi_4$, in this volume.

18. M. Kasaya, On the origin of an energy gap in semiconducting mixed-valent rare-earth compounds, *J. Phys. Soc. Jpn.* 61:3841 (1992).

19. A. Hasegawa and A. Yanase, Energy bandstructure and Fermi surface of LaB_6 by a self-consistent APW method, *J. Phys.* F7:1245 (1977).

20. M. Kasaya, T. Tani, H. Suzuki, K. Ohoyama, and M. Kohgi, Crossover from magnetic to non-magnetic ground state in the Kondo alloy system $Ce(Ni_{1-x}Pd_x)Sn$, *J. Phys. Soc. Jpn.* 60:2542 (1991).

21. A. Hasegawa, H. Yamagami, and H. Johbettoh, Electronic Structure of $CeSn_3$, *J. Phys. Soc. Jpn.* 59:2457 (1990).

22. I. Umehara, Y. Kurosawa, N. Nagai, M. Kikuchi, K. Satoh, and Y. Ônuki, High field magnetoresistance and de Haas-van Alphen effect in $CeSn_3$, *J. Phys. Soc. Jpn.* 59:2848 (1990).

THERMODYNAMICS AND TRANSPORT IN Ce$_3$Bi$_4$Pt$_3$ AND RELATED MATERIALS

J.D. Thompson[1], W.P. Beyermann[1,2], P.C. Canfield[1], Z. Fisk[1], M.F. Hundley[1], G.H. Kwei[1], R.S. Kwok[1,3], A. Lacerda[1], J.M. Lawrence[4], and A. Severing[5]

[1]Los Alamos National Laboratory, Los Alamos, NM 87545
[2]Department of Physics, University of California, Riverside, CA 92521
[3]Hughes Aircraft, Los Angeles, CA 90009
[4]Department of Physics, University of California, Irvine, CA 92717
[5]Institut Laue-Langevin, 38042 Grenoble, France

INTRODUCTION

The interplay between electronic and magnetic correlations in certain 4f and 5f intermetallic compounds has been shown[1] to lead to novel ground state properties, including strongly renormalized effective electron masses, homogeneous mixed valence of the f-configuration and unconventional superconductivity. Whereas the vast majority of these compounds are metallic at low temperatures, a few examples, such as SmB$_6$ (Ref. 2), SmS (Ref. 3), and YbB$_{12}$ (Ref. 4), have been known for some time to be small-gap semiconductors, with gaps of order 100K. Within the past two or three years several new examples have been discovered, notably in Ce[5-8] and U[9] compounds, suggesting that small-gap semiconductors may be a general consequence of strongly interacting electrons. With very few exceptions, the crystal structures of these materials are cubic which, as will be argued, is favorable for the appearance of a small gap in the electronic spectrum. Further common features are temperature variations in the cubic lattice parameter, in magnetic susceptibility and in L-edge absorption spectra that are consistent with an unstable f-configuration arising from hybridization between f and ligand electrons. The extent to which the physics of these small-gap semiconductors is similar to metals with otherwise similar properties remains an outstanding question but their analogous behavior to Kondo-lattice metals has led to the terminology Kondo insulators.[10,11] In the following we review transport and thermodynamic properties of the small-gap semiconductor Ce$_3$Bi$_4$Pt$_3$ and discuss them in relation to what is known about other Kondo insulators and metals.

Transport and Thermal Properties of f-Electron Systems
Edited by G. Oomi *et al.*, Plenum Press, New York, 1993

RESULTS

Ce3Bi4Pt3

Figure 1 compares the temperature-dependent resistivity $\rho(T)$ of single crystal Ce3Bi4Pt3, La3Bi4Pt3 and Pr3Bi4Pt3. Whereas $\rho(T)$ for the isostructural La and Pr analogues is typical of dirty intermetallic compounds, that of the Ce compound increases strongly with decreasing temperature. An Ahrrenius plot of the Ce3Bi4Pt3 data shows activated behavior above approximately 50K, with an activation energy Δ=50K or gap energy $E_g = 2\Delta$ of about 100K, if the Fermi level lies in the center of the gap.[*] Below 50K, $\rho(T)$ increases less rapidly than exponentially, with the deviation from activated behavior depending on sample quality. For crystals prepared with higher purity (Ames Laboratory) Ce, the Ahrrenius plot remains activated to lower temperatures but the magnitude of Δ is relatively insensitive to these effects, suggesting that small amounts (of order 0.1%) of impurities introduce electronic states in the gapped region. A crude measure of sample quality, therefore, is the resistivity ratio $\rho(2K)/\rho(300K)$ which approaches 1000 in the "best" crystals. Even in these, one should consider the low-temperature transport to be influenced by extrinsic effects.

Similar conclusions are drawn from Hall-coefficient R_H and thermoelectric power measurements[12] which show above about 50K temperature dependences characteristic of an electronic gap E_g=100K, but at lower temperatures both are dominated by parallel conduction channels arising from impurity states in the gap. Above the extrinsic carrier-dominated regime the Hall mobility,[13] $\mu = R_H/\rho$, saturates to a small, temperature-independent value of 10 cm^2/V-sec, suggesting strong scattering of carriers with enhanced effective mass.

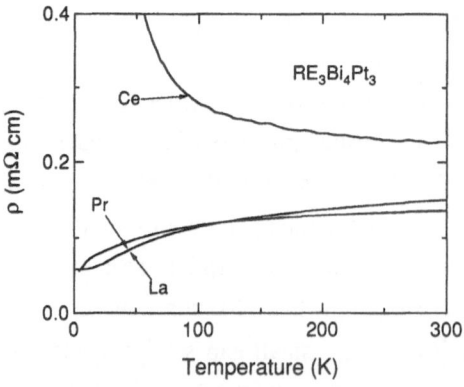

Fig. 1.   Resistivity, plotted on a linear scale, as a function of temperature for single crystals of Ce3Bi4Pt3, La3Bi4Pt3 and Pr3Bi4Pt3. Above approximately 350K, the resistivity of Ce3Bi4Pt3 begins to increase at a rate $\partial\rho/\partial T$ comparable to what is seen in La and Pr analogues below 300K. Measurements to 40 mK show no evidence for a phase transition in Ce3Bi4Pt3.

As to be expected for a small-gap semiconductor, the electronic contribution to the specific heat[6] of $Ce_3Bi_4Pt_3$ is small, $\gamma = 3mJ/mole \cdot Ce \cdot K^2$, a value about one-third that of $La_3Bi_4Pt_3$. That a measurable contribution is found at all reflects the presence of extrinsic carriers. Application of a 10T-magnetic field has no effect on the specific heat in the temperature interval $1.5 < T < 20K$ of the measurements. This result contrasts to observations[14] on the orthorhombic, small-gap ($\Delta = 5K$) semiconductor CeNiSn in which C/T at low temperatures increases by about a factor of two when an 8T field is applied along the a-axis. This could arise because the ratio of field-strength to gap-energy differs substantially in these two experiments and/or from the anisotropic nature of the energy gap in CeNiSn. Magnetoresistance behavior in these two materials will be discussed below.

Evidence for mixed-valence nature of the 4f configuration is found in the magnetic susceptibility χ and thermal expansion of $Ce_3Bi_4Pt_3$. Figure 2 shows χ versus temperature for a $Ce_3Bi_4Pt_3$ crystal produced from Ames Laboratory Ce. The temperature dependence is Curie-Weiss above ~100K, with an effective moment $\mu_{eff}=2.42\mu_B/Ce$ and a paramagnetic $\Theta_p=-125K$. The broad peak in χ, centered at $T_\chi=80K$, is characteristic of metallic mixed-valence compounds having a Kondo temperature $T_K \approx (3-4)T_\chi=240-320K$ (Ref. 6). In contrast to earlier measurements[6] on samples prepared with less-pure (99.9%) Ce that showed a pronounced "Curie-tail" at low temperatures, the data of Fig. 2 approach a constant value $\chi(0) \approx 2.3 \times 10^{-3}$ emu/moleCe. This large $\chi(0)$ most definitely is not Pauli-like and possibly arises from an unquenched orbital contribution of the 4f moment that gives a Van Vleck-type susceptibility with characteristic energy on the order of Δ.

Inelastic neutron scattering measurements[15] on powdered single crystals reproduce the temperature dependence of the static susceptibility, as also shown in Fig. 2. What is

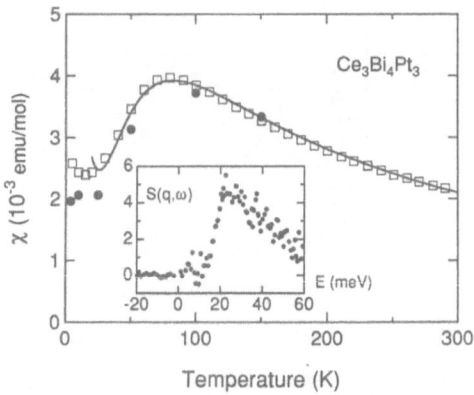

Fig. 2. Magnetic susceptibility χ versus temperature for $Ce_3Bi_4Pt_3$ measured in a field of 1T. Open circles denote χ obtained from the neutron scattering function $S(Q,\omega)$. The solid line is $\chi(T)$ calculated as described in the text. Inset shows the magnetic contribution to $S(Q,\omega)$ versus energy transfer at T=2K for an incident neutron energy of 69 meV. A gap in the spin-spin correlation function of 12meV is apparent.

not apparent in Fig. 2 but is revealed by neutron scattering is that below approximately 50K a well-defined gap in the spin-spin correlation function develops. (See Fig. 2 inset.) At 2K, this gap is 12 meV (140K), a value comparable to E_g inferred from transport measurements and below which magnetic intensity is zero. The gap remains well-defined to 25K but above 50K magnetic scattering at small energy transfer develops rapidly with increasing T. By 150K, the magnetic scattering is quasi-elastic like, consistent with a T_K inferred from susceptibility measurements. Similar conclusions are drawn from NQR studies.[16] Evidence for a spin-gap has also been reported in the small-gap semiconductor CeNiSn[17] and in the mixed-valent metal YbAl$_3$ (Ref. 18).

Temperature variations in the cubic lattice parameters a_0 of Ce$_3$Bi$_4$Pt$_3$ and La$_3$Bi$_4$Pt$_3$ are given in Fig. 3(a). An anomalous decrease in a_0 is seen below 100K for Ce$_3$Bi$_4$Pt$_3$, which for metallic Ce compounds would be argued as an indication of admixed $4f^0$ and $4f^1$ configurations. (Preliminary L_{III}-edge x-ray absorption experiments on Ce$_3$Bi$_4$Pt$_3$ indicate an f-occupancy $n_f \approx 0.9$).[19] The 4f-contribution to the volume-thermal expansion

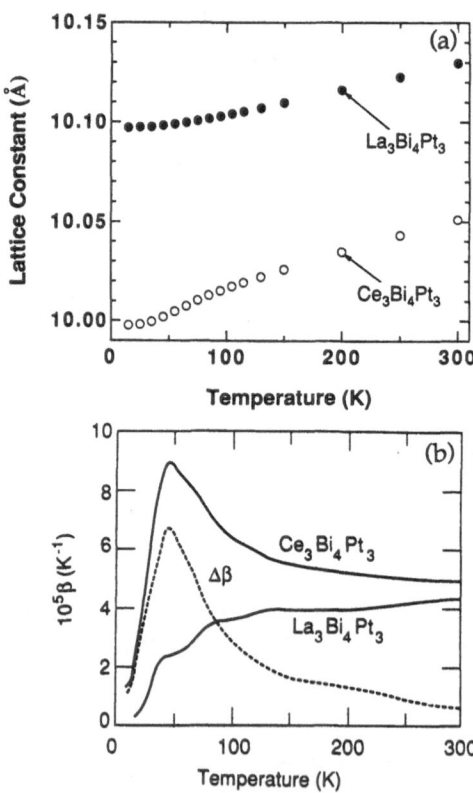

Fig. 3. (a) Cubic lattice parameter, determined by Rietveld analysis of neutron-diffraction spectra, of Ce$_3$Bi$_4$Pt$_3$ and La$_3$Bi$_4$Pt$_3$ as a function of temperature. (b) Volume-thermal expansion coefficient for both compounds calculated from the data in (a) and their difference $\Delta\beta(T)$, which gives the 4f-derived contribution to the thermal expansion.

is displayed more obviously in Fig. 3(b) where the volume thermal expansion coefficient β = $3\partial(\ln a_0)/\partial T$ of both compounds and their difference $\Delta\beta = \beta_{Ce}-\beta_{La}$ are plotted as a function of temperature. The difference $\Delta\beta$ peaks at $T_{\Delta\beta} \approx 50K$. By Maxwell's relations, then, the pressure derivative of the 4f entropy also peaks at $T_{\Delta\beta}$. Similar measurements[20] performed with the samples subjected to an applied pressure of 17.7 kbar give $T_{\Delta\beta} \approx 85K$ and, from a comparison of the ambient and high pressure data for $Ce_3Bi_4Pt_3$, a $T\rightarrow0$ bulk modulus of about 950 kbar. An analysis of these results yields a Gruneisen parameter Ω =36 comparable to that expected of a metallic-mixed valent compound with a T_K of 200-300K.

A remarkable observation[13] is that the difference in lattice parameters Δa_0 between that of $Ce_3Bi_4Pt_3$ and $La_3Bi_4Pt_3$ is linear in the product χT, as shown in Fig. 4(a). In the absence of interactions, χT is just the volume density of moments contributing to the susceptibility. A linear fit to these data gives $\chi T=-27.31\Delta a_0+2.76$ (emu·K/mole·Ce). A second interesting observation[+] is that, as shown in Fig. 4(b), the function $1/(1+\exp(\Delta/T))$ linearizes $\Delta a_0(T)$ to a good approximation. More precisely $\Delta a_0=-0.529/(1+\exp(120/T))+.099$ (Å). We note that the numerator in the exponential is very close in

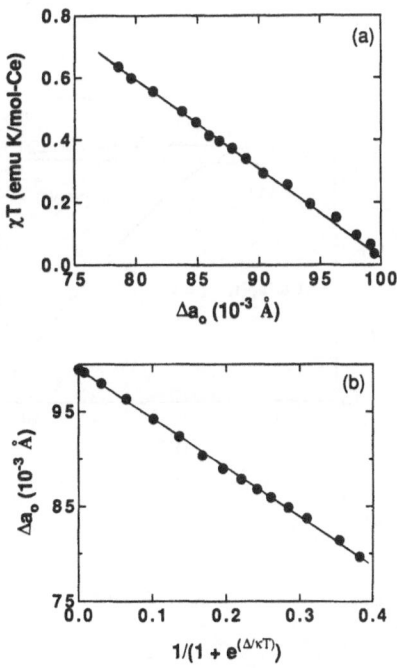

Fig. 4.   (a) Product of static susceptibility times temperature versus the lattice-parameter difference $\Delta a_0 = a_0(La_3Bi_4Pt_3)-a_0(Ce_3Bi_4Pt_3)$, with temperature as the implicit variable. (b) Δa_0 versus $1/[1+\exp(\Delta/k_BT)]$. A linear relationship is found for $\Delta/k_B = 120 \pm 5K$.

magnitude to the value of the gap in the spin-spin correlation function measured at low temperatures by inelastic neutron scattering[15] and to the paramagnetic Θp found from fitting a Curie-Weiss-form to the high-temperature susceptibility (Fig. 2). Together these two relationships allow the temperature-dependent susceptibility to be calculated directly, the results of which are shown in Fig. 2 as the solid line. The relatively good agreement between measured and calculated values of χ above ~30K confirms consistency in the parameterizations of $\Delta a_0(T)$; but perhaps more significant is that when $\chi(T)$ is viewed this way no Curie-Weiss Θp, ie., no interactions in the conventional sense, is required to understand the temperature variations in χ at low T.[21]

To probe the ground state of $Ce_3Bi_4Pt_3$ in more detail, the response of the electrical resistivity to applied magnetic fields and pressures has been measured. Fig. 5(a) shows the magnetoresistance $\Delta\rho/\rho=[\rho(H)-\rho(H=0)]/\rho(H=0)$ of $Ce_3Bi_4Pt_3$ at selected temperatures. At low temperatures, $\Delta\rho/\rho$ is strongly negative at high fields but exhibits a weak positive contribution at low fields. (Preliminary measurements[22] at 4K in fields to 50T find an approximately two-order-of-magnitude decrease in the resistivity, so that the sample is nearly metallized by a field comparable to Δ. Similar results have been reported[23] for CeNiSn with H=11T parallel to the orthorhombic a-axis.) With increasing temperature,

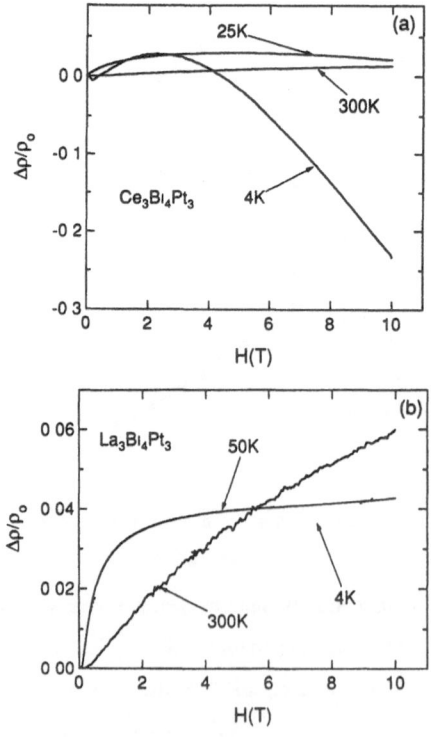

Fig 5   Normalized magnetoresistance $\Delta\rho=[\rho(H)-\rho(H=0)]/\rho(H=0)$ as a function of magnetic field for (a) $Ce_3Bi_4Pt_3$ and (b) $La_3Bi_4Pt_3$ at selected temperatures

the positive contribution dominates at all fields below 10T. An unusually large, positive magnetoresistance is also found in the non-magnetic analogue La$_3$Bi$_4$Pt$_3$ (Fig. 5(b)) that saturates at progressively lower fields with decreasing temperature and can be scaled to fit the positive $\Delta\rho/\rho$ in Ce$_3$Bi$_4$Pt$_3$ (Ref. 24). Given the low mobility in Ce$_3$Bi$_4$Pt$_3$ and comparable values of room-temperature resistivity in the Ce and La compounds, we believe this field and temperature dependence reflects the condition $\omega_c\tau \approx 1$ at about 1T, where ω_c is the cyclotron frequency. Comparison of Figs. 5(a) and (b) then suggests that the room-temperature band structures of both materials are similar and that the 4f interaction with the common underlying electronics results in the appearance of the gap in Ce$_3$Bi$_4$Pt$_3$ and its large negative $\Delta\rho/\rho$ at low temperatures.

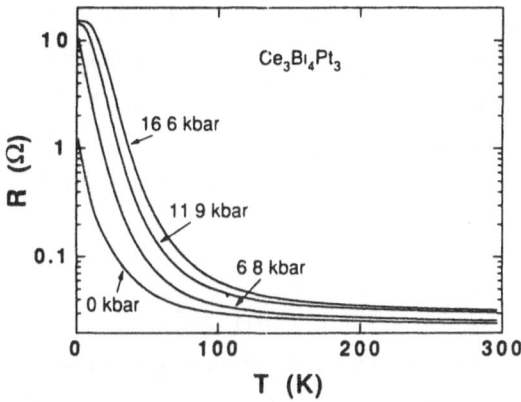

Fig. 6.   Resistance as a function of temperature for Ce3Bi4Pt3 subject to various applied hydrostatic pressures.

From the linear relationship found in Fig. 4(b) and the observation[20] that at low temperatures Δa_0 increases with pressure, it is straightforward to show that $\partial\Delta/\partial P > 0$, if Δ in Fig. 4(b) is associated with the spin or charge gap. That is, decreasing the cell volume, which favors stronger admixture of the $4f^0$ configuration, should enhance the gap. Figure 6 gives resistance versus temperature measurements for Ce$_3$Bi$_4$Pt$_3$ at pressures to over 16 kbar. These experiments were performed on an early sample in which $\rho(2K)/\rho$ (300K) was only 30. In spite of this, the data clearly show trends consistent with $\partial\Delta/\partial P > 0$. At the highest pressures, the low-temperature resistance saturates, most likely reflecting parallel conduction by impurity states in the gap. Depending on details of the analysis, we find that $\partial ln\Delta/\partial P$ ranges from 0.05 to 0.16 kbar^{-1} consistent with the Gruneisen interpretation of the pressure-dependent thermal expansion.[20] This result contrasts to observations[25,26] on SmB$_6$ and YbB$_{12}$ in which pressure suppresses the electronic gap. In the case of SmB$_6$, $\partial ln\Delta/\partial P$ varies from -0.02 to about -0.03 kbar^{-1} depending on the sample. (The sign difference between Ce$_3$Bi$_4$Pt$_3$ and SmB$_6$ in their logarithmic derivatives of Δ are reflected as well in sign differences in their 4f-derived thermal expansion.)[20,27]

At pressures of 55 to 70 kbar the electronic gap in SmB$_6$ is closed and the temperature-dependent resistance becomes that of a typical Kondo-lattice metal.[10,25] X-ray diffraction at room temperature indicates[28] the valence of Sm changes from 2.8 at P=0 to 2.9 at 60 kbar. Thus, with applied pressure the magnetic 4f^5 configuration of Sm is favored over the 4f^6 (J=0), leading to a decrease in charge hybridization and an approach to the Kondo-limit. Similar arguments apply to YbB$_{12}$. On the other hand, the 4f^0 configuration in Ce is favored at high pressure and we expect a more strongly mixed-valent, less magnetic ground state. In these materials, then, Δ tracks the expected change[29] with pressure in charge/spin hybridization and not band filling because in all cases the 4f^{n-1} configuration is stabilized relative to the 4f^n with decreasing volume. The small-gap semiconductor CeNiSn does not follow the expected response to pressure; instead of $\partial\Delta/\partial P$ being positive, the gap closes at a rate $\partial ln\Delta/\partial P \approx - 0.03$ kbar^{-1} and extrapolates to $\Delta=0$ at a critical pressure of about 30 kbar.[30] A possible interpretation is that the anisotropic gap[31], not observed in these measurements on a polycrystalline sample[30], is shunted by conduction through non-gapped regions of the Fermi surface. Pressure studies on single crystals of CeNiSn should prove valuable in resolving this possibility.

Substitution Studies

Lutetium substitutions[32] for Yb in YbB$_{12}$ and La substitutions[33] for Sm in SmB$_6$ rapidly metallize these Kondo insulators which have gaps comparable[2,4,6] to Ce$_3$Bi$_4$Pt$_3$. Because Lu is smaller than Yb but La is larger than Sm, this immediately suggests that the primary role of the dopant is not to suppress the gap by chemical pressure. However, in both cases the 4f-sublattice periodicity is broken and nonmagnetic, trivalent atoms replace those with some divalent character.

In the case of (Ce$_{1-x}$La$_x$)$_3$Bi$_4$Pt$_3$, La substitutions also induce a metallic, Kondo-like state, shift the maximum in χ to lower temperatures and produce a contribution to the electronic specific heat γ that is consistent with a Kondo temperature of about 300K.[6] Previously we have established[12] that, for La concentrations above about x=0.20, $\gamma \propto 1/T\chi$. Experimentally, the proportionality agrees quantitatively with the prediction from a Bethe ansatz solution of the Coqblin-Schrieffer model. Figure 7 shows that within uncertainties in absolute values of x the electronic specific heat per mole Ce is also proportional to $\sqrt{x}$ over the range studied $0.015 \leq x \leq 0.5$. Such a relationship has been predicted[34] recently to arise from an impurity band of Kondo holes produced by breaking translational invariance of the Ce sublattice through non-magnetic substitutions. (We note that in Ce$_3$Bi$_4$(Pt$_{1-x}$Au$_x$)$_3$, gold substitutions for Pt also metallize the compound but in this case $\gamma \propto x$ for x=0.1 and 0.2.)

Similar studies have not been performed for other rare-earth (RE) substitutions. However, resistivity, susceptibility and specific heat have been measured[35] on a series of (Ce$_{.985}$RE$_{.015}$)$_3$Bi$_4$Pt$_3$ crystals. Detailed analysis of the specific heat is complicated by a low-temperature upturn in C/T that scales approximately with the spin of the rare-earth

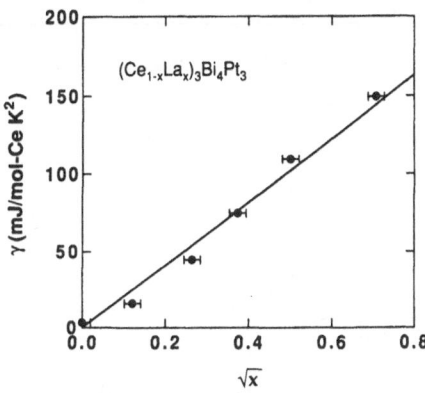

Fig. 7.   Linear contribution to the specific heat $\gamma \equiv C/T$ at low temperature for $(Ce_{1-x}La_x)_3Bi_4Pt_3$ versus the square root of La concentration x.

dopant. However, a linear extrapolation of C/T vs T^2 from above the upturn to T=0 gives $C/T|_{T=0}=16\pm6$mJ/mole·Ce·K^2 for all seven rare-earth dopants. This result argues again that chemical pressure is not a dominant effect and that the increase in C/T for small x is independent of the magnetic character of the substituted element as expected[36] in a Kondo-hole interpretation, provided the low-temperature tail in C/T is not intrinsic to the correlated-electron ground state.   This last point has not been resolved and deserves further attention, particularly in light of resistivity and susceptibility measurements that exhibit non-monotonic trends with RE substituents. Most pronounced is the resistivity that tends toward metallization for rare-earths lighter than Ho and toward more strongly semiconducting behavior for heavier rare earths.

Although arguments have been made that chemical pressure is not a dominant effect in substituted alloys, pressure does play a role. Shown in Fig. 8 is the resistance of $(Ce_{.985}La_{.015})_3Bi_4Pt_3$ at applied pressures to 18 kbar. The temperature-dependent resistance at ambient pressure is typical of $(Ce_{.985}RE_{.015})_3Bi_4Pt_3$ compounds for rare-earths lighter than Ho and of mixed-valent metals, eg. $CePd_3$, in which the Ce sublattice periodicity has been broken by non-magnetic or magnetic substitutions.[36] With increasing pressure (increasing hybridization) there is a clear trend for the resistance to approach that of the undoped compound.   A linear extrapolation between the lattice parameters of $La_3Bi_4Pt_3$ and $Ce_3Bi_4Pt_3$, combined with a bulk modulus of 950 kbar, allows an estimate of -3 kbar for the negative chemical pressure induced by substituting the larger La atom for Ce.   Although local chemical pressure around the dopant must be larger than this estimate, the data of Fig. 8 suggest an applied pressure of 30 to 40 kbar would be required to reproduce $\rho(T)$ of undoped $Ce_3Bi_4Pt_3$. Comparison of data for P=12.4 and 18.0 kbar shows a qualitative change in the low-temperature dependence of $\rho(T)$ that could be associated with a metal-insulator transition in the Kondo-hole band that is formed by La substitutions. Higher pressure experiments on this material and on more heavily doped compounds would be helpful in clarifying this possibility.

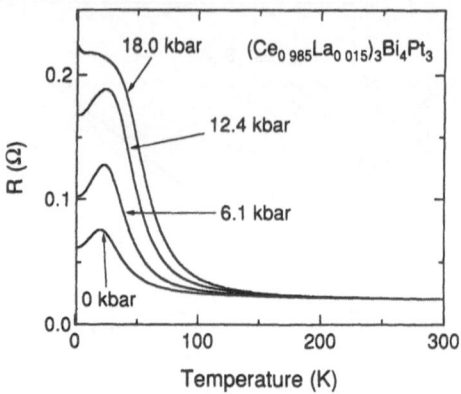

Fig. 8.   Resistance as a function of temperature for $(Ce_{.985}La_{.015})_3Bi_4Pt_3$ under various applied hydrostatic pressures.

Attempts to replace Ce with tetravalent ions, such as Zr and Th, have been unsuccessful in $Ce_3Bi_4Pt_3$. However, a pronounced asymmetry has been observed[37] in tri- versus tetravalent doping in the Kondo insulators, orthorhombic CeRbSb and cubic $U_3Sb_4Pt_3$. Substituting 10% La in CeRbSb destroys the small, ~7K, gap; whereas, the same amount of Zr makes the sample more resistive at all temperatures below 300K. Three percent Y or Lu in $U_3Sb_4Pt_3$ decreases the low-temperature resistivity and enhances the electronic specific heat by about one order of magnitude. On the other hand up to 10% Th addition causes no change in γ.

DISCUSSION

The body of data presented for $Ce_3Bi_4Pt_3$ and existing in the literature for related small-gap semiconductors is consistent with their being Kondo-like metals at temperatures $T \geq E_g/k_B$. Purely trivalent, isostructural analogues are normal metals at all temperatures; whereas, isostructural compounds formed with an element having the anomalous valence of the rare-earth it replaces, e.g., tetravalent elements replacing Ce or divalent elements replacing Sm or Yb, often are conventional semiconductors.[13,21] An example is EuB_6, in which Eu is divalent and which has a large gap, but SmB_6, in which Sm is nearly trivalent, has a small semiconducting gap. The existence of these isostructural conventional semiconductors indicates that the band structure near the Fermi level in Kondo insulators is relatively simple. We believe this is why most Kondo insulators form in cubic crystal structures, ie., very loosely, simple crystal and band structures are mutually compatible.

The modestly large electronic specific heat that develops when the Kondo insulator is metallized by doping[12,38] or by pressure[3], quasi-elastic neutron scattering at high temperatures that evolves into a well-defined gap in the spin-spin correlation function at low temperature[15], and the inter-relationship between 4f-derived thermal expansion, static susceptibility and the spin gap are incontrovertible evidence that the charge-excitation gap originates from spin/change hybridization between the 4f electron and the underlying s,p,d band structure. In mixed-valent metals this behavior is believed to be described by the Anderson Lattice Hamiltonian. In this model many-body interactions renormalize the bare f level to be degenerate with the conduction band, allowing hybridization between conduction electrons and the renormalized f level. Generally there will be more than one conduction band crossing the Fermi level and in this case it is easy to argue that the resulting hybridized-band structure will produce a metallic ground state.[21] However, if there is only a single conduction band cutting E_F and the electron count, which includes the strongly interacting f-electrons, is exactly two, the lower-hybridized band will be filled and the upper band empty. Under these conditions, a mean-field treatment of the Anderson Lattice Hamiltonian predicts[39,40] an indirect gap for excitations from the zone center to zone boundary proportional to $(1-n_f)V^2D$, where n_f is the f occupation, V is the hybridization matrix element and D is the conduction-band density of states. This interpretation** has a number of interesting consequences: at temperatures greater than E_g/k_B, the physics of Kondo insulators and metals is identical; the low-temperature transport and magnetic gaps have a common origin and should be of comparable magnitude, as experimentally observed; the existence of isostructural compounds having a conventional semiconducting gap, ie., not induced by electronic correlations, provides experimental proof that the Fermi- surface volume, in these cases zero, is independent of the Coulomb repulsion U, as expected from Luttinger's theorem[21,39]; an asymmetry in the thermodynamics is expected[39] between electron and hole doping, as observed, because Coulomb interactions forbid doping by more than one electron per f ion but there is no such restriction for hole doping; and the temperature dependence of the static susceptibility below T_χ arises from intraband processes allowed by thermal population of holes at finite T, i.e., interactions, characterized by a Curie-Weiss Θ_p are not required to explain $\chi(T)$ at low T.

CONCLUSIONS

Kondo insulators appear to be an unusually simple realization of the Anderson Lattice Hamiltonian in which the lower-hybridization band is exactly filled, or in Kondo language, the Abrikosov-Suhl resonance exactly fills a Brillouin zone. As such, this class of materials offers the possibility of detailed comparison between theory and experiment and the hope of a more complete understanding of both strongly-correlated insulators and metals.

Acknowledgments

We thank G. Aeppli, P.S. Riseborough and P. Schlottmann for helpful discussions. Work at Los Alamos was performed under the auspices of the U.S. Department of Energy.

Footnotes

* Because the temperature range over which the activation energy is evaluated is larger than or comparable to Δ, Δ and E_g may be underestimated by ~20%.

\+ We thank G. Aeppli for this suggestion.

** Strictly, the model is for a doubly degenerate ground state. Inelastic neutron scattering (Ref. 15) finds no evidence for crystal-field splitting of the J=5/2 manifold in $Ce_3Bi_4Pt_3$. In spite of this, the calculation should reflect qualitatively the essential physics at larger degeneracy.

REFERENCES

1. See, for example, N. Grewe and F. Steglich, Heavy Fermions, *in*: "Handbook on the Physics and Chemistry of Rare Earths", Vol. 14, K. A. Gschneidner and L. Eyring, ed. Elsevier Science Publishers, Amsterdam (1991).

2. S. von Molnar et al., Study of the energy gap in single crystal SmB_6, *in*: "Valence Instabilities," P. Wachter and H. Boppart, ed. North-Holland, Amsterdam (1982); A. Menth, E. Buehler and T. H. Geballe, Magnetic and semiconducting properties of SmB_6, Phys. Rev. Lett. 22: 295 (1969).

3. D. Bader, N. E. Phillips and D. B. McWhan, Heat capacity and resistivity of metallic SmS at high pressure, Phys. Rev. B 7: 4686 (1973).

4. M. Kasaya et al., Mixed valence properties of YbB_{12}, J. Magn. Magn. Mat. 47 & 48: 429 (1985).

5. T. Takabatake, Y. Nakazawa and M. Ishikawa, Gap formation in the valence fluctuation system CeNiSn, Jpn. J. Appl. Phys. 26, Suppl. 26-3: 547 (1987); F. G. Aliev et al., Transport and magnetic properties of intermetallic systems RNiM(R=U,Ce,Er,Ho,Tm,Yb,Sc,Ti,Zr,Hf; M=Sn,Sb), J. Magn. Magn. Mat. 76 & 77: 295 (1988).

6. M. F. Hundley et al., Hybridization gap in $Ce_3Bi_4Pt_3$, Phys. Rev. B 42: 6842 (1990).

7. K. Malik and D. T. Adroja, Evidence of pseudogap formation in a new valence-fluctuating compound: CeRhSb, Phys. Rev. B 43: 6267 (1991).

8. M. Kasaya, K. Katoh and K. Takegaraha, Semiconducting properties of the isomorphous compounds $Ce_3Au_3Sb_4$ and $Ce_3Pt_3Sb_4$, Solid State Commun. 78: 797 (1991).

9. T. Takabatake et al., Heavy-fermion and semiconducting properties of the ternary uranium compounds $U_3T_3Sn_4$ and $U_3T_3Sb_4$ (T=Ni,Cu,Pd,Pt and Au), J. Phys. Soc. Jpn. 59: 4412 (1990).

10. V. Moshchalkov et al., SmB_6 at high pressures: the transition from insulating to the metallic Kondo lattice, J. Magn. Magn. Mat. 47 & 48: 289 (1985).

11. A. J. Millis, Heavy electron metals and insulators, *in*: Physical Phenomena at High magnetic Fields, E. Manousakis, P. Schlottmann, P. Kumar, K. Bedell and F. M. Mueller, ed., Addison-Wesley, Redwood (1991).

12. M. F. Hundley et al., Evidence for a coherence gap in $Ce_3Bi_4Pt_3$, Physica B 171: 254 (1991).

13. Z. Fisk et al., $Ce_3Bi_4Pt_3$ and hybridization gap physics, J. Alloys Cmpds. 181: 369 (1992).

14. T. Takabatake et al., Anisotropic suppression of the energy gap in CeNiSn by high magnetic fields, Phys. Rev. B 45: 5740 (1992-II).

15. A. Severing et al., Gap in the magnetic excitation spectrum of $Ce_3Bi_4Pt_3$, Phys. Rev. B 44: 6832 (1991-I).

16. A. P. Reyes et al., (unpublished).

17. T. E. Mason et al., Spin gap and antiferromagnetic correlations in the Kondo insulator CeNiSn, Phys. Rev. Lett. 69: 490 (1992).

18. A. P. Murani, Observation of f-band hybridization gap in the anomalous rare-earth compound $YbAl_3$, Phys. Rev. Lett. 54: 1444 (1985).

19. G. H. Kwei et al., (unpublished).

20. G. H. Kwei et al., Thermal expansion of $Ce_3Bi_4Pt_3$ at ambient and high pressures, Phys. Rev. B (in press).

21. G. Aeppli and Z. Fisk, Kondo insulators, Comm. Cond. Mat. Phys. (in press).

22. G. Boebinger et al., (unpublished).

23. T. Takabatake et al., Magnetoresistance and Hall effect in the Kondo-lattice system CeNiSn with an anisotropic energy gap, J. Magn. Magn, Mat. 108: 155 (1992).

24. M. F. Hundley et al., Magnetoresistance of the Kondo insulator $Ce_3Bi_4Pt_3$, Physica B (in press).

25. J. Beille et al., Suppression of the energy gap in SmB_6 under pressure, Phys. Rev. B 28: 7397 (1983).

26. F. Iga, "Experimental Study of Internmediate Valence Compound YbB_{12}", Ph.D. thesis, Tohoku University (1988).

27. T. Kasuya et al., Anomalous properties of valence fluctuating CeB_6 and SmB_6 in: "Valence Fluctuations in Solids", L. M. Falicov, W. Hanke and M. B. Maple, ed., North-Holland, Amsterdam (1981).

28. H. E. King et al., Effects of valence and intermediate valence on the compressibility of the rare-earth hexaborides, in: "Valence Fluctuations in Solids," L. M. Falicov, W. Hanke and M. B. Maple, ed., North-Holland, Amsterdam (1981).

29. J. D. Thompson, Magnetic interactions in correlated electron systems: high pressure investigations, in: "Frontiers in Solid State Sciences," L. C. Gupta and M. S. Multani, ed., World Scientific, Singapore (in press).

30. M. Kurisu, T. Takabatake and H. Fujiwara, Gap suppression in CeNiSn under hydrostatic pressure, Solid State Commun. 68: 595 (1988).

31. T. Takabatake et al., Formation of an anisotropic energy gap in the valence-fluctuating system in CeNiSn, Phys. Rev. B 41: 9607 (1990).

32. F. Iga, M. Kasaya and T. Kasuya, Kondo state in the alloy system $Lu_{1-x}Yb_xB_{12}$, J. Magn. Magn. Mat. 52: 279 (1985).

33. M. Kasaya et al., Valence instabilities and electrical properties of the La- and Yb-substituted SmB_6,, in: "Valence Fluctuations in Solids," L. M. Falicov, W. Hanke and M. B. Maple, ed., North-Holland, Amsterdam (1981).

34. P. Schlottmann, Impurity bands in Kondo insulators, Phys. Rev. B 46: 998 (1992-II).

35. P. C. Canfield et al., (unpublished).

36. J. M. Lawrence, J. D. Thompson and Y. Y. Chen, Two energy scales in $CePd_3$, Phys. Rev. Lett. 54: 2537 (1985).

37. P. C. Canfield et al., Effects of doping on hybridization gapped materials, J. Magn. Magn. Mat. 108: 217 (1992); P. C. Canfield et al., Doping and pressure study of $U_3Sb_4Pt_3$, J. Alloys Compds. 181: 77 (1992).

38. F. Iga, M. Kasaya and T. Kasuya, Specific heat measurements of YbB_{12} and $Yb_xLu_{1-x}B_{12}$, J. Magn. Magn. Mat. 76 & 77: 156 (1988).

39. P. S. Riseborough, Theory of the dynamic magnetic response of Ce3Bi4Pt3: a heavy-fermion semiconductor, Phys. Rev. B 45: 13984 (1992-II).

40. R. M. Martin and J. W. Allen, Classification of states at the Fermi energy in mixed valence systems, *in*: "Valence Fluctuations in Solids," L. M. Falicov, W. Hanke and M. B. Maple, ed., North-Holland, Amsterdam (1981).

ORIGIN OF GAP FORMATION IN $Ce_3Pt_3Sb_4$ AND $Ce_3Pt_3Bi_4$

Katsuhiko Takegahara,[1] Hisatomo Harima,[2] Yasunori Kaneta,[3]and
Akira Yanase[2]

[1]Education Center for Information Processing, Tohoku University
Sendai 980, Japan
[2]College of Integrated Arts and Sciences, University of Osaka Prefecture
Sakai 593, Japan
[3]Department of Physics, Tohoku University, Sendai 980, Japan

INTRODUCTION

Valence fluctuation systems with an energy gap of activation type have been attracting much attention to investigate the origin of gap formation. SmB_6, SmS, and YbB_{12} are well known to have clear gap.[1,2] The existence of gap has been experimentally suggested in CeNiSn, $Ce(Pd_{1-x}Cu_x)_3$, CeRhSb, and TmSe.[1,2,3]

Recently, new groups of ternary compounds containing Ce or U have been found to show semiconductor-like behavior.[2,3,4,5] The first group is UNiSn, UPtSn and URhSb, the second $U_3T_3Sb_4$ (T = Ni, Pd, Pt) and $Ce_3Au_3Sb_4$, and the third $Ce_3Pt_3X_4$ (X = Sb, Bi)[3,4,5] that is the subject of this report. All of these ternary compounds have a characteristic crystal structure; a transition or a noble metal atom is placed at the largest empty site of corresponding binary compound.[2] In the first and second groups, as f electrons are well localized and have a large magnetic moment, an origin of gap formation has been explained from band structures of non-f reference Th- or La-compounds.[2] Furthermore the band structures of the ternary compounds are understood on the basis of those of the corresponding binary compounds because of the characteristic crystal structure. The third group compounds belong to the valence fluctuation regime because the magnetic susceptibility of these compounds is typical for a valence fluctuating Ce compound.[3,4,5] Thus the ground state is considered to be different from the first and second groups.

In recent years, there is notable progress in the Fermi surface study for Ce compounds.[6,7] For the materials belonging to the valence fluctuation regime, the Fermi surfaces can be explained very well by conventional band calculations. This implies that the $4f$ electrons are itinerant in the ground state and contribute directly to the formation of the Fermi surface. However, cyclotron effective masses cannot be explained by the band theory alone and the large mass enhancement factor of the $4f$ electrons is ascribed to correlation effects. For the valence fluctuation systems with an energy gap, the electronic band structures of CeNiSn, SmB_6, and YbB_{12} have been calculated within the local density approximation.[1] The calculated results show an energy gap or pseudo-gap at the Fermi level. Therefore, we consider that this picture is applicable to the gap formation in the valence fluctuating Ce compounds as a straight extension. To confirm this picture, we have carried out self-consistent relativistic LAPW band calculations for $Ce_3Pt_3X_4$. Details of the method of calculation are almost the same as those described in the previous paper on $LaCu_6$ (ref. 8) and will be presented elsewhere.

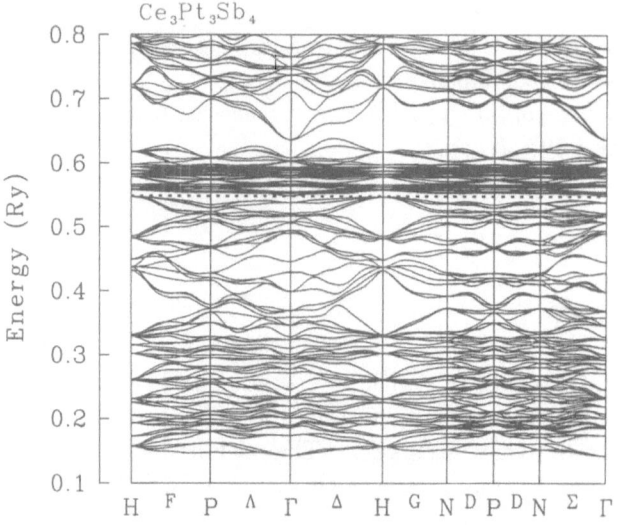

Figure 1. Self-consistent LAPW energy band structure for $Ce_3Pt_3Sb_4$. Dashed lines show the energy gap at the Fermi level.

ENERGY BAND STRUCTURES

At first, we summarize our recent results[2] on band structures for Th_3X_4 (X = P, As, and Sb), $Th_3Ni_3Sb_4$, and $La_3Au_3Sb_4$ because these are instructive to understand the band structures for the Ce compounds.

As the stable ionic state is Th^{4+} and X^{3-}, the band structures of Th_3X_4 are expected to be that the filled valence bands are derived from the X p states and the empty conduction bands from the Th $6d$ states. The calculated results show that the narrow gap appears at the Fermi level for Th_3P_4 and Th_3As_4 but the valence and conduction bands overlap slightly for Th_3Sb_4.

In $Th_3Ni_3Sb_4$, where the Ni atom is embedded in Th_3Sb_4, the $3d$ states on Ni are fully occupied by electrons. Therefore Ni exists as neutral atom. Then, as compared with the band structure of Th_3Sb_4, the Th $6d$ bands are shifted up by the hybridization with the Ni $3d$ bands to open a gap between the Sb $5p$ valence and Th $6d$ conduction bands because of the short distance between Th and Ni atoms. Therefore the system becomes an insulator.

In $La_3Au_3Sb_4$, the Sb $5p$ valence bands are filled by the transfer of three electrons from La atom and one electron from Au atom, leading to the insulator with an energy gap at the Fermi level.

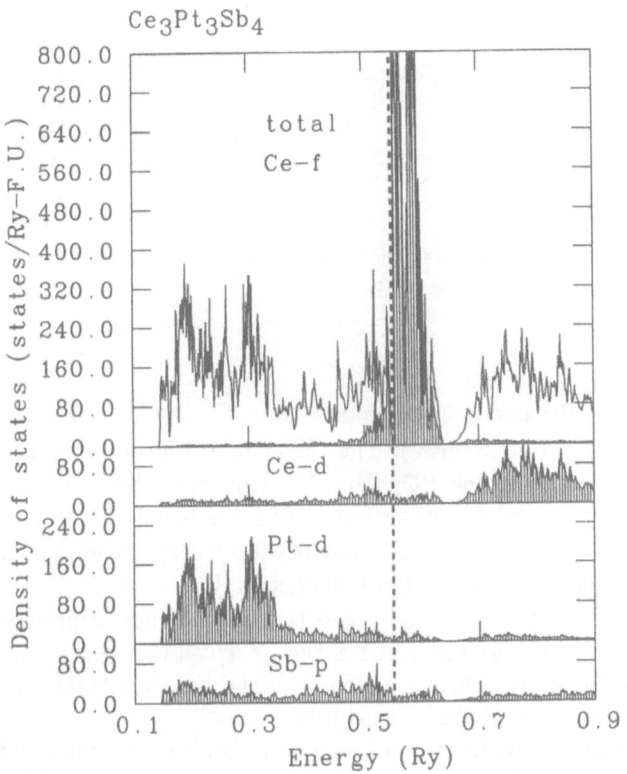

Figure 2. Calculated densities of states for $Ce_3Pt_3Sb_4$. Solid curves show the total density of states and hatched parts show the partial densities of states. Dashed lines show the energy gap at the Fermi level.

Figures 1 and 2 show the calculated energy band structure and the density of states for $Ce_3Pt_3Sb_4$, respectively. The tightly bound Ce $5p$ and Sb $5s$ bands are not shown in the figures. Similar to the band structure for $Th_3Ni_3Sb_4$, the $5d$ states on Pt are fully occupied by electrons. Then Pt remains as neutral atom.

The number of the valence electrons amounts to 108, $5d^14f^16s^2$ for each Ce, $5d^96s^1$ for each Pt, and $5p^3$ for each Sb and two formula units (F.U.) in the primitive unit cell. Thus, in Fig. 1, the lower 108 bands (from 0.1431 Ry to 0.5390 Ry at the Γ point) are the occupied valence bands and the empty bands lie above them. Note that, on the D and Σ axes, a degeneracy is lifted by the spin-orbit interactions and thus one band is occupied by one electron. As compared with the band structures of $Th_3Ni_3Sb_4$

and $La_3Au_3Sb_4$, the Ce $4f$ bands are shifted down and are located just in the energy gap between the Sb $5p$ valence and the Ce $5d$ conduction bands. Furthermore, a very narrow gap appears at the Fermi level because the top of valence bands is depressed by the Ce $4f$ bands around the H point. The valence bands consist of the Sb $5p$ and the Pt $5d$ states but the Pt $5d$ state is dominant at lower part. At upper part, the Ce $4f$ state mixes substantially as shown in Fig. 2. Then the number of $4f$ electron in the Ce muffin-tin sphere is 1.26 per Ce.

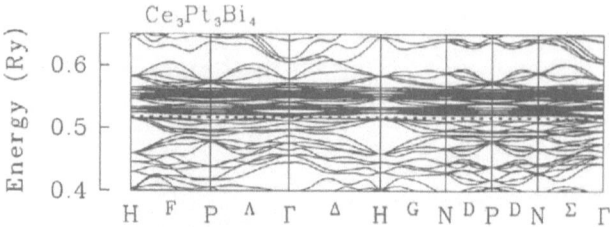

Figure 3. Self-consistent LAPW energy band structure in the vicinity of the Fermi level for $Ce_3Pt_3Bi_4$. Dashed lines show the energy gap at the Fermi level.

Between the 108th and 109th bands, there is the narrow gap with the indirect gap width of 0.0020 Ry (27 meV). The Ce $4f$ states form 84 bands (= 14 states × 6 atoms) from the 109th to the 192nd bands. The spin-orbit splitting is seen clearly in Fig. 1. The Ce $4f$ component in almost the 84 bands is more than 90 %. However, in several bands at upper part of the 84 bands the $4f$ component decreases to 45 ~ 70 % while the Ce $5d$ and Sb $5p$ components increase. Note that, at the Γ and H points, the symmetry label of 108th and 192nd bands is the same, indicating that these two bands repel each other due to mixings between the Ce $4f$ and $5d$ states and the Sb $5p$ states. Above the 193rd band (0.6369 Ry at the Γ point), the Ce $5d$ state is dominant but the Ce $4f$, Pt $5d$, and Sb $5p$ states mix slightly.

The electronic band structure of $Ce_3Pt_3Bi_4$ is also calculated and shown in Fig. 3. Overall feature is very similar to that of $Ce_3Pt_3Sb_4$ except some differences as follows. 1) As the spin-orbit interactions are larger for heavier atom Bi, the spin-orbit splitting of Bi $6p$ valence bands is about two times as large as that of Sb $5p$ bands.
2) The Ce $5d$ conduction bands are located closer to the valence bands. The first reason is that the lattice constant of $Ce_3Pt_3Bi_4$ is larger than that of $Ce_3Pt_3Sb_4$. Thus the hybridization between the Ce $5d$ and the Pt $5d$ states is weaker in $Ce_3Pt_3Bi_4$ thus results small repulsion between these states. The second reason is well known fact that the pnictogen p bands and the rare-earth $5d$ and $4f$ bands become closer as the atomic number of pnictogen increasing in many pnictides.
3) The number of $4f$ electron in Ce muffin-tin sphere increases slightly but the energy gap at the Fermi level is nearly same as that of $Ce_3Pt_3Sb_4$, 0.0022 Ry (30 meV).

DISCUSSION

For $Ce_3Pt_3Sb_4$, an energy gap is experimentally observed as 78 meV (ref. 4), but the calculated value is rather small, 27 meV. It is well known that the energy gap in the band calculation is much smaller than the actual value due to the inadequate treatment of the exchange-correlation potential in the local density approximation. As mentioned above, this problem also makes the disagreement of cyclotron effective mass between experimentally observed one and calculated one by band theory in the Fermi surface study.

In $Ce_3Pt_3Bi_4$, the resistivity shows activated behavior above 100 K with the energy gap of 70 K (6 meV).[5] The calculated energy gap is 30 meV, factor of 5 larger than the observed value. We rather speculate that the observed energy gap is not the intrinsic gap but due to impurity state, because it is common that the gap in the band calculation is smaller than the actual one as mentioned above.

The number of $4f$ electron in the Ce muffin-tin sphere is 1.26 per Ce for $Ce_3Pt_3Sb_4$ and 1.29 for $Ce_3Pt_3Bi_4$. These values are very close to the calculated ones for many other Ce compounds belonging to either the valence fluctuation regime or the Kondo regime. As the $4f$ bands are located near the Fermi level, the $4f$ components are induced in the valence bands by the hybridization effects between the $4f$ states and other states and thus the number of occupied $4f$ electron is about one. Thus the magnetic susceptibility at sufficiently high temperature has Curie-Weiss behavior.[5]

To conclude, $Ce_3Pt_3X_4$ is a very interesting material, one of a few examples belonging to the valence fluctuation regime with an energy gap. The origin of gap formation is clarified reasonably well by the band theory based on the itinerant electron model for the $4f$ electrons.

ACKNOWLEDGMENTS

The numerical computations were performed at the Computer Center of the Institute for Molecular Science and at the Computer Center of Tohoku University.

REFERENCES

1. A. Yanase and H. Harima, Band calculations on YbB_{12}, SmB_6 and CeNiSn, *Prog. Theor. Phys. Suppl.* No. 108:19 (1992), and references therein.
2. K. Takegahara and Y. Kaneta, Electronic band structures of f-electron ternary compounds with an energy gap, *Prog. Theor. Phys. Suppl.* No. 108:55 (1992), and references therein.
3. P. C. Canfield, J. D. Thompson, Z. Fisk, M. F. Hundly, and A. Lacerda, Effects of doping on hybridization gapped materials, *J. Magn. Magn. Mater.* 108:217 (1992), and references therein.
4. M. Kasaya, K. Katoh, and K. Takegahara, Semiconducting properties of the isomorphous compounds, $Ce_3Au_3Sb_4$ and $Ce_3Pt_3Sb_4$, *Solid State Commun.* 78:797 (1991).
5. M. F. Hundly, P. C. Canfield, J. D. Thompson, Z. Fisk, and J. M. Lawrence, Evidence for a 'coherence' gap in $Ce_3Bi_4Pt_3$, *Physica* B171:254 (1991).
6. Y. Ōnuki, T. Goto, and T. Kasuya, Fermi surfaces in strongly correlated electron systems, *in*: "Materials Science and Technology," R. W. Cahn, P. Haasen, and E. J. Kramer, ed., VCH, Weinheim (1992), Vol. 3A, Part I.
7. A. Hasegawa and H. Yamagami, Band theory of itinerant f-electron compounds, *Prog. Theor. Phys. Suppl.* No. 108:27 (1992).
8. H. Harima, A. Yanase and A. Hasegawa, Electronic structure and Fermi surface of $LaCu_6$, *J. Phys. Soc. Jpn.* 59:4054 (1990).

MAGNETIC AND TRANSPORT STUDIES ON Ce-BASED COMPOUNDS

S.K. Malik,[1] and D.T. Adroja[2]

[1]Tata Institute of Fundamental Research
Bombay 400 005, India

[2]Department of Physics
University of Southampton, Southampton, S09 5NH, U.K.

INTRODUCTION

In recent years, there has been a great deal of interest in the structural, magnetic and transport properties of intermetallic compounds containing rare earths and actinides. Among the rare earths, the compounds containing cerium have attracted a great deal of attention. Cerium is the first element in the lanthanide series having a non zero 4f occupation. Therefore, the 4f shell is rather unstable in cerium and can lead to very interesting properties.

The differing behaviour in many Ce-based compounds can be understood in terms of the hybridization width $\Delta \approx \pi <V_{sf}>^2 \rho(E_F)$ (where V_{sf} is the hybridization strength and $\rho(E_F)$ is the density of states at the Fermi level, E_F) and the position of the 4f level (E_f) relative to the Fermi level given by $E_0 = |E_f - E_F|$. Though Δ and E_0 may vary continuously, one can visualize three extreme cases.

(i) $\Delta << E_0$: In this case the system shows a stable valence state and in some cases it may exhibit a long range magnetic order below a certain temperature. The 4f electrons can be described as localized which interact with the neighbouring 4f electrons through RKKY interaction. Such compounds can be described by the free ion behaviour - at best influenced by the electrostatic fields, the so called crystal fields arising from the neighbouring ions.

(ii) $\Delta < E_0$: The system exhibits Kondo type behaviour and the 4f occupancy may slightly deviate from the integral value. If the intersite interactions between the ions are small, the system gains energy by losing the magnetic moment and forming a singlet ground state. The energy gain by such a process is $k_B T_K$ where T_K is the Kondo temperature which can vary from milli Kelvins to few tens of degree Kelvins. At temperatures well below T_K, the electrical resistivity shows a drop which is due to the coherence between the local moments. In the coherence regime, the single impurity Anderson model no longer holds [1]. If the intersite interaction is strong in the Kondo Lattice system, it may order magnetically with reduced moment.

(iii) $\Delta \geq E_0$: The system shows intermediate valence or valence fluctuating behaviour. Such compounds exhibit Curie-Weiss behaviour in their susceptibility at high temperatures with an effective moment intermediate between the values of the integral valent state. Most of the Ce and Yb based compounds show a maximum in the susceptibility at some temperature. At low temperatures the susceptibility becomes nearly temperature independent. The ground state properties of such systems can be described by a Fermi liquid model.

The degree of delocalization and the extent of hybridization strongly influences the magnetic and transport behaviour of these compounds and may lead to large values of the electronic specific heat coefficient γ, large value of the coefficient of the T^2 term in the resistivity, enhanced Pauli paramagnetism, etc. The interplay of the RKKY type interaction and the Kondo effect leads to further interesting features.

In this article we present a brief review of the magnetic and transport studies on some of the Ce-based compounds primarily carried out by the authors. The systems discussed are the equiatomic ternary compounds of the type RTX where R is a rare earth ion -primarily Ce, T is a transition metal and X is a metalloid atom, and compounds of the type $CeXPt_4$ where X=Cu, Ga, Rh, Pd, Ir, Sn and In.

EQUIATOMIC TERNARY COMPOUNDS

The equiatomic ternary compounds are represented by the general formula RTX where R is a lanthanide or an actinide ion, T is a transition metal such as Cu, Ag, Au, Pd, Pt, Rh, Ru, etc., and X is one of the sp metals or metalloid such as Si, Ge, Al, In, Ga, Sn, Sb etc. The RTX compounds crystallize in a variety of structure types ranging from the cubic (LaIrSi, MgAgAs type), the hexagonal ($CaIn_2$, Fe_2P type), the orthorhombic (TiNiSi, $CeCu_2$ type), the tetragonal (LaPtSi type), etc. Therefore, this family of compounds offers the opportunity of a systematic investigation of the effect of crystal structure, transition metals and metalloids on the electronic behaviour of the rare earth ions. With this in mind, we initiated a systematic investigation of such equiatomic ternary compounds. In what follows, we discuss the results on some selected cerium-based compounds which show rather unusual behaviour.

Antiferromagnetic Kondo Lattice System: CePdSn

The compound CePdSn crystallizes in the orthorhombic TiNiSi-type structure (space group Pnma). Its unit cell volume follows the lanthanide contraction observed in the RPdSn compounds. This suggests that the cerium ions are in the 3+ or nearly 3+ state in this compound. The magnetic susceptibility of CePdSn (Fig. 1) exhibits Curie-Weiss (CW) behaviour in the temperature range of 50-300K. Below 50K, deviations from CW behaviour are observed [2]. These have been analyzed in terms of the crystalline electric-field effects. The compound orders antiferromagnetically with Néel temperature (T_N) of about 7.5K. This is confirmed by neutron, ^{119}Sn Mössbauer, heat capacity and resistivity measurements. The isostructural Gd compound also orders antiferromagnetically at 14.5K [3]. On the basis of de Gennes scaling, the T_N of CePdSn should have been 91 times smaller than that of the Gd compound assuming that the exchange constant J_{sf} for the interaction between the conduction electron spins and the Ce-4f spins is the same. This is not the case experimentally. This already indicates the presence of strong exchange interaction between the conduction-electron spins and the Ce-4f spins presumably caused by the hybridization effects.

The resistivity of CePdSn is shown in Fig. 2. It decreases linearly with decreasing temperature from 300K but shows a downward curvature starting at about 80K. It exhibits a broad minimum at about 20K, a small rise and then a sudden drop at about 7.5K, the latter of which corresponds to the antiferromagnetic ordering of the Ce moments. The minima in the resistivity, shown clearly as an inset in Fig. 2 suggests the presence of Kondo type interactions in this compound. Thermoelectric power of CePdSn has also been measured and analyzed [4] in terms of the Bhattacharjee-Coqblin model [5] which takes into account the Kondo effect in the presence of crystalline electric fields. Thus CePdSn appears to be an antiferromagnetically ordered Kondo lattice system.

Ferromagnetic Kondo Lattice System: CePdSb

The occurrence of antiferromagnetic state is often observed in Ce-based ternary compounds and is in accord with such an ordering predicted by the single impurity Kondo

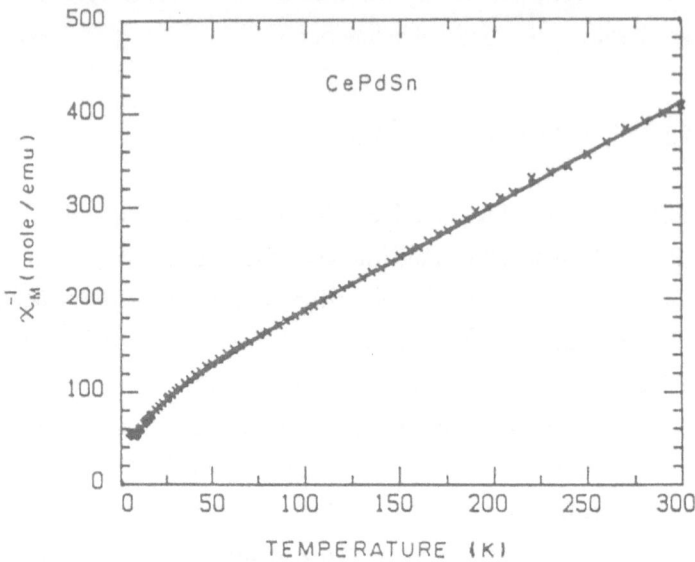

Figure. 1   Inverse magnetic susceptibility vs temperature for CePdSn.

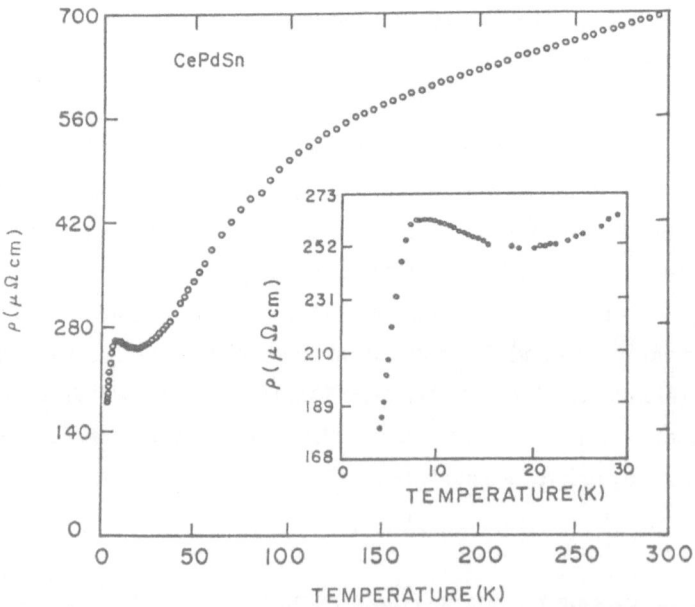

Figure 2   Electrical resistivity of CePdSn as a function of temperature. Inset shows expanded data at low temperatures.

models due to the negative exchange interaction inherent in such models. Of late ferromagnetic ordering has also been observed in Ce-based compounds many of which show features characteristic of Kondo lattice systems. Here we present the results on CePdSb which suggest that this compound may represent a ferromagnetic Kondo lattice system [6].

The RPdSb compounds with R=La to Sm crystallize in the hexagonal structure. Originally, based on the x-ray data, the hexagonal $CaIn_2$ type structure (space group P63/mmc) was proposed for these compounds. In this structure Pd and Sb atoms randomly occupy the In sites. More recent neutron diffraction studies carried out by us on CePdSb have shown that the diffraction pattern can also be fitted to an ordered structure (space group P63mc) in which the Pd and the Sb atoms occupy distinct sites. However, the Pd and Sb scattering lengths are almost similar, and from the neutron data alone one cannot distinguish between the ordered and the disordered structure. The inelastic scattering measurements [7] show well defined peaks corresponding to transitions from the ground state doublet and two excited state crystal field doublets. This may be taken as an indirect evidence for the ordered P63mc space group. It is likely that the other RPdSb compounds mentioned above also crystallize in this structure type. The unit cell volume of CePdSb does not show any pronounced anomaly in comparison to that expected on the basis of lanthanide contraction suggesting that the Ce ions are in a nearly trivalent state in this compound.

Among the RPdSb compound, those with Nd, Sm and Gd order antiferromagnetically with Néel temperature of 11K, 18K and 15.5K, respectively [8]. There are indications of another transition at lower temperatures, presumably of magnetic origin, in most of these compounds. Moreover, Nd compound shows a metamagnetic or spin flop transition in applied fields starting at about 2kOe at 4.2K. In contrast to these compounds, CePdSb orders ferromagnetically with a relatively high ordering temperature of about 17K (Fig. 3). The magnetization-field isotherms at 4.2K yield a value of $1.2\mu_B$ as the ordered state moment per Ce ion. Not only is the magnetic ordering so different in CePdSb and GdPdSb, but also the magnetic ordering temperature (T_M) of Ce compound is rather high and does not follow de Gennes scaling. The large observed T_M in Ce compound suggests the presence of stronger exchange interaction between the conduction electron spins and the Ce spins (compared to that in the Gd compound), possibly arising due to the hybridization effects.

The susceptibility of CePdSb follows Curie-Weiss behaviour above 50K with $\mu_{eff}=2.64\mu_B$ and $\theta_p=10K$. The deviation of susceptibility from CW behaviour below 50K is attributed to the effect of crystalline electric fields. This is consistent with the Ce ordered moment being less than the free ion value.

The resistivity behaviour of CePdSb is rather unusual. Figure 4 shows the plot of resistivity versus temperature for CePdSb, GdPdSb and LaPdSb. The phonon contribution to the resistivity in various RPdSb compounds may be taken to be the same as in isostructural nonmagnetic LaPdSb. The 4f contribution to the resistivity of CePdSb, obtained by subtracting the resistivity of LaPdSb, increases as the temperature is lowered from 300K; passes through a broad maximum at about 150K and then starts decreasing. It

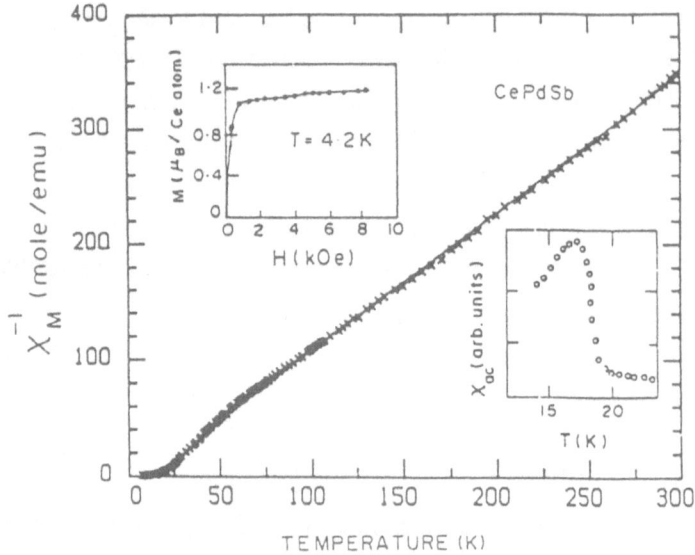

Figure. 3   Inverse magnetic susceptibility of CePdSb (in 5 kOe field) as a function of temperature. Insets show the magnetization-field isotherm at 4.2K and ac susceptibility as a function of temperature.

shows a precipitous drop at the magnetic ordering temperature. The high temperature resistivity can be fitted to a ln T behaviour which is one of the characteristic features of the dense Kondo systems. According to the Cornut-Coqblin model [9], the broad maxima in resistivity arises due to a combined influence of CEF on the 4f moments and the presence of Kondo type interactions. This theory predicts different ln T behaviour for different temperature regimes corresponding to different crystal field split levels. In the CePdSb resistivity data, the low temperature ln T behaviour is not observed, the absence of which may be due to the onset of ferromagnetic ordering and the ferromagnetic correlations which may set in at temperatures above T_M.

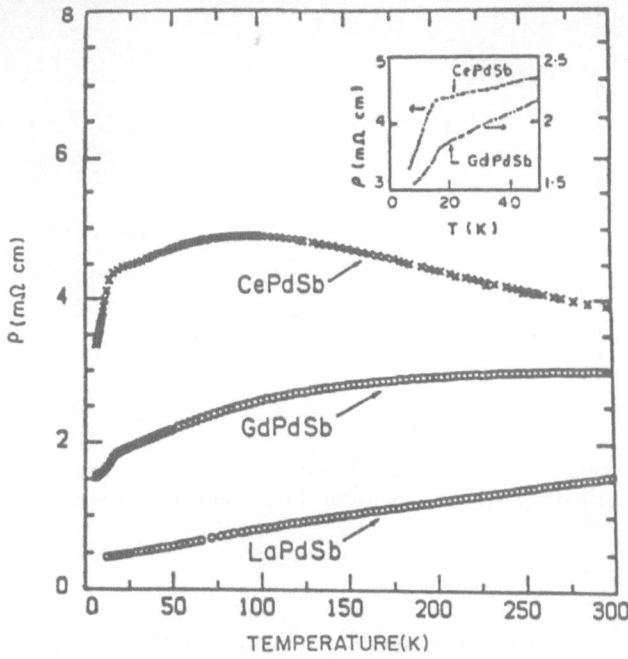

Figure. 4   Electrical resistivity of RPdSb (R=La, Ce and Gd) as a function of temperature. Inset shows the resistivity behaviour at low temperatures.

The nearest neighbour Ce-Ce distance in CePdSb is 3.96Å along the c axis, which is greater than 3.25-3.4Å, the Hill limit, for the direct overlap between the 4f electrons at adjacent sites to occur. Therefore, the mechanism for the relatively high magnetic ordering temperature does not involve direct 4f-4f overlap. The results can be understood on the basis of the interplay between two competing processes, namely, the indirect exchange interaction of Ce^{3+} ions via the RKKY interaction and the effective suppression of the Ce magnetic moments due to Kondo effect. This is the so called Kondo necklace model proposed by Doniach [10]. By considering a one dimensional lattice of localized spins coupled to a system of conduction electrons he showed that, when the local spin - conduction electron coupling is weak, an antiferromagnetic ground state can exist. For strong J_{sf} the local moments are quenched forming a Kondo like state. In the very weak hybridization limit (i.e. when the 4f level is well separated from the Fermi level and the 4f occupation is nearly unity), the Anderson single ion Hamiltonian [11] can be transformed in to a simple Kondo model by the Schreiffer-Wolf transformation [12]. In this case J_{sf} depends on the hybridization strength V_{sf}, the position of the 4f level, E_f, relative to the Fermi energy given by $E_0 = |E_F - E_f|$, and the inter-site Coulomb repulsion U of the two electrons with opposite spin directions at the same ion. In the limit of both V_{sf} and E_0 much smaller than U, the exchange constant is given by $J_{sf} = -V_{sf}^2/E_0$.

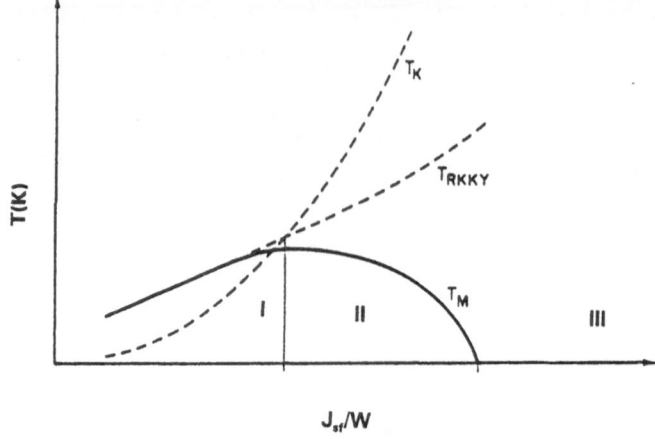

Figure. 5   Schematic diagram showing T_K, T_{RKKY} and T_M as a function of J_{sf}/W.
I: Pure magnetic system, II: Magnetic Kondo lattice, III: Nonmagnetic Kondo lattice.

The RKKY interaction temperature T_{RKKY} varies as J_{sf}^2 while the Kondo temperature T_K varies as $\exp(-1/|J_{sf}|)$. Thus in the limit when E_0 decreases significantly, J_{sf} increases appreciably in magnitude and so do T_{RKKY} and T_K. However because of their different dependences on J_{sf}, the system shows a crossover from a magnetically ordered ground state to a spin compensated ground state at a critical value of Jsf. This is also noted in the Kondo necklace model. The magnetic ordering temperature calculated by Doniach exhibits a maximum at certain value of J_{sf} [13] and this has been observed experimentally in some intermetallic compounds [14]. The schematic dependence of the magnetic transition temperature T_M on the normalized Kondo coupling constant J_{sf}/W (where W is the width of the conduction band) is shown in Fig. 5. Thus the high ordering temperatures of CePdSn and CePdSb may be understood on the basis of the interplay between the Kondo and the RKKY type interactions. However, it should be pointed out that CePdSb orders ferromagnetically unlike most other Kondo lattice systems which show antiferromagnetic ordering.

Pseudo-gap in the Density of States in CeRhSb

The study of gap formation in the valence fluctuating compounds has evinced a great deal of interest in recent times. The valence fluctuating compounds SmB_6 [15], SmS [16] and YbB_{12} [17] are known to develop an energy gap in their density of states. Very recently this phenomenon has been observed in CeNiSn [18], $Ce_3Bi_3Pt_4$ [19] and CeRhSb [20]. These compounds exhibit an insulating ground state with a small energy gap in the electronic density of states at the Fermi level which clearly manifests itself in transport properties such as resistivity, thermoelectric power and also in inelastic neutron scattering experiments. Here we present detailed results on CeRhSb.

The RRhSb (R=La to Sm) compounds form in the orthorhombic structure (TiNiSi type). The unit cell volume of CeRhSb shows pronounced deviation from the lanthanide contraction suggesting the mixed valent nature of cerium ions in this compound. This is also borne out by the magnetic susceptibility measurements. The susceptibility of CeRhSb (Fig. 6) is only weakly temperature dependent, shows a broad maximum at about 113K and a rise at low temperatures. The latter may be due to the stabilization of a fraction of cerium ions in the 3+ state which carry a magnetic moment. The susceptibility behaviour observed in CeRhSb, in particular the broad maximum, is typical of mixed valent Ce and Yb compounds. The susceptibility results on CeRhSb have been analyzed on the basis of Coqblin-Schrieffer model [21]. The solid line in Fig. 6 represents a fit to this model.

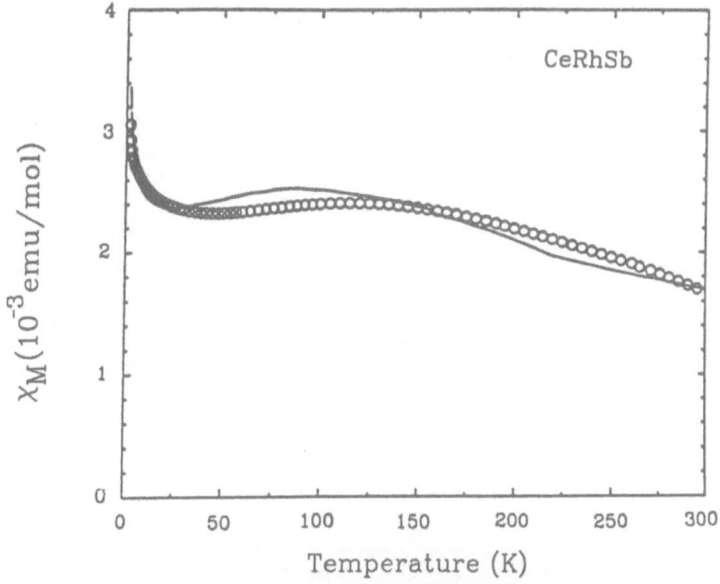

Figure. 6   Magnetic susceptibility of CeRhSb as a function of temperature.

The resistivity of CeRhSb shows very interesting behaviour (Fig. 7).   As temperature is lowered from 300K, the resistivity increases slightly and reaches a broad maximum at 113K - the same temperature where susceptibility also shows a broad maximum. It then decreases with decreasing temperature and exhibits a minimum at about 21K. However, below 21K, the resistivity increases rapidly with decreasing temperature. In contrast, the resistivity of LaRhSb (Fig. 7) shows the behaviour typical of metallic systems throughout the temperature range investigated (4.2-300K). The rapid rise in the resistivity of CeRhSb below 21K may be interpreted in terms of the opening of a gap in the density of states. A plot of $\ln \rho$ vs T^{-1} (Fig. 8) is a straight line from which the gap energy

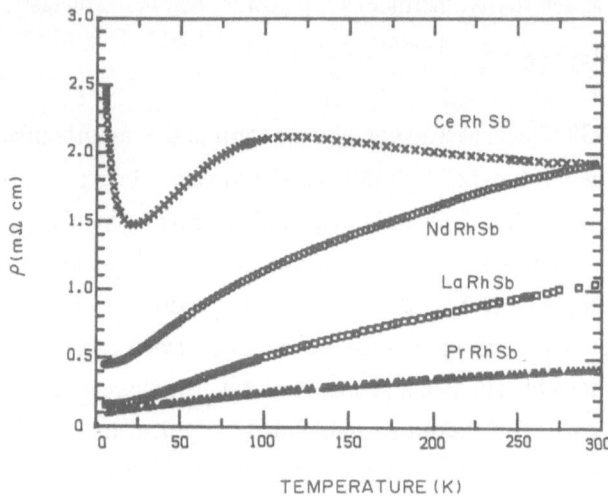

Figure. 7   Temperature dependence of the electrical resistivity of RRhSb (R = La, Ce, Pr and Nd).

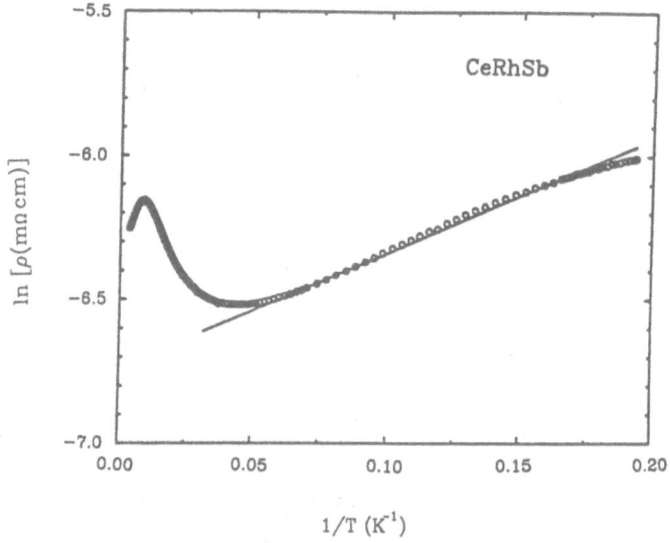

Figure. 8   Ln ρ versus 1/T for CeRhsb. The solid line represents the fit ($\rho = \rho_0 \exp(-E_g/kT)$) with E_g of about 4K.

is estimated to be about 4K. It is interesting to note that a 20% replacement of La by Ce suppresses the rise in resistivity at low temperatures though the susceptibility continues to exhibit a broad maximum. This suggests that a crystallographically well-ordered Ce-lattice may be essential for the formation of a gap in such compounds. The magnetoresistance measurements on CeRhSb and gapless $Ce_{0.8}La_{0.2}RhSb$ compound [22] reveal that the gap in CeRhSb decreases with increasing applied field (Fig. 9). For comparison purpose, data on CeNiSn taken from ref. 23 are also shown in the same figure.

Although, the study of the origin of the energy gap in the mixed valent insulating ground state systems is an area of active research at present, not much progress has been made in understanding all the factors leading to the formation of such a gap. It is believed

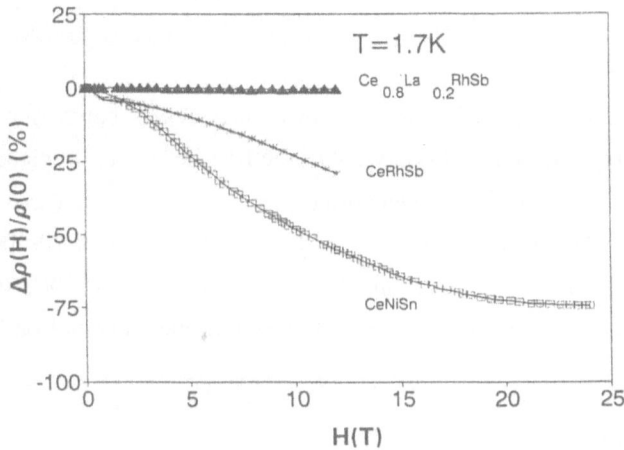

Figure. 9 Normalized change in resistance vs applied field at 1.7K for various compounds.

that lattice coherence and strong hybridization between Ce-4f electrons and the conduction electrons are responsible for the stability of the gap at E_F [24,25]; consequently it is frequently referred to as hybridization or coherence gap.

COMPOUNDS OF THE TYPE CeXPt4 (X= Cu, Ga, Rh, Pd, Ir Sn and In)

Hexagonal Compounds

The compound $CePt_5$ crystallizes in the hexagonal $CaCu_5$ structure (space group P6/mmm) with one formula unit per unit cell and orders antiferromagnetically at low temperatures below 1K [26]. There are two crystallographically inequivalent Pt sites in this compound in the ratio of 2:3. Thus a selective replacement of one of the two Pt sites may be possible by other elements which may influence the magnetic and transport properties of these compounds. In many cerium-based compounds, it is often observed that the magnetic ordering is lost when T_K exceeds T_M, the temperature corresponding to the RKKY interaction. In this respect CePt$_5$ offers the opportunity to study the possible crossover from magnetic to nonmagnetic regime. With this mind, we have investigated the structural, magnetic and transport properties of the compounds CeXPt$_4$ with X=Cu, Ga, Rh, Pd, Ir, Sn and In.

Powder X-ray diffraction studies on CeXPt$_4$ compounds with X=Cu, Ga, Rh, Pd, Ir and Sn show that these compounds also crystallize in the hexagonal structure [27]. However, it can not be ascertained whether the X atoms selectively occupy one of the two Pt sites or are randomly distributed on both the sites. The lattice parameters of CeXPt$_4$ compounds are similar to those of CePt$_5$ from which these are derived.

The susceptibility of CePt$_5$ follows Curie-Weiss behaviour with an effective moment of 2.48μ_B which is close to that of the free Ce^{3+} ion. Replacement of one Pt by the X atom does not alter the behaviour substantially. Curie-Weiss behaviour in the susceptibility is also observed in CeXPt$_4$ compounds between 20-300K with moments again close to the free ion value. A slight deviation from the Curie-Weiss behaviour is seen below 20K and is attributed to the effects of crystalline electric fields on Ce^{3+} ions.

Cubic Compounds

In contrast to the above, the compound CeInPt$_4$ is found to crystallize in a cubic structure - either the Laves phase MgCu$_2$ type or the cubic MgSnCu$_4$ type structure. One may visualize that CeInPt$_4$ is derived from the cubic CePt$_2$ (MgCu$_2$ type) by partial replacement of the Ce atoms by the In atoms. Such a substitution will lead to a distribution of the Ce and In atoms at the rare earth (8a) site. On the other hand MgSnCu$_4$ type of structure is an ordered structure with all the atoms occupying distinct sites. Further x-ray and/or neutron diffraction measurements are needed to determine the atomic ordering.

The CeInPt$_4$ compound behaves very differently [28] compared to the other CeXPt$_4$ compounds mentioned above. A comparison of the unit cell parameters of the RInPt$_4$ (R=rare earth) compounds reveals that the unit cell volume of CeInPt$_4$ does not deviate significantly from that expected on the basis of lanthanide contraction. From this it may be inferred that Ce is in a trivalent or nearly trivalent state in this compound. Figure 10 shows the plot of inverse susceptibility versus temperature for CeInPt$_4$. A Curie-Weiss behaviour is observed in the temperature range of 100-300K. The effective paramagnetic moment and the paramagnetic Curie temperatures are 2.54μ_B and -225K respectively. The large negative paramagnetic Curie temperature is indicative of the presence of Kondo type interactions in this compound. In fact, the susceptibility behaviour of CeInPt$_4$ is similar to those of heavy fermion systems CeAl$_3$ [29] and CeCu$_2$Si$_2$ [30]. Pronounced deviation of the susceptibility from Curie-Weiss behaviour is observed below 100K.

The results of the heat capacity (C) measurements on CeInPt$_4$ down to 100mK [28] are shown in Fig. 11 where C/T is plotted as a function of T^2. This compound does not appear to order magnetically in the temperature range investigated. The C/T (or γ) continues to increase below 2K and attains a value of 1750 mj/mole K^2 at 100mK. The value of γ extrapolated to T=0 is 2500 mj/mole-K^2 suggesting the formation of a quasiparticle state with very large effective mass. This value of γ is among the highest reported in the literature. According to the single ion Kondo model, the Kondo temperature T$_K$ is related to the maximum value of γ by γ_{max}=0.68R/T$_K$ where R is the gas constant [31]. The value of T$_K$ obtained by using γ=2500 mj/mol-K^2 is 2.2K.

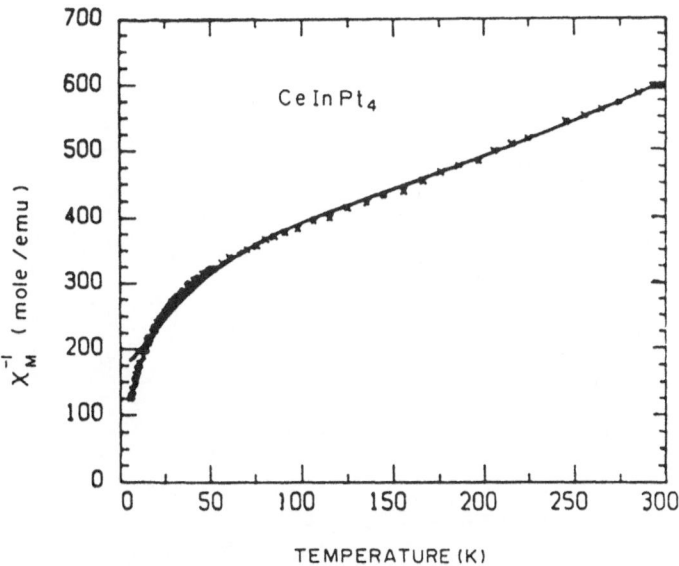

Figure. 10   Inverse magnetic susceptibility vs temperature for CeInPt$_4$.

The electrical resistivity of CeInPt$_4$ and that of the nonmagnetic LaInPt$_4$ is shown in Fig. 12. If the phonon contribution to the resistivity is assumed to be the same in La and Ce compounds then the magnetic contribution to the resistivity (ρ_m) due to Ce can be obtained by subtracting the resistivity of La compound from that of the Ce compound. The most prominent feature of ρ_m is a broad maximum at about 70K followed by a shallow minimum at 20K. Below 20K, ρ_m increases with decreasing temperatures down to 4.2K. This resistivity behaviour is similar to that observed in other heavy fermion systems.

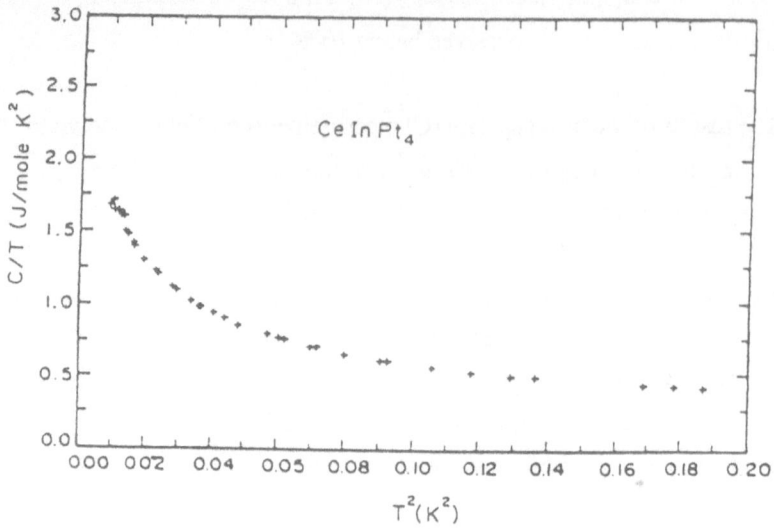

Figure. 11   Specific heat divided by temperature (C/T) as a function of T^2.

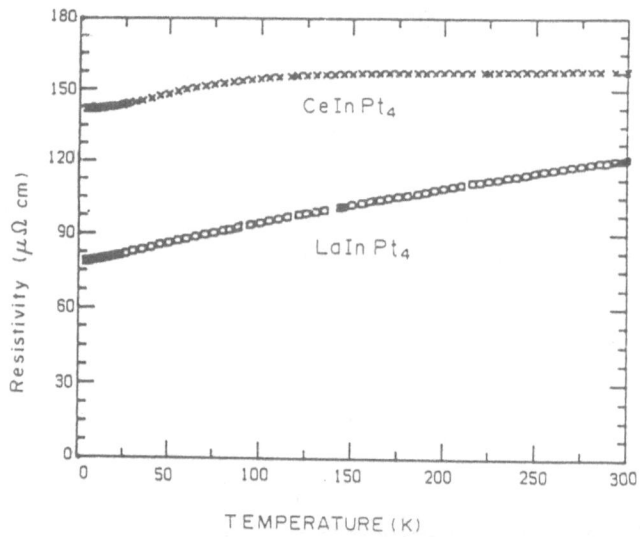

Figure. 12   Electrical resistivity of LaInPt$_4$ and CeInPt$_4$ as a function of temperature.

The magnetic susceptibility of CeInPt$_4$ has been analyzed on the basis of a model which takes into account the valence fluctuation as well as the crystal field effects on the J=5/2 state of Ce^{3+} ion. Solid line in Figure 10 shows one of the fits using this model. Details of the analysis are given elsewhere [28].

In conclusion, we have shown that Ce-based compounds show a variety of behaviour in their magnetic and transport properties. In the family of equiatomic compounds discussed above, CePdSn represents an antiferromagnetic Kondo lattice system while CePdSb is ordered ferromagnetically and shows features characteristic of Kondo lattice systems. The compound CeRhSb exhibits nearly temperature independent susceptibility typical of Ce-based mixed valent systems. Its resistivity shows a rapid rise at low temperatures which is attributed to the opening of an energy gap in the density of states. The compounds of the type $CeXPt_4$ crystallize in the hexagonal structure for X=Pt, Cu, Ga, Rh, Pd, Ir and Sn in which Ce behaves like a normal trivalent rare earth ion. However, $CeInPt_4$ forms in the cubic structure and shows one of the highest reported values of the electronic specific heat coefficient .

ACKNOWLEDGEMENT

One of us (SKM) is thankful to JSPS for the award of their fellowship which enabled him to attend this Conference and spend some time in Japan. He also thanks Profs. H. Fujii and T. Takabatake of Hiroshima University for their kind hospitality at Hiroshima during which period this manuscript was written.

REFERENCES

[1]   See, for instance, "*Theory of Heavy Fermions and Valence Fluctuations* ", edited by T. Kasuya and T. Saso, Springer Series in Solid State Science, No. 62, (Springer-Verlag 1985).

[2]   S.K. Malik, D.T. Adroja, S.K. Dhar, R. Vijayaraghavan and B.D. Padalia, Phys. Rev. B **40**, 2414 (1989).

[3]   D.T. Adroja and S.K. Malik, Phys. Rev. B **45**, 779 (1992).

[4]   D.T. Adroja, B.D. Padalia, S.N. Bhatia and S.K. Malik, Phys. Rev. B **45**, 477 (1992).

[5]   A.K. Bhattacharjee and B. Coqblin, Phys. Rev. B **13**, 3441 (1976).

[6]   S.K. Malik and D.T. Adroja, Phys. Rev. B **43**, 6295 (1991).

[7]   B.D. Rainford and D.T. Adroja, ISIS, Rutherford Lab. U.K., Annual Report (1992), p. A243; and to be published.

[8]   S.K. Malik and D.T. Adroja, JMMM **102**, 42 (1991).

[9]   B. Cornut and B. Coqblin, Phys. Rev. B **5**, 4541 (1972).

[10]   S. Doniach, Physica B **91**, 231 (1977).

[11]   P.W. Anderson, Phys. Rev. **124**, 41 (1961).

[12]   J.R. Schrieffer and P.W. Wolf, Phys. Rev. **149**, 491 (1966).

[13] See, for instance, S. Doniach, in "*Valence Instability and Related Narrow-Band Phenomena* ", edited by R.D. Parks (Plenum Press, New York, 1977), p. 169.

[14] See, for instance, N.B. Brandt and V.V. Moschalkov, Adv. Phys. **33**, 373 (1984).

[15] F. Lapierre, M. Ribalt, F. Holtzberg and J. Flouquet, Solid State Commun. **40**, 347 (1981).

[16] A. Jayaraman, V. Narayanamurti, E. Bucher and R.G. Maines, Phys. Rev. Lett. **25**, 1430 (1970).

[17] M. Kasuya, F. Iga, M. Takigawa and T. Kasuya, JMM **47 & 48**, 429 (1985).

[18] T. Takabatake, F. Teshima, H. Fujii, S. Nishigori, T. Suzuki, T. Fujita, Y. Yamaguchi, J. Sakurai, and D. Jaccard, Phys. Rev. B **41**, 9607 (1990).

[19] M.F. Hundley, P.C. Canfield, J.D. Thompson, Z. Fisk, and J.M. Lawrence, Phys. Rev. B **42**, 6842 (1990).

[20] S.K. Malik and D.T. Adroja, Phys. Rev. B **43**, 6277 (1991).

[21] V.T. Rajan , Phys. Rev. Lett. **51**, 308 (1983).

[22] D.T. Adroja and B.D. Rainford, submitted to Z. Physik B (1992); S.K. Malik et al, to be published.

[23] T. Takabatake, Y. Nakazawa, M. Ishikawa, T. Sakakibara, K. Koga and I. Oguro, JMMM **76 & 77**, 87 (1988).

[24] T. Takabatake, M. Nagasano, H. Fujii, M. Nohara, T. Suzuki, T. Fujita, G. Kido and T. Hiraoka, JMMM **108**, 155 (1992).

[25] R.M. Martin, Phys. Rev. Lett. **48**, 362 (1982).

[26] P.K. Ikonomou, J. Less-Common Met. 71, 13 (1980);
K.S.V.L. Narasimhan, V.U.S. Rao and R.A. Butera, AIP Conf. Proc. **10**, 1081 (1973);
R. Vijayaraghavan, S.K. Malik and V.U.S. Rao, Phys. Rev. Lett. **20**, 106 (1968).

[27] D.T. Adroja, S.K. Malik, B.D. Padalia and R. Vijayaraghavan, Solid State Commun.**71**, 649 (1989).

[28] S.K. Malik, D.T. Adroja, M. Slaski, B.D. Dunlap and A. Umezawa, Phys. Rev. B **40**, 9378 (1989).

[29] B.C. Sales and R. Viswanathan, J. Low Temp. Phys. **23**, 449 (1976).

[30] K.H.J. Buschow and J.F. Fast, Z. Phys. Chem. **50**, 1 (1966).

[31] N. Anderi, R. Furuya and M. Lowenstein, Rev. Mod. Phys. **55**, 331 (1983).

LOW-TEMPERATURE MAGNETOTRANSPORT OF THE INTERMETALLIC
ACTINIDE COMPOUND NpPt$_3$

M. Amanowicz[a], C.Ayache[a], H.Kitazawa[a)d], S.Kwon[a)c], Y.Ohe[a)c],
J.Rebizant[b], J.C.Spirlet[b], J.Rossat-Mignod[a], T.Suzuki[c] and T.Kasuya[c]

[a]Centre d'Etudes Nucléaires, DRFMC/SPSMS, BP 85X
38041 Grenoble-céde, France
[b]T.U.I.Karlsruhe, 7500 Karlsruhe, Germany
[c]Tohoku University, Sendai 980, Japan
[d] The Institute of Physical and Chemical Research (RIKEN)
Wako Saitama 351-01, Japan

INTRODUCTION

The study of actinide intermetallics AnPt$_3$ is interesting in many respects. The
prototype in the series is the well-known uranium compound, UPt$_3$, characterized by the
high γ-value of its electronic specific heat (γ=420 mJ.mole^{-1}.K^{-2}) and which becomes
superconducting below T$_c$=0.5 K [1]. The superconducting state involves heavy mass
electrons resulting from the hybridization between the broad electronic band structure and
the localized 5f states. It is thus interesting to follow the evolution of the Fermi liquid
parameters of the heavy electrons while reducing the spatial extension of 5f electrons is
reduced without changing the matrix. This can be realized by spanning the actinide
sequence.

On the other hand, antiferromagnetic order with a highly reduced ordered momentum has
been evidenced below T$_N$=5 K in UPt$_3$ [2] and substituting a few percent of Pt ions by
other metallic ions like Th, Pd or Au, strongly reinforces magnetism. Understanding the
role of magnetic fluctuations is one of the essential goals in order to elucidate the nature of
the superconductivity in UPt$_3$ for which some authors have invoked an odd-coupling

mechanism. Comparison with magnetically ordered parent compounds thus appears usefull. Such is the case of $NpPt_3$ which has been reported to be antiferromagnetically ordered below $T_N=22$ K [3]. More recently, resistivity measurements indicated rather a value close to 17 K [4]. However, as will be shown below and in a companion article [5], it is now established that the A.F.-order develops below $T_N\#29$ K. This is clearly settled owing to ^{237}Np Mössbauer spectroscopy and to neutron diffraction. In addition, Mössbauer spectroscopy provides the valency 4+ for neptunium. This result is important for a good understanding of the band properties in $AnPt_3$ compounds while in the case of UPt_3 the valency of uranium is still controversial.

However, in order to compare consistently the electronic and magnetic properties of the $AnPt_3$ compounds, one must take care of their different crystallographic structures [6]. Both UPt_3 and $NpPt_3$ belong to the space group $P6_3/mmc$ but a major change concerns the stacking sequence of the $AnPt_3$ planes. In UPt_3 the stacking sequence is ABAB... and leads to the crystal structure of the $MgCd_3$-type. The unit crystallographic cell thus contains two unit formulas with U in hexagonal coordination.. The stacking sequence of $NpPt_3$ is of the ABACABAC...type which generates the $TiNi_3$ crystal structure. Each unit cell contains four unit formulas. That structure accomodates two distinct crystallographic sites for Np: one cubic and one hexagonal, and this predicts different magnetic ground states for the Np ions. On the other hand, the volume associated with each unit formula is slightly reduced from UPt_3 (70.42 $Å^3$) to $NpPt_3$ (69.85 $Å^3$) and $PuPt_3$ (cubic, 69.27 $Å^3$). These relatively small volume changes mask more important ones concerning the lattice parameters which vary in an opposite direction. The a-parameter of $NpPt_3$ (5.806 Å) is stretched relatively to that of UPt_3(5.760 Å) while the mean interlayer spacing is more contracted in $NpPt_3$ (4.785 Å= a/2 Å) than in UPt_3(4.902 Å). These opposite trends should have an effect on both the electronic and the magnetic properties.

In the present article we report measurements of the magnetoresistance and Hall effect of $NpPt_3$ obtained for temperatures ranging from 450 mK to 50 K and for magnetic fields up to 12 T. These measurements are interestingly discussed in terms of the $NpPt_3$ magnetism and by comparison with UPt_3. However, several of our present conclusions should be considered as provisional, waiting for a refined characterization of the material investigated.

SAMPLE PREPARATION AND EXPERIMENTAL CONDITIONS

$NpPt_3$ was prepared by arc melting, directly starting from pure metallic neptunium and platinum. Then it was annealed at 1500°C for one week. Metallography indicates that grains of 3-4 mm were grown in this way. Most probably, this preparation technique provides a highly textured polycrystalline material. Neutron diffraction studies are in project for a confirmation of this point.

Two samples were cut from the same ingot. They have the form of platelets whose dimensions are approximatly 3-4mm x 2mm x 1mm. Consequently, it is quite possible that each sample contains a few grain boundaries accross its length.

We have measured the galvanomagnetic effects using two different contact geometries. Sample 1 was first mounted with a square configuration of contacts adapted to the determination of the Hall coefficient. However the relative changes of the resistance with temperature and magnetic field were also deduced from the even part of the measured voltage on field reversal. The current flows parallely to the platelet and the magnetic field is directed perpendicularly to it. Then the same sample was mounted in a second geometry with contacts aligned, enabling a direct determination of the resistance. Current still flows parallely to the platelet but the magnetic field now is parallel to the platelet and perpendicular to the current direction. Sample 2 was mounted using this second geometry. A helium-3 insert was specially designed for receiving neptunium materials. The lowest temperature presently achievable is 450 mK. Samples are encapsulated under a glove box in a contaminant free container which can be pluged into the insert. Electrical contacts are insured by beryllium copper springs.Electrical measurements are done either with a dc or an ac technique. Magnetic fields up to 12 T are provided by a superconducting magnet.

EXPERIMENTAL RESULTS

The temperature dependence of the resistivity of $NpPt_3$ is shown between 2 K and room temperature for sample 2, together with similar data obtained for UPt3 in the hexagonal plane [7] and for the mixed compound, $U_{0.5}Np_{0.5}Pt_3$, normalized to $NpPt_3$ at R.T. The RRR for $NpPt_3$ is of the order of 20, i.e. the same order as that measured previously by Van Sprang et al [8]. However, we presently obtain ϱ(R.T.)# 220 $\mu\Omega$.cm very similar to the value of UPt3 for the hexagonal plane. It is possible that the value of 80 $\mu\Omega$.cm obtained by Van Sprang et al results both of the textured nature of the sample and of their experimental setup [4] which could favour the c-axis component. The temperature dependences of $NpPt_3$ and UPt3 are very similar with the characteristic rounding due to spin fluctuation scatering whereas the alloy exhibits a much flatter resistance in the higher temperature interval.

Figure 2 shows the relative variation of the resistance of sample 1, first geometry between 1.3 K and 50 K for different applied magnetic fields. In this temperature interval, the resistance increases nearly by a factor ten for all fields.This change is fairly consitent with that of sample 2 reported above and indicates that the deconvolution presently made is correct. At H=0 T, the R(T) curve is mainly characterized by two slightly marked anomalies. At $T_N \approx 29$ K, a kink separates the higher temperature region

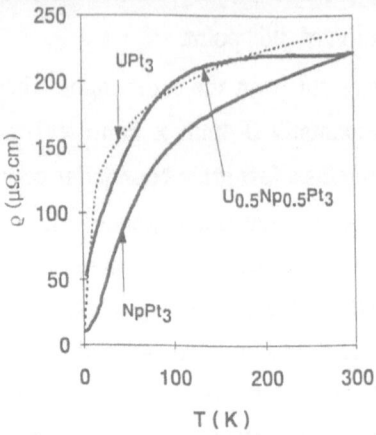

Figure 1. Electrical resistivity of NpPt$_3$, UPt$_3$ [7] and U$_{0.5}$Np$_{0.5}$Pt$_3$.

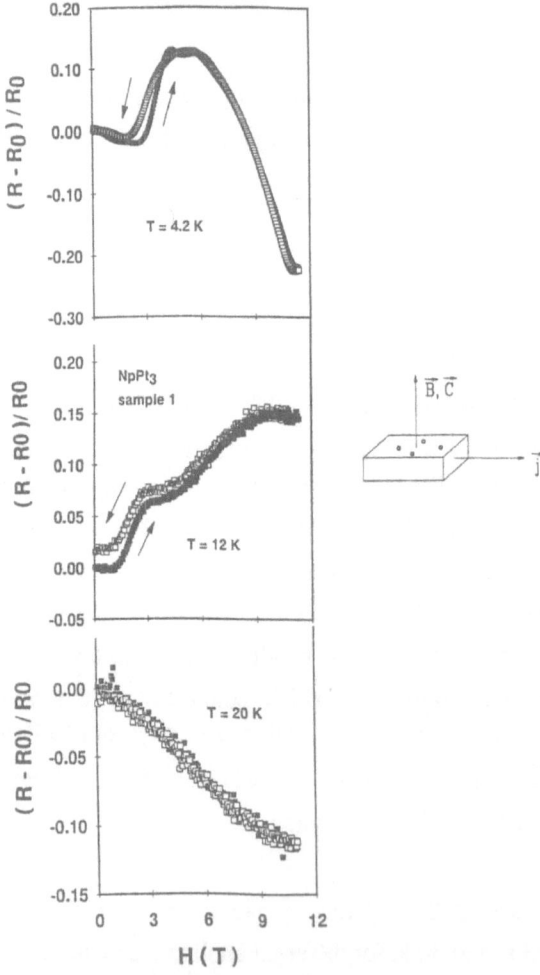

Figure 2. Temperature variations of the resistance of sample 1 at different applied fields. For clarity, the different curves are shifted by +2 relatively to each other. For higher fields, the H=0 T-curve is also reproduced by a thin line for comparison. The assumed experimental geometry is indicated.

with a negative curvature from the lower temperature one in which R increases more rapidly with T. Mössbauer spectroscopy and neutron diffraction clearly observe the development of magnetic order below the same temperature [5]. Between $T \approx 15$ K and $T \approx 18$ K, an excess of resistance, like a plateau, is observed. This could correspond to a similar excess observed in the magnetic susceptibility [9]. The anomaly at 29 K is not at all affected by the application of a magnetic field but the plateau is progressively changed into a peak shape. Despite this effect, the magnetoresitance remains moderate especially far from this temperature interval. However the R(H) curves at constant temperature reveal a more complex behaviour as shown in Figure 3. At lower temperatures, typically for T< 9-10 K, the magnetoresistance is non monotonous. The curve at T=4.2K for instance, shows a step-like increase around H=3T followed by a maximum and a negative contribution which approximately varies like H^2 at higher fields In the higher field range, R(H) is reversible upon increasing and decreasing field while the step is not. For $T \geq 10$ K , the resistance is always increasing. In Figure 3 the result for T=12 K is reported. The irreversible step is still present but it is less intense and located at lower field. At higher field, the negative contribution is replaced by a shoulder which could indicate the presence of a second step. Such a positive magnetoresistance is maintained up to the upper part of the peak in the resistance namely up to $T \cong 18$ K. However the step progressively broadens and disappears, and it loses irreversibility. For T>18 K, the magnetoresistance is monotonous and negative. At T=20 K, an inflexion is still observed but at higher temperatures a H^2- law dominates all the experimental field range.

It is possible to consider the magnetoresistance results for the first geometry, as the superposition of two contributions: 1/ a negative contribution which varies as H^2 at all temperatures and 2/ a positive contribution which varies rapidly with one or eventually two steps. Hysteresis clearly affects the lower step. This first step is easily associated with the metamagnetic effect observed in magnetization measurements [9]. For H=11 T, the negative contribution decreases from about 35% at T=2 K to 12% at T=20 K. In the intermediate region it is masked by the different structures of the positive part. The decrease is very strong between T=20K and T=30K where it reaches 2%. At T=50K it is limited to about 1%. Thus clearly the strong magentoresistance for this first geometry is a character of the ordered state as is the positive one.

The variations of the resistivity of sample 2 between 2 K and 50 K are shown in Figure 4a for different applied fields ranging from H=0 T to H=11.4 T. The agreement also holds as concerns the kink at $T_N \approx 30K$ which separates the upper range with negative curvature from the lower range. However, the other anomalies appear to be different from those observed in sample 1. At a temperature $T \approx 19.5K$, $\varrho(T)$ exhibits a drop of about 20% instead of the small plateau excess observed in sample1 with an onset at 18 K. A bump follows this drop on decreasing temperature; this has a lower inflexion point at $T \approx 16$ K close to the inflexion point at $T \approx 15$ K in sample 2. Another broad bump located around 10 K is also shown much better than in sample 1. Measurements performed down to

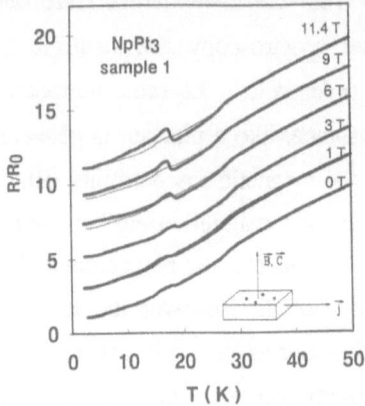

Figure 3. Magnetoresistance of sample 1, shown at three characteristic temperatures in the ordered phase: T= 4.2 K, 12 K and 20 K . The experimental geometry is the same as in Figure 2.

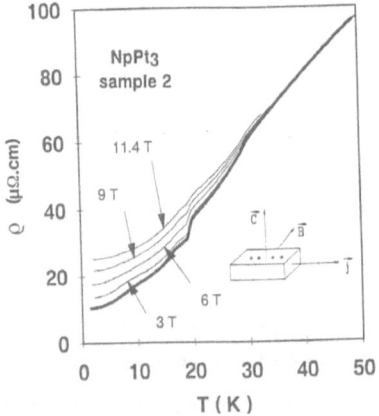

Figure 4a Temperature dependence of the resistivity of sample 2 at several applied fields. The thick line represents the H=0 T-curve. The assumed geometry is shown in insert.

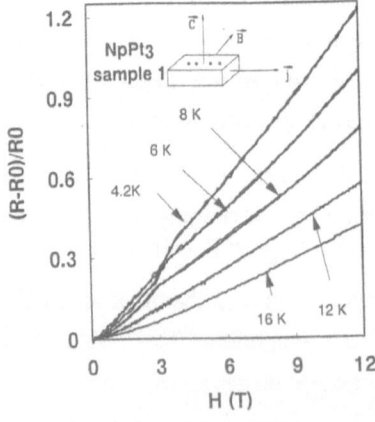

Figure 4b. Magnetoresistance of sample 1 with the same geometry as in Figure 4a. Similar data where obtained for sample 2.

450 mK showed no further irregularity. The effect of the magnetic field is more dramatic than in the case of sample 1. In the lower temperature range , e.g. at T=2 K, the magnetoresistance is positive and very stong, of the order of 150% at H=11.4 T. The other dramatic effect of the field is to wipe out all the anomalies in the $\varrho(T)$ curve. One is left only with a T^2-law for all the temperatures below T_N. Even the anomaly at $T_N\approx30K$ is progressively smoothened. However the position of the different anomalies do not shift significantly with field.

In order to clarify the above differences in $\varrho(T)$, sample 1 was remounted in the same geometry as sample 2. The result is surprising as it fully agrees with that obtained with sample 2 as well in magnitude and as regards the details of the anomalies.

This agreement is further comforted by the field dependence.In the present case it is always positive and monotonous. Corresponding results are shown for sample 1 in Figure 4b. At T=4.2 K, ϱ varies first like H^2 only up to about H≤1 T. A kink is observed about 3 T but without hysteresis. Above this kink, $\varrho(H)$ seems to have a linear contribution whereas the quadratic one is considerably reduced.

The preceeding results clearly establish than the observed differences in the $\varrho(t)$ (fig4a vs fig 2) and $\varrho(H)$ (fig 4b vs fig 3) dependences are due to differences between the measuring geometry rather than between samples. This should be related to the certainly high-textured samples expected from the synthesis method. As stated above, the value of the resistivity measured along the platelets is close to that reported for UPt_3 in the hexagonal plane.

Thus it is fairly reasonable to assume that current is flowing along the hexagonal plane also for our samples. The differences in the anomalies of the ordered region can be attributed to several reasons:1/the measuring method for which the deconvolution procedure could be deffective though correct in a first approximation, 2/one or more grain boundaries could be located between the contacts and bias the results following the contact geometry, 3/ a possible domain structure or even an intrinsic   anisotropy within the hexagonal plane could lead to different effects on the R(T) curve depending on the orientation within the plane. This last point seems to be excluded as the observed differences are stable as shown from thermal cycling and from the similar behaviours of the two samples when mounted similarly.

Figure 5a shows the temperature variations of the Hall coefficient R_H at different fields. Above T_N, only marked by a change in slope, $R_H(T)$ presents a positive curvature indicating an anomalous contribution. Data must be extended to higher temperatures in order to separate normal and anomalous contributions in a confident way. A shoulder is observed around T≈25 K and a maximum around T≈15K. The latter slightly shifts when field increases.

Below T≈20 K, R_H is no more linear in magnetic field (Figure 5b). There is a general reducing trend but, in addition, one has several maximums or kinks corresponding to the step anomalies observed in the magnetoresistance. In particular hysteresis is observed for the lowest field anomalies.

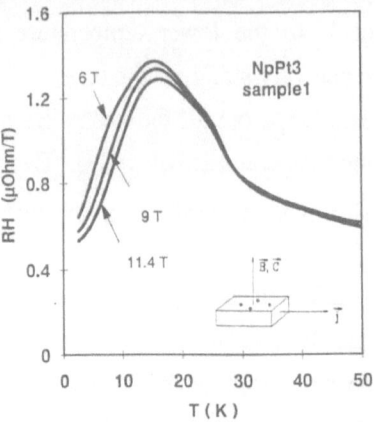

Figure 5a. Temperature dependence of Hall coefficient at different applied fields. The differences at low temperature reveal the non-linear field dependence.

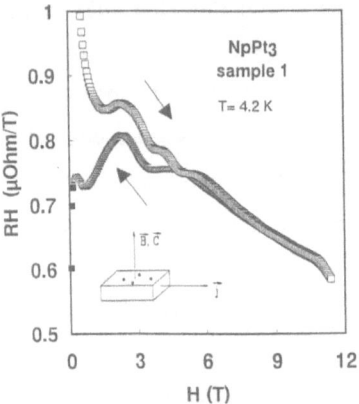

Figure 5b. Hall coefficient at T=4.2 K as a function of magnetic field. This curve shows hysteretic behaviour, field dependence and ondulations which correspond to the steps in magnetoresistance.

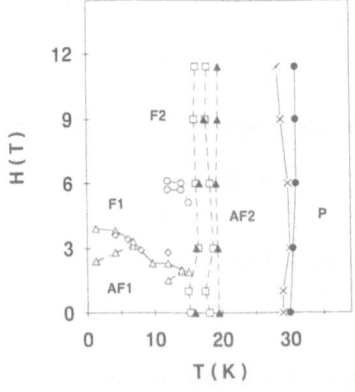

Figure 6. Magnetic phase diagram of NpPt3. This diagram is deduced from the magnetoresistance data of samples 1 (open symbols) and 2 (full symbols). H is assumed perpendicular and parallel to the hexagonal plane for the first and second geometry, respectively.

DISCUSSION

The above results are interestingly discussed in terms of the presence of an antiferromagnetic ordering below $T_N=30K$.

Provided one retains the hypothesis concerning the two geometries used, when H is parrallel to c, one has a complicated effect on transport due to the combined effect of a positive contribution connected with a first order spin-flop transition and of a strong negative contribution in the ordered state. This could indicate that applying the field along c decreases the magnetic disorder in the plane. But the spin-flop transition has just an opposite effect. The higher field step indicate that the transition from the AF zero field state to the induced ferrimagnetic state could be rather complex both as a function of T and H. Preliminary results with H//ab plane seem indicative of a spin-flop transition also for this orientation. But the main result concerns the very strong positive MR and could indicate a strongly increased disorder when H is applied in that direction and it smoothes all the anomlies connected with different transitions.

The position of the anomalies at T_N and in the range 15-19K, appear less sensitive to orientation and intensity of the magnetic field.They could possibly involve some structural transitions though Mössbauer and neutrons show that T_N corresponds to a true magnetic order.All these characteristics are summarized in the H-T phase diagram where vertical lines separate the paramagnetic state from The AF2, possibly modulated phase and the latter from a second low temperature one:AF1. The separation between the two AF phases looks complex. At low T, an equaly complex transition is predicted from AF1 to the strong field ferrimagnetic state.

Finally, let us remark that some similarities exist with the magnetoresistance of UPt$_3$ [7,9,10]. However strong differences also exist and further work should aim at a refined comparison on the basis of a better description of the microscopic magnetic order.

REFERENCES

[1] G.R. STEWART , Z. FISK , J.O. WILLIS and J.L. SMITH, Phys.Rev.Lett. 52 (1984) 679.

[2] G. AEPPLI, E. BUCHER, C. BROHELM, J.K. KJEMS, J. BAUMANN and J. HUFNAGE, Phys.Rev.Lett.60 (1988) 615.

[3] B. ERDMANN and C. KELLER, J.Solid State Chemistry, 7 (1973) 40

[4] M. van SPRANG, PhD Thesis, University of Amsterdam (1989)

[5] J.P. SANCHEZ et al., SCES' 92, Sendai sept. 1992.

[6] J. REBIZANT, unpublished.

[7] A. de VISSER, A.MENOVSKY and J.J.M.FRANSE, Physica 147B (1987) 81.

[8] M.Van SPRANG, J.J.M.FRANSE, J.M.ROSSAT-MIGNOD,J.M.FOURNIER,J.CHIAPUSIO and J.C.SPIRLET, J.Physique 49 (1988) C8-975

[9] K.BEHNIA, thesis, Université de Paris-Sud, Orsay (1990).

[10]G.REMENYI, U.WELP, J.FLOUQUET, J.J.M.FRANSE, and A.MENOVSKY, J.Magn.Magn.Mat.63-64(1987) 391.

SPECIFIC HEAT OF SOME URANIUM–BASED TERNARY COMPOUNDS

Toshizo Fujita,[1] Shin-ichi Ikeda,[1] Shijo Nishigori,[1]
Yuji Aoki,[2] Toshiro Takabatake,[3] and Hironobu Fujii[3]

[1]Department of Physics, Hiroshima University
 Higashi–Hiroshima 724, Japan
[2]Department of Physics, Tokyo Metropolitan University
 Hachioji, Tokyo 129–03, Japan
[3]Faculty of Integrated Arts and Sciences, Hiroshima University
 Hiroshima 730, Japan

INTRODUCTION

Unusual low–temperature properties of uranium–based intermetallic compounds originate from $5f$ electrons of uranium atoms, which have an intermediate nature between itinerant and localized characters. The $5f$ states are more extended in space than $4f$ orbits in rare earth compounds. Usually the $4f$ electrons of rare earth atoms are well localized inside the $5s$ and $5p$ shells. On the other hand, the $5f$ electrons are not so delocalized as compared with d electrons in transition metals, for which an itinerant description is suitable.

The degree of itinerancy is frequently discussed using the Hill criterion,[1] in which the atomic spacing d_{U-U} between neighboring uranium sites is an important parameter. For $d_{U-U} < 3.5$ Å, the $5f$ wave functions of neighboring uranium atoms overlap with one another and the $5f$ electrons have an itinerant character, whereas, for $d_{U-U} > 3.5$ Å, the $5f$ electrons are expected to be localized as the $4f$ electrons in rare earth compounds. In uranium–based intermetallic compounds, however, the $5f$ states hybridize with s, p or d bands of ligand elements in the vicinity of Fermi level. Such hybridization mainly determines the degree of delocalization of the $5f$ states for large d_{U-U} and gives rise to a wide variety of unusual magnetic and transport properties in uranium compounds at low temperatures.

In binary systems of $U_k X_m$, the hybridization of the $5f$ electrons with p electrons derived from a metalloid or non–transition metal element X is essential, while the hybridization with d electrons from transition metal element T plays an important role in compounds $U_k T_l$. In the case of weak hybridization, the Hill criterion holds true and a localized character of the $5f$ electrons is observed for large

$d_{\text{U-U}}$. For strong hybridization, the $5f$ electrons are delocalized even for large $d_{\text{U-U}}$ and the systems exhibit metallic conduction as well as Pauli paramagnetism. The most interesting is the case of intermediate hybridization, in which the systems display various exotic phenomena including unusual magnetic ordering, heavy–fermion behavior, spin–fluctuating state and unconventional superconductivity. In addition, these consequences of hybridization delicately depend on the type of ligand elements X or T. In ternary compounds $U_kT_lX_m$, the $5f$ electrons hybridize with both of the p electrons from X and the d electrons from T, and the hybridization can be controlled systematically if we choose a proper series of compounds.

Specific heat measurement provides a useful parameter to experimentally characterize the effects of hybridization, since the Sommerfeld coefficient γ or the low–temperature ratio C/T of electronic specific heat C to temperature T is closely related to the density of states $N(E_F)$ at the Fermi level E_F. Metallic systems with widely extended conduction bands show a small γ value, typically smaller than 30 mJ/K^{2}mol. In strongly localized systems or insulators, the γ value should be zero. In the intermediate regime, however, uranium compounds exhibit a remarkably enhanced γ value, in some cases larger than 100 mJ/K^{2}mol, suggesting a renormalized narrow band of quasiparticles with heavy effective mass at E_F.

In this article, we review recent studies on the specific heat of uranium–based ternary compounds $U_kT_lX_m$. A large collection of the γ values is given in the next section to show an outline of the general trend as well as the individual characters which the ternaries exhibit. To demonstrate some consequences of hybridization, we present the specific heat studies which we have recently performed on three series of compounds.

URANIUM–BASED TERNARY COMPOUNDS

In Table 1, we list crystal structures and Sommerfeld coefficients of various uranium–based ternary compounds $U_kT_lX_m$ which have been reported in recent literatures. The uranium–uranium spacings $d_{\text{U-U}}$ of the ternaries are usually greater than the Hill limit of $d_H = 3.5$ Å. Of equiatomic systems UTX ($k=l=m=1$), hexagonal compounds have $d_{\text{U-U}}$ ranging between 3.5 and 4.0 Å, whereas cubic compounds, including UNiSn, UPtSn and URhSb, have a larger $d_{\text{U-U}}$ value than 4.0 Å. Most compounds of this class undergo ferromagnetic or antiferromagnetic transition. UPdIn exhibits double magnetic transitions and has a large C/T value of 280 mJ/K^{2}mol at 1.5 K in the lowest–temperature phase. An antiferromagnetic transition is accompanied by a semiconductor to metal transition in UNiSn.

In all other ternaries $U_kT_lX_m$, the nearest neighbor uranium–uranium distance $d_{\text{U-U}}$ is larger than 4.0 Å. It is noted that cubic compounds have particularly large $d_{\text{U-U}}$ values. Without hybridization, $5f$ electrons would be well localized in these compounds. However, not a few compounds exhibit metallic conduction. Furthermore, some of them become superconducting at low temperatures. There are many antiferromagnets also in this class of ternaries. The C/T ratios at low temperatures are widely distributed. UPt$_4$Au is reported to have an exceptionally large value of $C/T = 725$ mJ/K^{2}mol. An interesting change has been found in low–temperature transport properties of UCu$_{3+x}$Ga$_{2-x}$ with varying x, although no apparent anomaly is seen in $d_{\text{U-U}}$ or C/T.

Table 1. Crystal structures, uranium–uranium spacings d_{U-U}(Å) and Sommerfeld coefficients C/T(mJ/K^{2}mol) of various ternaries $U_kT_lX_m$. Symbols (c), (t), (o) and (h) denote the cubic, tetragonal, orthorhombic and hexagonal structures, respectively. Most of C/T are evaluated at the lowest temperature of measurements below 2 K or by extrapolation to 0 K. Superconducting(SC), antiferromagnetic(AF), ferromagnetic(F), structural(CS) transition temparatures and Kondo(K) temperature are given in Kelvin in Remarks. P means paramagnetic, PP Pauli–paramagnetic, CW Curie–Weiss paramagnetic and NM non–magnetic behavior.

Compounds $U_kT_lX_m$	Structure	d_{U-U} (Å)	C/T (mJ/K^{2}mol)	Remarks	Reference
UCoAl	ZrNiAl(h)	3.48	68	AF16K	2
UNiAl	ZrNiAl(h)	3.51	164	AF19K	3,4,5
URuAl	ZrNiAl(h)	3.61	45	P	2
UCoGa	ZrNiAl(h)	3.5	40	F47	2
UNiGa	ZrNiAl(h)	3.51	59	F36	2,3
URuGa	ZrNiAl(h)	3.73	52	P	2
URhGa	ZrNiAl(h)	3.67	40	F40	6
UIrGa	ZrNiAl(h)	3.68	41	F60	6
UPtGa	ZrNiAl(h)	3.66	72	F68	6
UPdIn	ZrNiAl(h)	3.87	280	AF21,8.5	7
UCoSn	ZrNiAl(h)	3.86	61	F80	2,3
UNiSn	MgAgAs(c)	4.53	20	AF43	8,9
UCuSn	CaIn$_2$ (h)	3.62	53	AF60	9
URuSn	ZrNiAl(h)	3.98	50	F53	2
UPdSn	CaIn$_2$ (h)	3.65	4.3	AF29	3
UPtSn	MgAgAs(c)	4.63	11	K75	3
URhSb	MgAgAs(c)	4.62	2.1	K40	3
UPdSb	CaIn$_2$ (h)	3.61	62	F65	3
UAu$_2$Al	YPd$_2$Si(o)	4.03	102	AF25	10
UNi$_2$Ga	ZrPt$_2$Al(h)	4.21	62	P	ours[*]
UPd$_2$Ga	YPd$_2$Si(o)	3.83	172	AF6.5	ours[*]
UAu$_2$In	MnCu$_2$Al(c)	4.60	60	AF61,K40	11
UNi$_2$In	MnCu$_2$Al(c)	4.60	45		12
UPd$_2$In	MnCu$_2$Al(c)	4.81	200	AF20,CT180	13
UNi$_2$Sn	MnCu$_2$Al(c)	4.58	52		12
UPd$_2$Sn	YPd$_2$Si(o)	>4.0	80	K81	14
UCu$_2$Sn	ZrPt$_2$Al(h)	4.46	60	AF16.6	ours[*]
UPt$_2$Sn	ZrPt$_2$Al(h)	4.55	17	AF60	ours[*]
UAl$_2$Si$_2$	Cu$_3$Au(c)	>4.0	27.9	PP,SC1.3	15
UPt$_2$Si$_2$	CaBe$_2$Ge$_2$(t)	4.19	35	AF35	16
URu$_2$Si$_2$	ThCr$_2$Si$_2$(t)	4.13	75	AF17,SC1.31	17 (cont.)

Table 1 (Cont.)

UPt_2Ge_2	$CaBe_2Ge_2$(t)	4.33	14	AF72	16
UNi_2Al_3	$PrNi_2Al_3$(h)	4.02	125	AF4.6,SC1	18
UPd_2Al_3	$PrNi_2Al_3$(h)	4.20	150	AF14,SC2	19
UPt_4Au	$AuBe_5$(c)	5.29	725~900	NM	15,20,21,22
UPt_4Ir	$AuBe_5$(c)	5.23	45	NM	20
$UCu_{3.1}Ga_{1.9}$	$CaCu_5$(h)	4.13	370	AF9.4	ours[*]
$UCu_{3.3}Ga_{1.7}$	$CaCu_5$(h)	4.14	205	AF17.4	ours[*]
$UCu_{3.8}Ga_{1.2}$	$CaCu_5$(h)	4.15	285	CW	ours[*]
U_2PtSi_3	AlB_2(h)	3.97	200	WF?	23,24,25
$U_2Pt_{15}Si_7$	(c)	5.9	100	CW	25,26
$U_3Ni_3Sn_4$	$Y_3Au_3Sb_4$(c)	4.38	92	CW	12,27
$U_3Cu_3Sn_4$	$Y_3Au_3Sb_4$(c)	4.45	380	AF12	27
$U_3Pt_3Sn_4$	$Y_3Au_3Sb_4$(c)	4.52	94	CW	27
$U_3Au_3Sn_4$	$Y_3Au_3Sb_4$(c)	4.59	280	CW	27
$U_3Ni_3Sb_4$	$Y_3Au_3Sb_4$(c)	4.39	2	F?	27

[*] The details will be reported elsewhere.

It is difficult to find any simple systematics concerning the relation between d_{U-U} and low–temperature properties listed in Table 1. This suggests importance of hybridization depending on ligand elements. Different consequences of hybridization are described in the following sections.

DOUBLE MAGNETIC TRANSITIONS IN UPdIn

Brück et al.[28] have predicted that the ternary compound UPdIn is a heavy–fermion compound with a spontaneous ferromagnetic moment of 0.30 μ_B per U atom in the ground state. The crystal structure of hexagonal ZrNiAl–type gives rise to a strong anisotropy in various properties. The nearest–neighboring uranium atoms in the c plane form a two–dimensional network with the uranium–uranium distance of d_{U-U} = 3.87 Å. Unusual anisotropy was found to become clear in the temperature dependence of resistivity[7] ρ below 50 K. As temperature decreases, the resistivity ρ_c along the c axis decreases initially, but it exhibits a rapid upturn below 50 K, and finally tends to saturation below 10 K. In contrast, the a–axis resistivity ρ_a rapidly decreases below 21 K and shows a T^2–dependence in the lowest temperature range below 4 K.

In Fig. 1, the ratio of specific heat[7] to temperature C/T is plotted as a function of T for UPdIn and the non–magnetic reference material ThPdIn. The 5f contribution to the Sommerfeld coefficient C_m/T is estimated by subtracting C/T of ThPdIn from the ratio of UPdIn. Two peaks are clearly found in C_m/T at 21 and 8.5 K, indicating

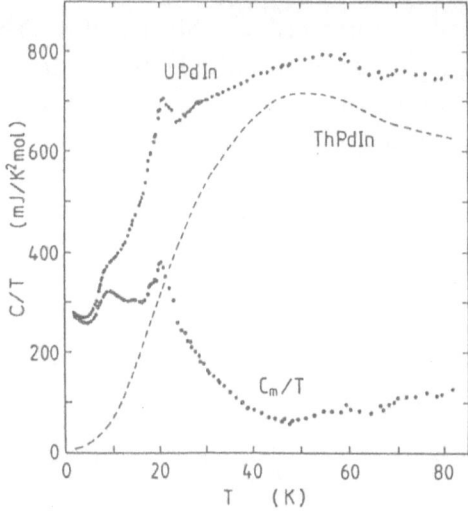

Fig. 1. Temperature dependence of C/T for UPdIn. The dotted line show the data for isostructural ThPdIn. The magnetic contribution C_m/T estimated by subtracting C/T of ThPdIn from C/T of UPdIn is also shown.

double phase transitions. With further decreasing temperature, C_m/T exhibits an upturn below 4 K and attains 280 mJ/K^{2}mol at 1.5 K. Thus the specific heat measurements not only confirmed the prediction by Brück et al.[28] for the realization of a magnetic heavy–fermion state, but also revealed double phase transitions. Magnetic entropy S_m is roughly estimated to attain $R\ln2$ at 20 K and is saturated at $R\ln3$ around 30 K. This temperature dependence of entropy suggests that a triplet ground state is responsible for the low–temperature properties.

A recent neutron diffraction study on a single crystal[29] revealed that a c–axis incommensurate structure of the uranium moments coupled ferromagnetically within a c plane with a propagation vector $k = (0, 0, 0.405)$ is realized below 21 K. The magnetic structure changes at 8.5 K into a square–up structure with the stacking sequence +−++− of the ferromagnetic c–plane sheets along the c–axis with $k = (0, 0, 0.400)$. This arrangement of uranium moments yields a net moment of $0.3\mu_B$ per U atom, whereas each uranium atom has a moment of ~$1.5\mu_B$. The large value of $C_m = 280$ mJ/K^{2}mol at 1.5 K indicates that the heavy–fermion nature survives in the ferromagnetic phase below 8.5 K.

PARAMAGNETIC SEMICONDUCTOR TO ANTIFERROMAGNETIC METAL TRANSITION IN UNiSn

A cubic compound UNiSn attracts a special interest since a semiconductor to metal transition[3] occurs in concurrence with an antiferromagnetic ordering. In this compound, the uranium–uranium interatomic distance, $d_{U-U} = 4.53$ Å, is larger than the critical value of $d_H = 3.5$ Å. Hence, the 5f electrons are expected to be localized. In fact, the electrical resistivity exhibits a semiconductor–like temperature dependence at high temperatures and a pronounced peak around 50 K. The activation energy E_g in the semiconducting region depends critically on sample quality; E_g = 67 meV is reported above 200 K for a well annealed sample.[30] At low temperatures below 50 K, however, the resistivity decreases substantially and shows a metallic temperature-dependence.

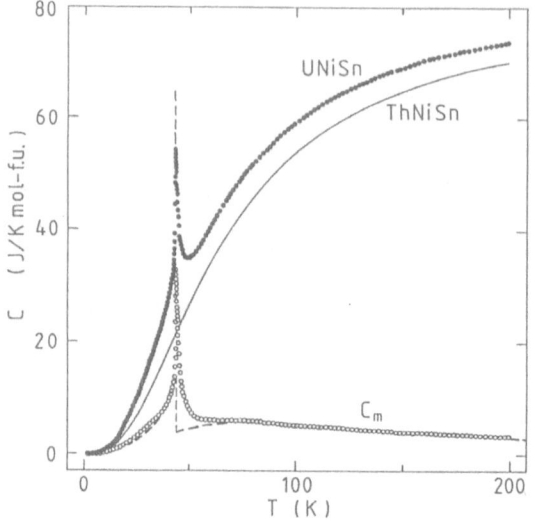

Fig. 2. Specific heat C of UNiSn and ThNiSn. The $5f$ contribution C_m is eatimated by subtracting C of ThNiSn from C of UNiSn. The broken curve is a calculated one for C_m.

The powder neutron diffraction[31] revealed that an antiferromagnetic order of type-I with single k was established in the low-temperature phase. The Néel temperature T_N was determined to be 45±2 K. However, the $M(110)$ magnetic intensity diminishes at T_N more rapidly than that expected for a typical antiferromagnetic transition. Similar behavior was reported for the ^{119}Sn Mössbauer spectra[32] of UNiSn, in which the hyperfine field discontinuously vanishes around T_N with increasing temperature, suggesting a first-order transition. It is also curious that the magnetic susceptibility[9,30] $\chi(T)$ exhibits a clear upturn below 43 K instead of a decrease following a cusp-like anomaly at T_N, although $\chi(T)$ follows the Curie–Weiss law in the paramagnetic high temperature phase.

In Fig. 2, the specific heat C is plotted as a function of temperature for UNiSn and ThNiSn. A sharp peak is found at 43 K in $C(T)$ of UNiSn, indicating a single transition from the paramagnetic semiconducting phase to the antiferromagnetic metallic phase. A low-temperature C/T vs T^2 plot for ThNiSn shows a linear variation up to 10 K with the Sommerfeld coefficient of γ = 0.9 mJ/K^{2}mol. The small γ value is ascribed to impurities or crystal imperfection since the resistivity and thermoelectric power[30] as well as the band calculation[33] indicate semiconducting characters for ThNiSn. To estimate the $5f$ contribution C_m to the specific heat of UNiSn, we subtracted C of ThNiSn from C of UNiSn; $C_m = C_{UNiSn} - C_{ThNiSn}$. The C_m/T vs T^2

curve is nearly linear below 7 K and the extrapolated value is 19 mJ/K^{2}mol at 0 K, which we adopt as the Sommerfeld coefficient. This is consistent with the metallic conduction induced in the antiferromagnetic phase. The Sommerfeld coefficient is substantially reduced below 7 mJ/K^{2}mol if 5% uranium is replaced by thorium.[34] On the other hand, both $C_m(T)$ and $\chi(T)$ in the paramagnetic semiconducting phase exhibit essentially the same values for UNiSn as the corresponding values per molar U for the diluted compound $(U_{1-x}Th_x)NiSn$.[34] This also supports the localized picture for the 5f electrons in UNiSn.

The 5f contribution $S_m(T)$ to the entropy calculated from $C_m(T)$ reaches $R\ln2$ around T_N = 43 K and tends to saturation approaching $R\ln5$ at 200 K. This fact suggests that five levels of the 5f electron lie in a energy range of 200 K and the lowest two of them split in the antiferromagnetic phase. As shown in Fig. 3, the low–temperature susceptibility $\chi(T)$ of the diluted compound $(U_{1-x}Th_x)NiSn$ tends to be temperature–independent in the paramagnetic phase below 50 K, suggesting that the ground state is a non–magnetic doublet, although $\chi(T)$ of UNiSn exhibits an unusual sudden increase below T_N = 43 K.

Assuming the tetravalent 5f^2 configuration for the uranium ion, we tried to reproduce the observed behavior within the framework of a simple localized model and a molecular field approximation. In a cubic lattice, a ground multiplet $^3H_4$ for the 5f^2 splits into a singlet Γ_1, a doublet Γ_3, and two triplets Γ_4 and Γ_5. We calculated the specific heat and magnetic susceptibility and chose the parameters involved so as to yield the best fit to the experimental data for the paramagnetic region. For the antiferromagnetic region, we introduced a molecular field H^{MF} = $-\lambda_{AF}M$, where λ_{AF} is the molecular field coefficient and the sublattice magnetization M was determined self–consistently. The detail will be described elsewhere.[34] As a result, we obtained the level scheme and wave functions proposed for the paramagnetic state, which are given in Table 2. The molecular field coefficient λ_{AF} = $0.1\Delta/g^2\mu_B^2$ reproduce a first–order antiferromagnetic transition at T_N = 43 K, where Δ is the energy separation between the ground state Γ_3 and the first excited state Γ_4. The estimated entropy jump ΔS is as small as 1.0

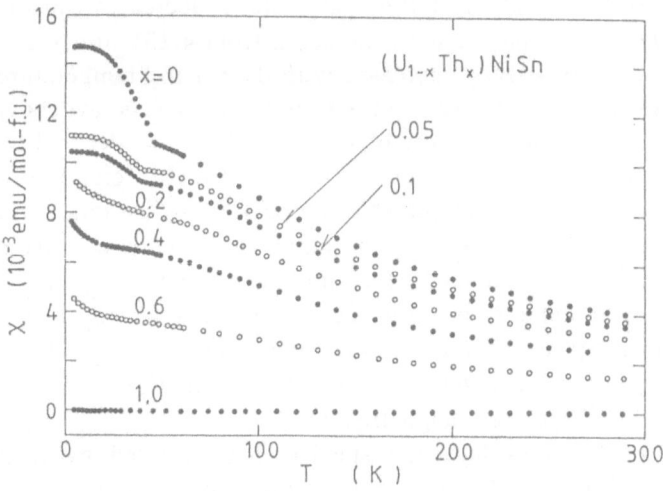

Fig. 3. Magnetic Susceptibility $\chi(T)$ of $(U_{1-x}Th_x)NiSn$.

Table 2. A proposed level scheme of U^{4+} in the cubic crystalline electric field of UNiSn.

State	Multiplicity	Energy
Γ_5	triplet	(~3000 K)
$\Gamma_1{}'$	singlet	430 K
Γ_4	triplet	180 K
Γ_3	doublet	0 K

J/Kmol, which is consistent again with our experimental result. The broken curve in Fig. 2 is the calculated specific heat using the parameters obtained above. Pauli paramagnetism due to the induced carriers in the metallic phase is one of possibilities explaining the unusual increase in $\chi(T)$ below 43 K, because the observed C_m/T value of 19 mJ/K^{2}mol leads to the Pauli susceptibility of ~10^{-3} emu/mol.

CHANGE OF HYBRIDIZATION IN $UCu_{3+x}Ga_{2-x}$

The off–stoichiometric compound $UCu_{3+x}Ga_{2-x}$, which crystallizes in the hexagonal CaCu$_5$–type structure, is available only in a limited range of $0.1 \leq x \leq 0.8$. The uranium–uranium distance is estimated to be $d_{U-U} = 4.13 \sim 4.15$ Å. As x increases from 0.1 to 0.8, the lattice parameter a decreases linearly from 5.144 to 5.068 Å, whereas the parameter c increases from 4.131 to 4.153 Å.

For all x, the resistivity increases with decreasing temperature down to 30 K, and the magnetic susceptibility follows the Curie–Weiss law above 150 K with the paramagnetic Curie temperature of $\Theta = -45$ K for $x = 0.1$ and $\Theta = -30$ K for $x = 0.8$. These results appear to suggest the localized character of 5f electrons. At low temperatures, the $x=0.1$ sample shows a peak in $\chi(T)$ around 10 K, indicating an antiferromagnetic transition. On the other hand, the resistivity $\rho(T)$ shows a broad peak around 30 K followed by a rapid upturn below T_N. As x increases to 0.3, T_N increases to about 18 K. With further increasing x, T_N begins to decrease and the antiferromagnetic ordering is missing at least down to 2 K for $x \geq 0.6$. At the limiting region of x close to 0.8, a pronounced reduction is observed in ρ below 20 K in the Curie–Weiss paramagnetic state.

Figure 4 illustrates how the specific heat divided by temperature C/T changes with varying x. At high temperatures above 20 K, C/T is essentially identical for $0.3 \leq x \leq 0.6$. A triangular peak is clearly observed for $x \leq 0.5$. The midpoint of the specific heat jump corresponds to the antiferromagnetic peak in $\chi(T)$. The Néel temperature T_N determined from C/T sifts from 9.4 K for $x = 0.1$ to 17.4 K for $x = 0.3$. With further increasing x, T_N decreases and the peak height of C/T diminishes substantially. In the case of $x = 0.6$, C/T shows an appreciable upturn below 10 K. The C/T value at the lowest temperature of 1.3 K initially decreases from 370 mJ/K^{2}mol for $x = 0.1$ to 205 mJ/K^{2}mol for $x = 0.3$ as T_N rises. As T_N is lowered by further doping Cu but the contribution of antiferromagnetic ordering to C/T reduces, the low temperature C/T value grows up to 435 mJ/K^{2}mol for $x = 0.6$. Moreover the $x=0.8$ sample has a smaller C/T of 285 mJ/K^{2}mol at 1.3 K.

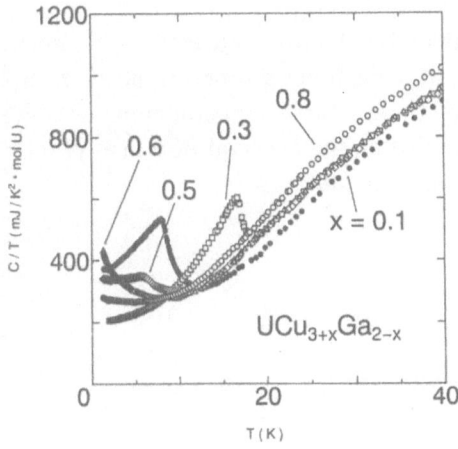

Fig. 4. The ratio C/T of specific heat to temperature of UCu$_{3+x}$Ga$_{2-x}$ for $x = 0.1$, 0.3, 0.5, 0.6 and 0.8.

The variation of C/T is not accidental but can be interpreted as a consequence of competition between delocalization and magnetic ordering. As x increases, the compound changes from a localized magnetic system to a heavy–fermion system. The initial growth of hybridization promotes the indirect magnetic coupling between localized magnetic moments of uranium ions, resulting in the antiferromagnetic ordering and depression of C/T. The increasing hybridization reduces the magnetic moment of uranium ions and forms an antiferromagnetic heavy–fermion state. Finally the delocalization prevails over the magnetic ordering and a moderate heavy–fermion state is realized due to the enhanced hybridization.

Plotted in Fig. 5 is the entropy which is calculated from the specific heat after the subtraction of the non–magnetic contribution estimated from the specific heat of the corresponding thorium compound. For most of x, the $5f$ contribution to the entropy $S_m(T)$ tends to approach a saturated value of $R\ln 3$ around 60 K, although S_m for $x = 0.8$ is somewhat larger than $R\ln 3$. The saturation appears to indicate a triplet ground state for uranium ions. The amount of entropy released by the antiferromagnetic ordering is remarkably reduced for $x = 0.5$.

CONCLUSION

We collected experimental data on specific heat of various uranium–based ternary compounds and discussed the low–temperature properties in terms of the degree of hybridization. The degree of hybridization, which depends delicately on ligand elements, governs the low–temperature properties of uranium–based ternary compounds with larger d_{U-U} than d_H.

The specific heat of UPdIn revealed that the compound undergoes double magnetic transitions; antiferromagnetic ordering at 21 K and magnetic structural change to the ground state with a ferromagnetic moment at 8.5 K. The observation of the large C/T value indicates the survival of a heavy–fermion state even in the low–temperature ferromagnetic state.

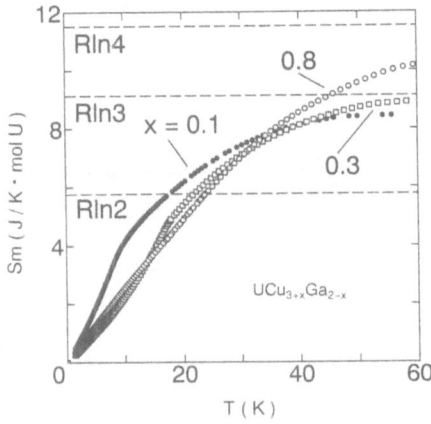

Fig. 5. The $5f$ contribution to the entropy S_m of $UCu_{3+x}Ga_{2-x}$ for $x = 0.1, 0.3$ and 0.8.

A sharp single peak was observed at 43 K in the specific heat of UNiSn, suggesting a first–order phase transition. This transition is an antiferromagnetic ordering accompanied by a semiconductor to metal transition. We analyzed the specific heat data of $(U_{1-x}Th_x)NiSn$ in conjunction with the magnetic susceptibility data. Assuming a simple localized model in the cubic crystalline field and introducing a molecular field approximation, we proposed a energy level scheme and wave functions for uranium ions so as to reproduce the temperature dependence of specific heat and susceptibility as well as the first–order transition. The small C/T value also implies that the $5f$ electrons are well localized in these compounds as suggested by a large d_{U-U}.

An off–stoichiometric series $UCu_{3+x}Ga_{2-x}$ shows a wide variety of low-temperature properties by varying x from 0.1 to 0.8. As x increases, the competition between antiferromagnetic ordering and delocalization give rise to a drastic change in the ground state from a localized magnetic state to a non-magnetic heavy–fermion state. The entropy consideration suggests a triplet ground state for the $5f$ electrons of uranium ions.

REFERENCES

1. H.H. Hill, in "Plutonium and Other Actinides," W.M.Miner, eds., AIME, New York (1970) p.2.
2. V. Sechovsky, L. Havela, L. Neuzil, A.V. Andreev, G. Hilscher and C. Schmitzer, J. Less Common Metals 121, 169 (1986).
3. T.T.M. Palstra, G.J. Nieuwenhuys, R.F.M. Vlastuin, J. van den Berg, J.A. Mydosh and K.H.J. Buschow, J. Magn. Magn. Mater. 67, 331 (1987).
4. L. Havela, V. Sechovsky, P. Nozar, E. Brück, F.R. de Boer, J.C.P. Klaasse, A.A. Menovsky, J.M. Fournier, M. Wulff, E. Sugiura, M. Ono, M. Date and A. Yamagishi, Physica B163, 313 (1990).
5. E. Brück, H.P. Van der Meulen, A.A. Menovsky, F.R. de Boer, P.F. de Chatel, J.J.M. Franse, J.A.A.J. Perenboom, T.T.J.M. Berendschot, H. Van Kempen, L. Havela and V. Sechovsky, J. Magn. Magn. Mater. 104&107, 17 (1992).
6. V. Sechovsky, L. Havela, N. Pillmayr, G. Hilscher and A.V. Andreev, J. Magn. Magn. Mater. 63&64, 199 (1987).
7. H. Fujii, H. Kawanaka, M. Nagasawa, T. Takabatake, Y. Aoki, T. Suzuki, T. Fujita, E. Sugiura, K. Sugiyama and M. Date, J. Magn. Magn. Mater. 90&91, 507 (1990).
8. Y. Aoki, T. Suzuki, T. Fujita, H. Kawanaka, T. Takabatake and H. Fujii, J. Magn. Magn. Mater. 90&91, 96 (1990).
9. H. Fujii, H. Kawanaka, T. Takabatake, E. Sugiura, K. Sugiyama and M. Date, J. Magn. Magn. Mater. 87, 235 (1990).
10. T. Takabatake, H. Iwasaki, H. Fujii, S. Ikeda, S. Nishigori, Y. Aoki, T. Suzuki and T. Fujita, J. Phys. Soc. Jpn. 61, 778 (1992).
11. M.J. Besnus, M. Benakki, J.P. Kappler, P. Lehmann, A. Meyer and P. Panissod, J. Less Common Metals 141, 121 (1988).
12. T. Takabatake, H. Fujii, S. Miyata, H. Kawanaka, Y. Aoki, T. Suzuki, T. Fujita, Y.Yamagushi and J. Sakurai, J. Phys. Soc. Jpn. 59, 16 (1990).
13. T. Takabatake, H. Kawanaka, H. Fujii, Y. Yamaguchi, J. Sakurai, Y. Aoki and T. Fujita, J. Phys. Soc. Jpn. 58, 1918 (1989).
14. C. Rossel, M.S. Torikachvili, J.W. Chen and M.B. Maple, Solid State Com. 60, 563 (1986).
15. H.R. Ott, F. Hulliger, H. Rudigier and Z. Fisk, Phys. Rev. B31, 1329 (1985).
16. T. Endstra, G.J. Nieuwenhuys, A.A. Menovsky and J.A. Mydosh, J. Magn. Magn. Mater. 108, 67 (1992).
17. W. Schlabitz, J. Baumann, B. Pollit, U. Rauchschwalbe, H.M. Mayer, U. Ahlheim and C.D. Bredl, Z. Phys. B62, 171 (1986).
18. C. Geibel, S. Thies, D. Kaczorowski, A. Mehner, A. Grauel, B. Seidel, U. Ahlheim, R. Helfrich, K. Petersen, C.D. Bredl and F. Steglich, Z. Phys. B83, 305 (1991).

19. C. Geibel, C. Schank, S. Thies, H. Kitazawa, C.D. Bredl, A. Bohm, M. Rau, A. Grauel, R.Caspary, R. Helfrich, U. Ahlheim, G. Weber and F. Steglich, Z. Phys. **B84**, 1 (1991).

20. C. Quitmann, B. Andraka, J.S. Kim, B. Treadway, G. Frauenberger, G.R. Stewart and J. Sticht, J. Magn. Magn. Mater. **76&77**, 91 (1988).

21. Z. Fisk, H.R. Ott and J.L. Smith, J. Less Common Metals **133**, 99 (1987).

22. H.R. Ott, H. Rudigier, E. Felder, Z. Fisk and J.D. Thompson, Phys. Rev. **B35**, 1452 (1987).

23. N. Sato, M. Kagawa, N. Tanaka, N. Takeda, T. Satoh and T. Komatsubara, J. Magn. Magn. Mater. **108**, 115 (1992).

24. N. Sato, M. Kagawa, K. Tanaka, N. Takeda, T. Sato, S. Sakatsune and T. Komatsubara, J. Phys. Soc. Jpn. **60**, 757 (1991).

25. C.G. Geibel, C. Kammerer, E. Goring, R. Moog, G. Sparn, R. Henseleit, G. Cordier, S. Horn and F. Steglich, J. Magn. Magn. Mater. **90&91**, 435 (1990).

26. C. Geibel, R. Kohler, A. Bohm, J. Diehl, B. Seidel, C. Kammerer, A. Grauel, C.D. Bredl, S. Horn, G. Weber and F. Steglich, J. Magn. Magn. Mater. **108**, 209 (1992).

27. T. Takabatake, S. Miyata, H. Fujii, Y. Aoki, T. Suzuki, T. Fujita, J. Sakurai and T.Hiraoka, J. Phys. Soc. Jpn. **59**, 4412 (1990).

28. E. Brück, F.R. de Boer, V. Sechovský and L. Havela, Europhys. Lett. **7**, 177 (1988).

29. T. Ekino, H. Fujii, M. Nagasawa, H. Kawanaka, T. Takabatake, M. Nishi, K. Motoya and Y. Ito, preprint.

30. H. Fujii, H. Kawanaka, T. Takabatake, M. Kurisu, Y. Yamaguchi, J. Sakurai, H. Fujiwara, T. Fujita and I.Oguro, J. Phys. Soc. Jpn. **58**, 2495 (1989).

31. H. Kawanaka, H. Fujii, M. Nishi, T. Takabatake, K. Motoya, Y. Uwatoko and Y. Ito, J. Phys. Soc. Jpn. **58**, 3481 (1989).

32. N. Bykovetz, Warren N.Herman, T. Yuen, Chan-Soo Jee, C.L. Lin and J.E. Crow, J. Appl. Phys. **63**, 4127 (1988).

33. K. Takegahara and T. Kasuya, Solid State Commun. **74**, 243 (1990).

34. Y.Aoki, T.Suzuki, T.Fujita, H.Kawanaka, T.Takabatake and H.Fujii, preprint.

MAGNETIC ORDERING OF 1-2-2 U AND Ce INTERMETALLIC

COMPOUNDS DESCRIBED VIA AN f-d HYBRIDIZATION MODEL

J.A. Mydosh, T. Endstra, and G.J. Nieuwenhuys

Kamerlingh Onnes Laboratorium
der Rijksuniversiteit Leiden
2300 RA Leiden, The Netherlands

INTRODUCTION

There exist a large number of well-characterized MT_2X_2 intermetallic compounds which form in two slightly different tetragonal crystal structures. For the purposes of this overview M is restricted to uranium or cerium, T denotes a transition metal (3d, 4d or 5d), and X is silicon or germanium. The crystal structures are shown in Fig. 1 and consist of the $ThCr_2Si_2$ and $CaBe_2Ge_2$ types. Note the symmetry change due to the different stacking sequences of T and X, while the U or Ce sites remain approximately fixed.

Here a rich variety of electronic ground-state properties were observed which ranged from simple Pauli paramagnetism to long-range (anti-) ferromagnetism, and also, extended into the more exotic phenomena associated with "heavy-fermion" behavior. The latter includes coherency effects resulting in strongly correlated Fermi liquids, superconductivity, reduced-moment magnetism, and the coexistence of superconductivity and antiferromagnetism. The complete experimental properties of these compounds are reviewed in Refs. [1-3].

In this article we summarize our efforts at collecting and interpreting the available data concerning the appearance and strength of the magnetic ordering. We first present, as a retrospect of the experimental behavior, the essential physical properties of UT_2Si_2 and UT_2Ge_2. Then we offer a brief outline of the theoretical model which, based upon a semi-quantitative band-structure approach, relates the magnetic-ordering temperature T_c (or disappearance of it) to the strength of the f-d hybridization V_{df}. The systematics of V_{df} is employed to determine the ordering trends in the phase diagram of a Kondo lattice. Using this model satisfactory agreement with the experimental data for T_c is attained.

BASIC PROPERTIES

In Tables I and II we show the basic physical properties of UT_2Si_2 and UT_2Ge_2, respectively. The appropriate references to these data can be found in Ref. [3]. The table captions describe the various symbols. Note especially the different magnetic features which appear as the T=3d, 4d

Transport and Thermal Properties of f-Electron Systems
Edited by G. Oomi *et al.*, Plenum Press, New York, 1993

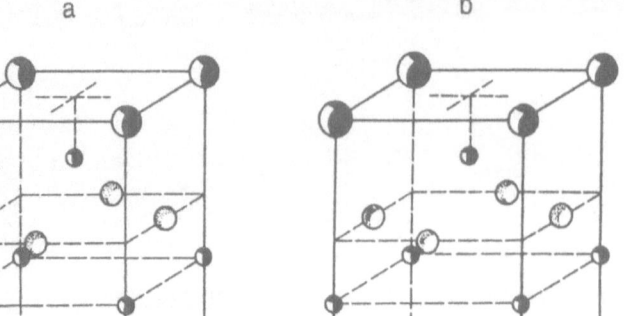

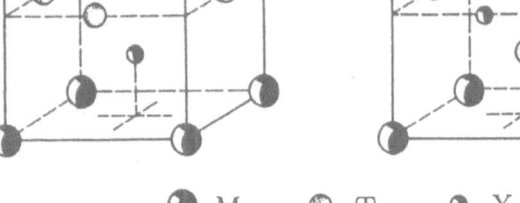

◐ M   ◉ T   ◑ X

Figure 1. A schematic drawing of (a) the $ThCr_2Si_2$ and (b) the $CaBe_2Ge_2$ crystal structure. The origin of the $CaBe_2Ge_2$ unit cell is shifted by $(3/4, 3/4, -z_M)$ to facilitate a comparison with the $ThCr_2Si_2$ unit cell. M denote lanthanide or actinide atoms, T are transition-metal atoms and X metalloids Si, Ge etc.

and 5d rows are traversed. The magnetic-ordering temperature, denoted by T_c for both ferro and antiferromagnetism, can vary between 0 and 174 K with the exception of the Mn compounds. This special case is due to the transition metal Mn (S=5/2) possessing a full local moment which orders at a much higher temperature than the U-moments. For all the other compounds (the situation of Cr in UCr_2Si_2 is still uncertain), the T-elements do not exhibit any magnetism. Based upon this accumulation of data we can ask about possible trends or systematics leading to a theoretical interpretation.

THEORETICAL MODEL: KONDO LATTICE AND HYBRIDIZATION

We consider these UT_2X_2 compounds as a periodic array of 5f spins in which a conduction-electron-mediated (RKKY) magnetic-exchange interaction is in competition with a Kondo-type spin-compensation. By comparing the binding energy of a Kondo singlet

$$k_B T_K \propto [N(0)]^{-1} \exp[-1/N(0)\mathcal{J}]$$

with that of a RKKY ordered magnetic state

$$k_B T_{RKKY} \propto \mathcal{J}^2 N(0)$$

where N(0) is the conduction-electron density of states at the Fermi

Table I. Physical properties of UT$_2$Si$_2$. Key to the abbreviations: AF denotes antiferromagnetism, FM means ferromagnetism, T_N and T_C are the Néel and Curie temperatures, PP means Pauli paramagnetism and MULT indicates that multiple magnetic transitions were found at different temperatures (only the highest one is given). S denotes superconductivity. I and P indicate $I4/mmm$ (ThCr$_2$Si$_2$) and $P4/nmm$ (CaBe$_2$Ge$_2$) crystal structures, respectively. For single crystals, first the value measured parallel to the c axis, and then (in parentheses) the value perpendicular to it is given.

UT$_2$Si$_2$						
T=3d	Cr	Mn	Fe	Co	Ni	Cu
Cryst. struct.	I	I	I	I	I	I
a (Å)	3.911	3.922	3.951	3.917	3.958	3.984
c (Å)	10.503	10.284	9.530	9.614	9.514	9.946
Magn. ord.	AF	FM (Mn)	PP	AF	MULT	FM
T_C, T_N (K)	30	377	<1.8	85	124	104
θ_{CW} (K)	–6	388	–	–285	–15 (–530)	103
μ_{eff} (μ_B/U)	1.8	5.41 (Mn)	–	4.85	3.67 (3.55)	2.7
γ (mJ/mol K^2)			17.5		22	
T=4d	Mo	Tc	Ru	Rh	Pd	Ag
Cryst. struct.	not	not	I	I	I	not
a (Å)	repor-	repor-	4.128	4.012	4.121	repor-
c (Å)	ted	ted	9.592	10.06	10.19	ted
Magn. ord.			AF+S	AF	AF	
T_C, T_N (K)			17.5	130	97	
θ_{CW} (K)			–65 (//c)	–40	–10	
μ_{eff} (μ_B/U)			3.51 (//c)	2.65	2.88	
γ (mJ/mol K^2)			180			
T=5d	W	Re	Os	Ir	Pt	Au
Cryst. struct.	not	I	I	P	P	I
a (Å)	repor-		4.121	4.087	4.197	4.28
c (Å)	ted		9.681	9.829	9.691	10.29
Magn. ord.		PP	PP	AF	AF	MULT
T_C, T_N (K)		<0.33	<0.33	4.9	35	48
θ_{CW} (K)		–	–	1.84 (//c)	–31 (–98)	–37
μ_{eff} (μ_B/U)		–	–	1.03 (//c)	2.87 (3.39)	3.1
γ (mJ/mol K^2)				105	32	

Table II. Physical properties of UT_2Ge_2. Abbreviations as in the previous table. $P^{1)}$ denotes a primitive orthorhombic crystal structure. Ko means Kondo. * For the Co and Ir germanides LT and HT crystal-structure modifications exist with different physical properties. Only the properties of the LT phase are listed. UCo_2Ge_2 HT phase: $P4/nmm$ (?), $a = 4.043$ Å, $c = 9.295$ Å, paramagnetic down to $T = 0.35$ K, $\theta_{CW} = -51$ K, $\mu_{eff} = 1.58$ μ_B/U, $\gamma = 62$ mJ/mol K^2. UIr_2Ge_2 HT phase: $P4/nmm$, $a = 4.156$ Å, $c = 9.773$ Å, AF with $T_N = 19$ K, $\theta_{CW} = -230$ K, $\mu_{eff} = 3.3$ μ_B/U. "does not exist" means that the compound cannot be stabilized in 1-2-2 composition, using standard metallurgical techniques.

UT_2Ge_2						
T=3d	Cr	Mn	Fe	Co (LT)	Ni	Cu
Cryst. struct.	does	I	I	I*	I	I
a (Å)	not	3.993	4.024	4.010	4.095	4.063
c (Å)	exist	10.809	9.964	9.878	9.478	10.229
Magn. ord.		FM (Mn)	PP	AF	AF	FM/AF
T_C, T_N (K)		380	<0.35	174	77	100/43
θ_{CW} (K)		400	–	-262	-70.7	70
μ_{eff} (μ_B/U)		5.46 (Mn)	–	4.0	3.08	3.02
γ (mJ/mol K^2)		30.3	24.4	34	39.5	26.4
T=4d	Mo	Tc	Ru	Rh	Pd	Ag
Cryst. struct.	not	not	does	?	I	not
a (Å)	repor-	repor-	not	4.154	4.200	repor-
c (Å)	ted	ted	exist	9.762	10.230	ted
Magn. ord.				Ko	AF	
T_C, T_N (K)					140	
θ_{CW} (K)				-10 (-19)	-81	
μ_{eff} (μ_B/U)				2.98 (1.25)	3.40	
γ (mJ/mol K^2)				305		
T=5d	W	Re	Os	Ir (LT)	Pt	Au
Cryst. struct.	not	not	does	$P^{1),*}$	P	not
a (Å)	repor-	repor-	not	4.054	4.330	repor-
b (Å)	ted	ted	exist	4.195		ted
c (Å)				10.25	9.752	
Magn. ord.				AF	AF	
T_C, T_N (K)				33	72	
θ_{CW} (K)				-240	-52	
μ_{eff} (μ_B/U)				3.6	2.93	
γ (mJ/mol K^2)					14	

level and J is an exchange coupling constant, Doniach [4,5] derived a phase diagram for this so-called Kondo lattice. Figure 2 illustrates the stability of the possible states in the T-J plane [6]. The dashed lines indicate the Kondo and RKKY temperatures; the thick line (T_M) denotes the effective magnetic-ordering temperature in the presence of the Kondo effect. Notice the three regimes along the abscissa: a conventional magnetic 4f or 5f metal at low J, a magnetically concentrated Kondo system at intermediate J where the magnetic-ordering temperature (T_M in the figure) begins to decrease with increasing J, and finally, a non-magnetic regime, i.e. $T_M=0$ at $J \geq J_c$.

In order to proceed we must determine the conduction-electron/f-electron exchange parameter J_{cf}. Here we invoke hybridization according to the proportionality

$$J_{cf} \propto V_{cf}^2 / (E_F - E_f)$$

where V_{cf} is the hybridization matrix element and E_f is the location of the f-level relative to the Fermi level E_F. This relation is the form given by the Schrieffer-Wolff transformation. If we take $E_F - E_f$ to be constant in a given transition-metal series and also assume the conduction-electron bandwidth W to be constant, we can, by calculating the c-f hybridization of the compound obtain an estimate of J_{cf}/W.

It is commonly believed that in these 1-2-2 compounds the hybridization between the f states and the conduction electrons is mainly governed by f-d hybridization. Our key parameter is the variable number of d electrons for a given series. Thus, we assume that the f-d hybridization depends, firstly, on the spatial extent of the d-orbitals resulting in an "effective radius" for each T-metal, and secondly, on the distance between the d and f atoms in the crystal structure. According to the first assumption with increasing number of d electrons the hybridization decreases since the d band is pulled down in energy away from E_F. The second assumption is analogous to the Hill criterion which states that when U-U interatomic distances are too large there will be no direct overlap of the 5f wave functions. We propose a similar criterion for the U-T interatomic distances, i.e., if d_{U-T} is too large no f-d hybridization can occur. Consequently with the above relation between J_{df} and V_{df}, J_{df} will increase (move to the right in Fig. 2) both with decreasing number of d electrons and with decreasing U-T separation.

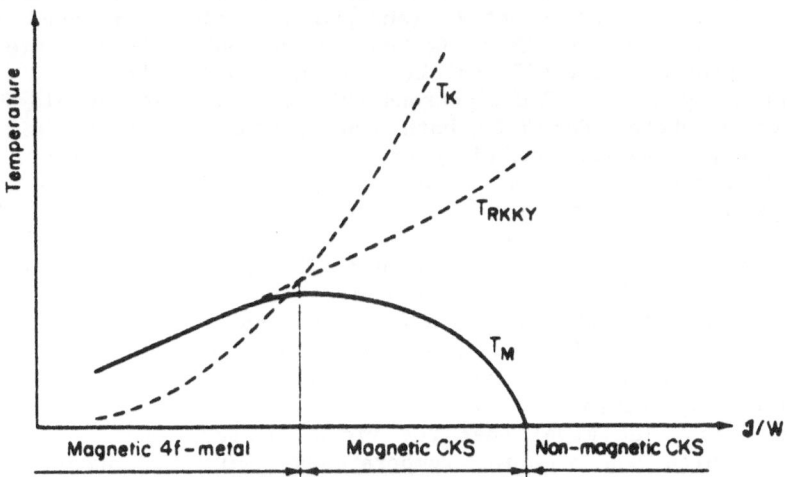

Figure 2. Phase diagram for the Kondo-lattice model after Brandt and Moshchalkov Ref. [6]

In order to put these ideas on more quantitative basis we follow the band-structure approach of Harrison and co-workers [7-11]. This method was initially used to calculate the d-band structure of solids, but was latter adapted to compute the coupling between atomic orbitals of s, p, d, f symmetries in different compounds. The formalism combines muffin-tin-orbital theory with transition-metal pseudo-potentials to obtain a general hybridization matrix element $V_{\ell\ell'm}$. The resulting equation [11] is

$$V_{\ell\ell'm} = \frac{\eta_{\ell\ell'm}\hbar^2}{m_e} \left[(r_\ell^{2\ell-1} r_{\ell'}^{2\ell'-1})^{1/2} / d^{\ell+\ell'+1} \right]$$

The input parameters are the atomic radii of the respective atoms (r_ℓ and $r_{\ell'}$) the interatomic distance d, the angular momenta ℓ,ℓ' (ℓ=0,1,2,3 for s,p,d,f orbitals) and the symmetry of the bond m. For a given d-series, which keeps V_{pf} and V_{sf} relatively fixed, $\eta_{\ell\ell'm}$ becomes a simple constant. As indicated previously the important parameters in determining V_{df} are the U-T interatomic distance and the number of d electrons of the transition metal. The former directly enter into the above equation as d=d_{U-T} and the latter via the tabulated values [10] of the atomic radii r_d since r_f = constant.

In Table III we have collected the U-T interatomic distances, the calculated V_{df} values and the magnetic ordering temperatures of the UT_2Si_2 compounds. Table IV repeats the same quantities for the UT_2Ge_2 compounds. Here we have also included the results on a number of pseudo-ternary (2 different transition metal) compounds. We now can proceed with a direct comparison of the T_c magnitudes according to their calculated V_{df} values. Since the constant of proportionality is missing in the relation between J_{df} and V_{df} only the sequence along the horizontal axis of Fig. 2 can be determined. Nevertheless this then should establish the T_c trends according to the curve in Fig. 2.

COMPARISON OF EXPERIMENT WITH f-d HYBRIDIZATION MODEL

Let us first consider the $U(3d)_2Si_2$ series of compounds. From the crystal-structure data of Table III and the number of d electrons, the following order of increase of V_{df} (and thus J_{df}) can be expected: Cu → Ni → Co → Fe. Since in the T=Ni, Co and Fe compounds the d_{U-T} are almost equal, the hybridization will solely be determined by the d-band filling. With decreasing number of d-electrons this hybridization should increase as discussed above. For T=Cu both the increased distance and d-band filling imply a smaller hybridization, thus completing the trend. Because of its 3d magnetism we must exclude T=Mn, and the situation for T=Cr remains unclear. From the entries in the tables it can be seen that the calculated V_{df} values (using the Harrison model) indeed confirm this expected trend. The results for the $U(3d)_2Si_2$ system, interpreted within the framework of the Kondo-lattice model are shown in Fig. 3 together with those for $U(4d)_2Si_2$ and $U(5d)_2Si_2$ compounds. For the latter two d series a similar procedure was used from Table III as for the 3d series. The maximum in the curve corresponds to a $T_c \approx 190$ K which is the highest transition temperature observed in these silicides for pseudo-ternary $U(Ru_{0.3}Rh_{0.7})_2Si_2$. The model does not yield absolute value of J_{df}/W, nor does it predict the ordering temperatures, only the order of J, or sequence in which we put the systems of a given series into the phase diagram, is given. Yet it is not trivial that this will always work.

Figure 4 exhibits the results for UT_2Ge_2 according to Table IV. A

Table III · Crystal-structure data, hybridization matrix elements (V_{df}), and magnetic-ordering temperatures (T_c) for UT_2Si_2 compounds. The nature of the magnetic order is given in brackets after the ordering temperature, the abbreviations denote the type of magnetism as follows: AF antiferromagnetism, FM ferromagnetism, and P (Pauli) param-agnetism. For $CaBe_2Ge_2$ type compounds (indicated by superscript "a") only the shortest d_{U-T} is given, but an average V_{df} has been calculated by taking into account two different d_{U-T} distances.

Compound	d_{U-T} (Å)	V_{df} (eV)	T_c (K)
UCr_2Si_2	3.274	0.269	27 (AF)
UMn_2Si_2	3.234	0.273	80–100 (FM, U)
UFe_2Si_2	3.095	0.320	– (P)
UCo_2Si_2	3.100	0.290	90 (AF)
UNi_2Si_2	3.092	0.270	124 (AF)
UCu_2Si_2	3.189	0.204	103–107 (FM)
$UCoNiSi_2$	3.091	0.283	115 (AF)
$UNiCuSi_2$	3.136	0.237	162 (AF)
URu_2Si_2	3.164	0.418	17.5 (AF)
URh_2Si_2	3.209	0.354	130–137 (AF)
UPd_2Si_2	3.241	0.308	97, 150 (AF)
UOs_2Si_2	3.179	0.458	– (P)
UIr_2Si_2[a]	3.113	0.436	4.9 (AF)
UPt_2Si_2[a]	3.211	0.376	35 (AF)
UAu_2Si_2	3.322	0.293	48, 78 (AF ?)

[a] $CaBe_2Ge_2$ crystal structure, others adopt the $ThCr_2Si_2$ structure.

Table IV. Crystal-structure data, hybridization matrix elements and magnetic-ordering temperatures for UT_2Ge_2 compounds. Abbreviations as in Table III. "LT" and "HT" denote a low- and a high-temperature crystal-structure modification respectively.

Compound	d_{U-T} (Å)	V_{df} (eV)	T_c (K)
UMn_2Ge_2	3.364	0.215	100–150 (FM, U)
UFe_2Ge_2	3.199	0.263	– (P)
$UCo_2Ge_2{}^{LT}$	3.182	0.248	174 (AF)
$UCo_2Ge_2{}^{HT,a}$	3.080[b]	0.302	– (P)
UNi_2Ge_2	3.128	0.251	77 (AF)
UCu_2Ge_2	3.266	0.177	100–110 (FM)
$U(Co_{0.875}Ni_{0.125})_2Ge_2$	3.153	0.259	46 (AF)
$U(Co_{0.75}Ni_{0.25})_2Ge_2{}^a$	3.134[b]	0.266	19 (AF)
$U(Co_{0.5}Ni_{0.5})_2Ge_2{}^a$	3.135[b]	0.260	21 (AF)
$U(Co_{0.5}Ni_{0.5})_2Ge_2{}^a$	3.131[b]	0.262	– (<10)
$U(Co_{0.25}Ni_{0.75})_2Ge_2{}^a$	3.133[b]	0.255	51 (AF)
$U(Co_{0.75}Cu_{0.25})_2Ge_2$	3.199	0.230	130 (AF)
$U(Co_{0.5}Cu_{0.5})_2Ge_2$	3.213	0.215	100 (AF)
$U(Co_{0.25}Cu_{0.75})_2Ge_2$	3.235	0.197	102 (FM)
$U(Ni_{0.75}Cu_{0.25})_2Ge_2$	3.182	0.222	135 (AF)
$U(Ni_{0.5}Cu_{0.5})_2Ge_2$	3.213	0.205	140 (AF)
$U(Ni_{0.25}Cu_{0.75})_2Ge_2$	3.234	0.192	133 (FM)
$U(Ni_{0.1}Cu_{0.9})_2Ge_2$	3.250	0.184	115 (FM)
$URh_2Ge_2{}^c$	3.205[b]	0.356	–
UPd_2Ge_2	3.309	0.272	140 (AF)
$UIr_2Ge_2{}^{LT,d}$	3.289[b]	0.346	33 (AF)
$UIr_2Ge_2{}^{HT,a}$	3.194[b]	0.412	19 (AF)
$UPt_2Ge_2{}^a$	3.261[b]	0.344	72 (AF)

[a] $CaBe_2Ge_2$ ($P4/nmm$) crystal structure.

[b] estimated value, exact z parameters not known.

[c] exact crystal structure not known.

[d] $Pmmm$ crystal structure.

similarly good agreement results for the germanides. Of particular importance is the existence of two phases with different crystal structures for UCo_2Ge_2 and UIr_2Ge_2. Their T_c's are indeed consistent with their V_{df} values. In addition Table IV has various groups of pseudo ternary compounds which, if plotted in Fig. 4, would also fall on the "bell-shaped" curve [12].

We could extend our considerations to the Ce 1-2-2 compounds where T_c has been experimentally determined and calculate the hybridization values using the same procedure. Once again the dependence of T_c follows the Kondo-lattice curve according to the sequence of V_{df} values obtained from the band-structure model.

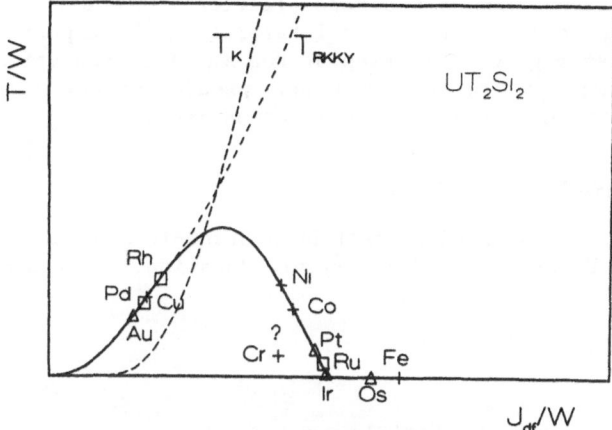

Figure 3. Schematic phase diagram for the Kondo-lattice along with the magnetic-ordering temperatures of UT_2Si_2.

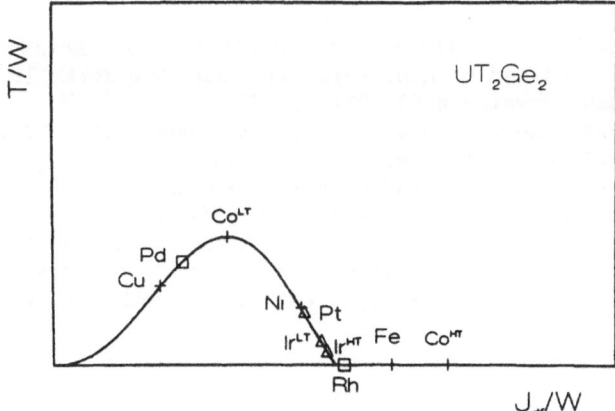

Figure 4. Schematic phase diagram for the Kondo-lattice with magnetic-ordering temperatures of UT_2Ge_2.

CONCLUSIONS

The systematic behavior of the magnetic properties of $(U, Ce)T_2X_2$ intermetallic compounds, with the transition metal as variable, has attracted considerable experimental attention in the past. However, little theoretical guidance was offered for the diversities in the magnetic ordering. Our hybridization model is the first attempt to clarify the mechanism underlying the absence or presence of magnetic ordering and the nonmonotonic variation of T_c for those compounds that do order magnetically.

We have shown that the magnetic-ordering characteristics of U and Ce 1-2-2 compounds are determined by the strength of the f-d hybridization. By means of a simple band-structure approach we have calculated V_{df}, the hybridization matrix element for four different series of compounds and used their relative values for a given d series to explain the observed trends within the Kondo-lattice phase diagram. The success of this f-d hybridization model in interpreting the experimental data may be taken as a strong justification of the initial assumptions. Additional experiments such as pressure and new pseudo-ternary combinations are needed to extend the experiment-model comparison.

ACKNOWLEDGEMENTS

This work was supported in part by a Grant-in-Aid for International-Joint Research Program from Ministry of Education, Science and Culture of Japan.

REFERENCES

[1]   A. Szytuła and J. Leciejewicz, in *Handbook on the Physics and Chemistry of Rare Earths, Vol. 12*, edited by K.A. Gschneidner, Jr. and L. Eyring (North-Holland, Amsterdam, 1989), p.133.

[2]   V. Sechovsky and L. Havela, in *Ferromagnetic Materials, Vol. 4*, edited by E.P. Wohlfarth and K.H.J. Buschow (North-Holland, Amsterdam, 1988), p. 309.

[3]   T. Endstra, S.A.M. Mentink, G.J. Nieuwenhuys, and J.A. Mydosh, to be published in *Frontiers in Solid State Sciences: Magnetism*, edited by L.C. Gupta and M.S. Multani (World Scientific, Singapore, 1992).

[4]   S. Doniach, in *Valence Instabilities and Related Narrow-Band Phenomena*, edited by R.D. Parks (Plenum, New York, 1977), p. 169.

[5]   S. Doniach, Physica B **91**, 231 (1977).

[6]   N.B. Brandt and V.V. Moshchalkov, Adv. Phys. **33**, 373 (1984).

[7]   W.A. Harrison, Phys. Rev. **181**, 1036 (1969).

[8]   W.A. Harrison and S. Froyen, Phys. Rev. B **21**, 3214 (1980).

[9]   W.A. Harrison, Phys. Rev. B **28**, 550 (1983).

[10] G.K. Straub and W.A. Harrison, Phys. Rev. B **31**, 7668 (1985).

[11] W.A. Harrison and G.K. Straub, Phys. Rev. B **36**, 2695 (1987).

[12] T. Endstra, G.J. Nieuwenhuys, and J.A. Mydosh, to be published.

ANOMALOUS HALL COEFFICIENT IN THE f ELECTRON SYSTEM

Y. Ōnuki,[1] S. W. Yun,[1] K. Satoh[1]
H. Sugawara[1] and H. Sato[2]

[1] Institute of Materials Science
University of Tsukuba
Tsukuba, Ibaraki 305, Japan
[2] Department of Physics
Tokyo Metropolitan University
Minami-osawa, Hachiōji, Tokyo 192-3, Japan

INTRODUCTION

The 4f(5f) electrons in the rare earth(uranium) atom are pushed deep into the interior of the closed 5s(6s) and 5p(6p) shells because of the strong centrifugal potential $1(1+1)/r^2$, where l=3 holds for the f electrons. This is a reason why the 4f(5f) electrons possess an atomic-like character even in the compound. On the other hand, the tail of their wave function spreads to the outside of the closed 5s(6s) and 5p(6p) shells, which is highly influenced by the potential energy, the relativistic effect, the distance between the rare earth(uranium) atoms, and hybridization of the 4f(5f) electrons with the conduction electrons. These cause the various phenomena such as valence and spin fluctuations, gap states, Kondo lattice, heavy electrons (fermion), metamagnetism and superconductivity for the f electron compounds.[1]

The Hall effect is a good experimental method to detect the f electron behavior. It is generally considered as a sum of an ordinary Hall effect which is related to the carrier concentration and an anomalous part dependent on the magnetization. The Hall coefficient in the paramagnetic state is phenomenologically expressed as

$$R_H = R_0 + 4\pi R_s \chi \qquad (1)$$

and

$$\chi = C / (T + \theta_p) \qquad (2)$$

where R_0 is the ordinary Hall coefficient, R_s is the temperature-independent anomalous Hall coefficient, χ is the magnetic susceptibility, C is the Curie constant and θ_p is the paramagnetic Curie temperature.

Theoretically the skew (asymmetric) scattering is the main mechanism of the anomalous Hall coefficient. The asymmetry comes from an interaction of conduction electrons with localized d or f electrons which possess orbital angular moments. In this case, the transition probability of the conduction electrons which are scattered from a state k to another state k' due to the localized d or f electrons, $W(k \rightarrow k')$ is not equal to the one from k' to k; $W(k' \rightarrow k) \neq W(k \rightarrow k')$. This leads to the anomalous Hall coefficient. Experimentally, the sign of R_s is positive in the ferromagnetic compound, while in the antiferromagnetic compound it becomes positive for one compound and negative for another one, dependent on the compound.

The Hall effect in the impurity Kondo system was calculated by Coleman et al.[2] and by Fert et al.[3] on the basis of the skew scattering by the Ce impurity;

$$R_H = R_0 + \gamma \bar{\chi} \rho_m \qquad (3)$$

where γ is $-(15/7)g\mu_B k_B^{-1}\sin\delta_2\cos\delta_2$, $\bar{\chi} = \chi/C$ and ρ_m is the magnetic resistivity. The skew scattering in this case is due to the d-wave (l=2) and f-wave (l=3) scattering. The phase shift δ_2 is associated with a spherical scattering in the channel l=2. The prominent scattering is due to the Kondo scattering in channel l=3. Therefore, the magnetic susceptibility originates from the Ce impurity, and ρ_m corresponds to the magnetic resistivity which is extremely dominant in the Ce compounds.

The conduction electrons in the Ce-based Kondo lattice compounds or the spin-fluctuating U compounds possess large cyclotron masses at low temperatures, forming the f-derived heavy bands. We present in this paper the characteristic behavior of the Hall coefficient in the heavy electron system.

ANOMALOUS HALL COEFFICIENT OF THE FERROMAGNETIC COMPOUND

As mentioned above, the Hall coefficient R_H indicates a large anomalous Hall coefficient with the positive value for the ferromagnetic compound. Figure 1 shows a typical example of the Hall coefficient in UGe_2 for current and field along the c- and b-axes, respectively under fields of 5kOe and 20kOe. UGe_2 is known to become ferromagnetic below 52K.[4] It increases in magnitude with decreasing the temperature, makes a maximum around the Curie temperature and finally decreases steeply with decreasing the temperature.

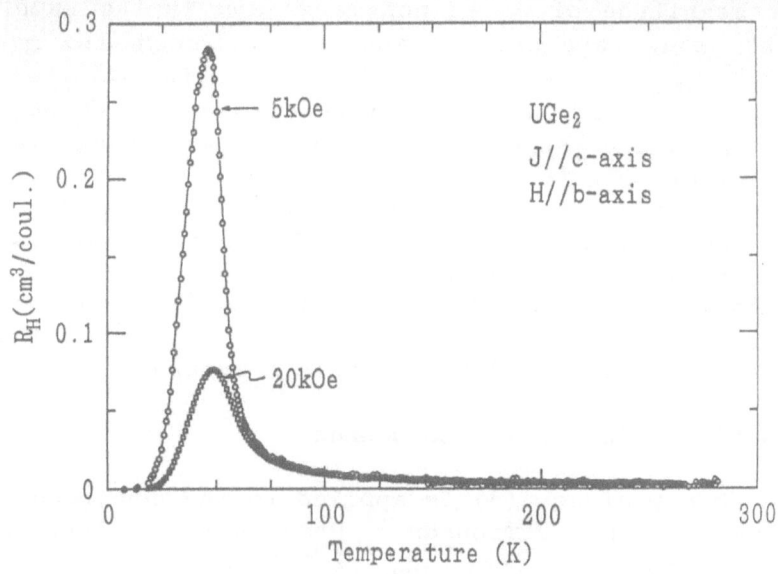

Fig. 1 Temperature dependence of R_H in UGe_2.

This is a characteristic behavior for the ferromagnetic compound. In the ferromagnetic state the magnitude of the Hall coefficient at 5kOe is larger than that at 20kOe. This is related to the saturated behavior of magnetization.

ANOMALOUS HALL COEFFICIENT IN THE HEAVY ELECTRON SYSTEM

Kondo Lattice Compounds with Clear Antiferromagnetic Ordering

First we show in Fig. 2 the Hall coefficient of the Kondo lattice compound with clear magnetic ordering such as $CeAl_2$ (Nèel temperature $T_N=3.8K$), CeB_6 (quadrupolar ordering temperature $T_Q=3.2K$ and $T_N=2.3K$) and $CeIn_3$ ($T_N=10.2K$).

The behavior of the Hall coefficient is highly different in the compounds. If we follow eq. (1), the sign of R_S is positive in $CeAl_2$, zero in CeB_6, and negative in $CeIn_3$.

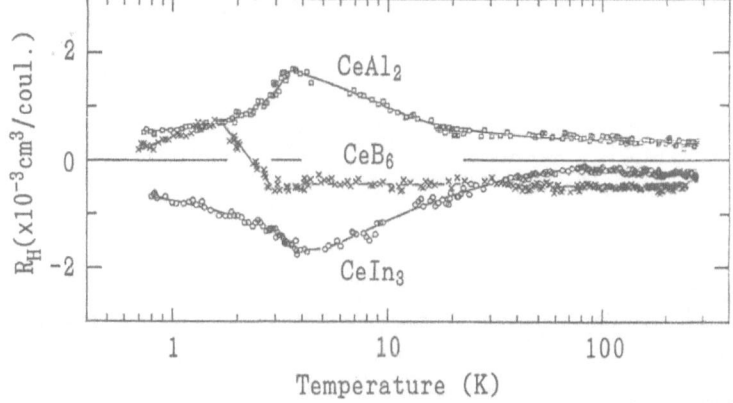

Fig. 2 Temperature dependence of R_H in $CeAl_2$, CeB_6 and $CeIn_3$.

The magnitude of R_H is not large and is the same order as in the usual rare earth compounds, although the magnetic resistivity ρ_m in eq. (3), which results in the Kondo effect, is extremely large for these compounds. We note that in the temperature range of 4K to room temperature, the Hall coefficient of CeB_6 is -4.5×10^{-4} cm^3/coul. in value, which is the same with that of LaB_6, indicating one conduction electron per primitive cell. Thus no anomalous Hall coefficient is present in CeB_6.

These results claim that the anomalous Hall coefficient for these compounds does not follow eq. (3) but rather eq. (1) as in the usual antiferromagnetic compounds.[5]

Non-magnetic Kondo Lattice Compounds

The theory of eq. (3) is applied to the non-magnetic Ce-based Kondo lattice compounds. Previously we analyzed the Hall coefficient of $Ce_xLa_{1-x}Cu_6$ (x=0-1) on the basis of eq. (3).[5] At higher temperatures than 40K, the Hall coefficient is explained well by eq. (3), as shown by the solid lines in Fig. 3. Here we used the magnetic susceptibility $\bar{\chi}=\chi/C$ and the magnetic resistivity $\rho_m=\rho(Ce_xLa_{1-x}Cu_6)-\rho(LaCu_6)$ for the same sample. We note that the fitting parameter of the phase shift δ_2 (=-0.02) is concentration-independent. This means that the f electron behavior in $Ce_xLa_{1-x}Cu_6$ is governed by the impurity Kondo effect. Thus obtained theoretical curves are in good agreement with the experimental data.

At low temperatures, the behavior of the Hall

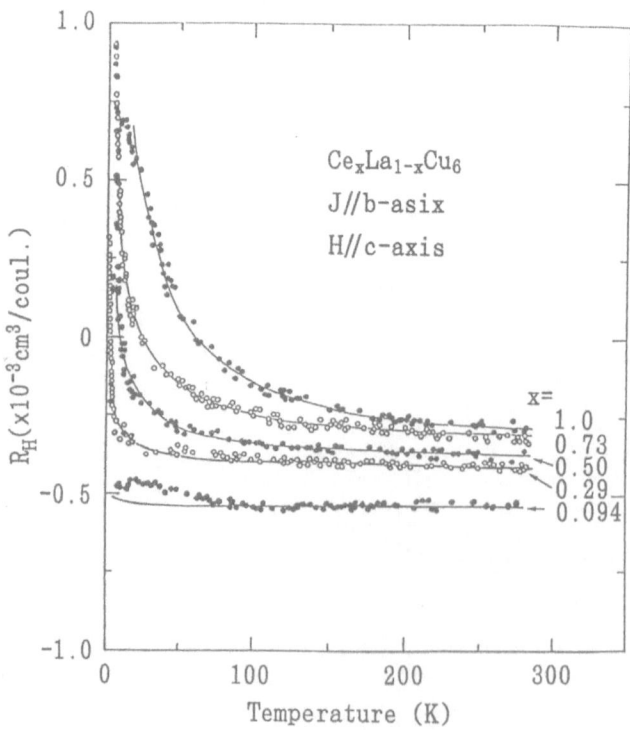

Fig. 3 Temperature dependence of R_H in $Ce_xLa_{1-x}Cu_6$.

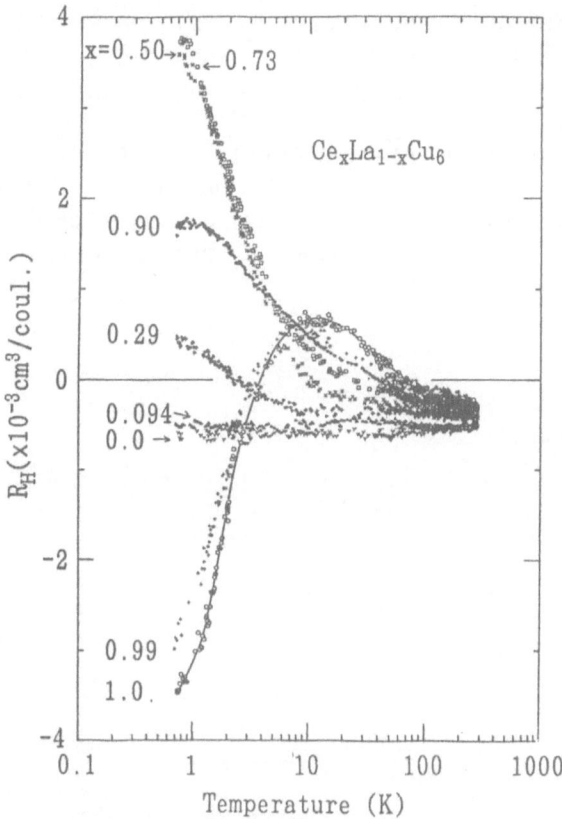

Fig. 4 Temperature dependence of R_H in $Ce_xLa_{1-x}Cu_6$.

coefficient is not simple in $Ce_xLa_{1-x}Cu_6$, as shown in Fig. 4. The Hall coefficient of $CeCu_6$ makes a maximum around 10K, which is highly different from the dilute Kondo system (x=0.094 or 0.29).

To know the overall behavior of the Hall coefficient we show in Fig. 5 the logarithmic temperature dependence of the Hall coefficient of the non-magnetic Kondo lattice compounds

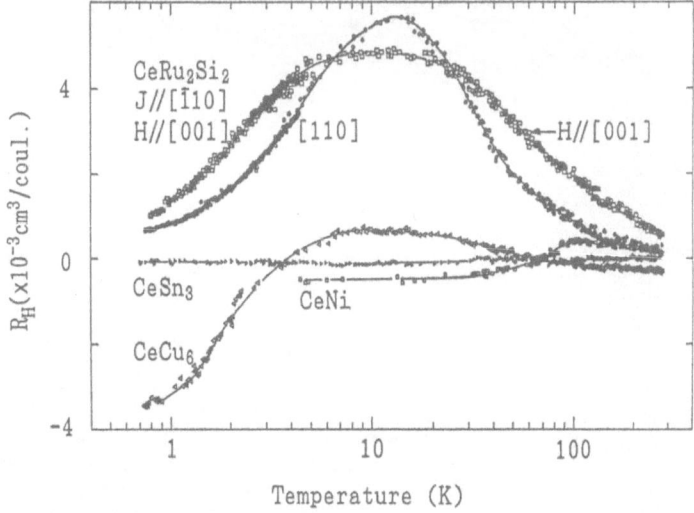

Fig. 5 Temperature dependence of R_H in $CeCu_6$, $CeRu_2Si_2$, CeNi and $CeSn_3$.

such as $CeCu_6$ (Kondo temperature $T_K=4K$), $CeRu_2Si_2$ ($T_K=20K$), CeNi ($T_K=150K$) and $CeSn_3$ ($T_K=200K$). The Hall coefficient of these compounds, except $CeSn_3$, indicates a maximum around the Kondo temperature.

We discuss the Hall coefficient of $CeRu_2Si_2$. This compound shows a highly anisotropic behavior, reflecting the tetragonal crystal structure. Figure 6 shows the temperature dependence of the electrical resistivity, magnetic

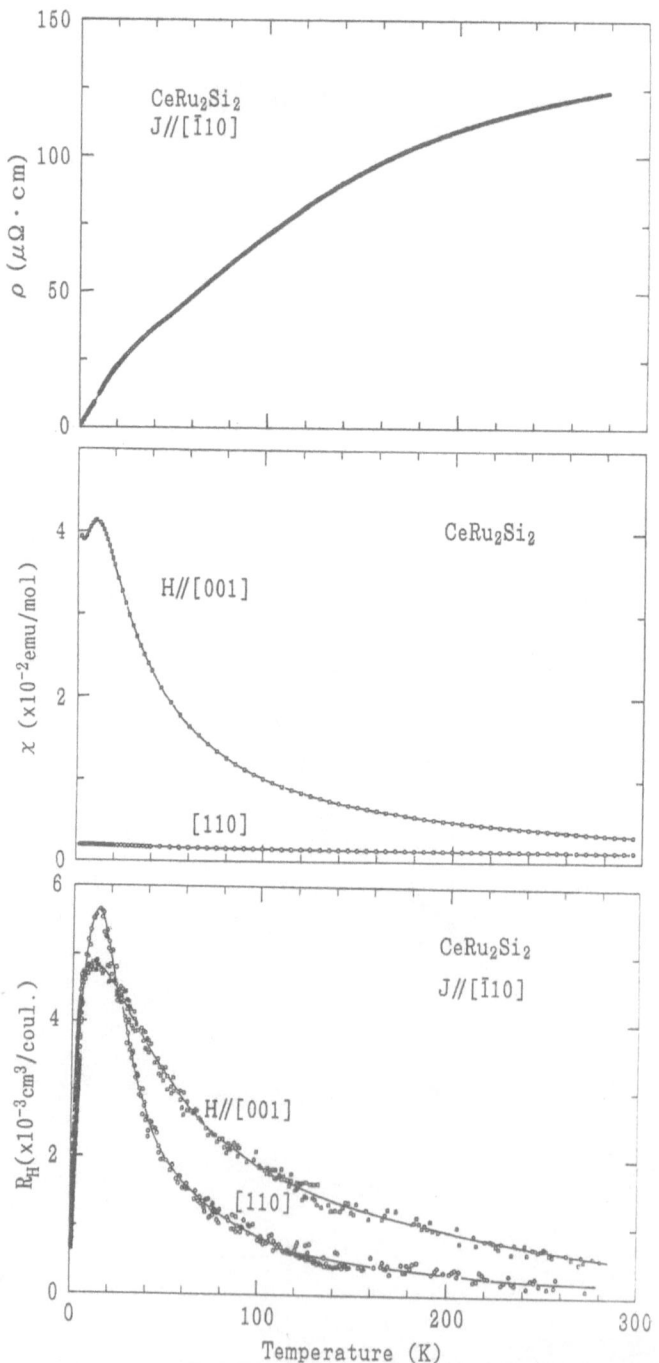

Fig. 6 Temperature dependence of ρ, χ and R_H in $CeRu_2Si_2$.

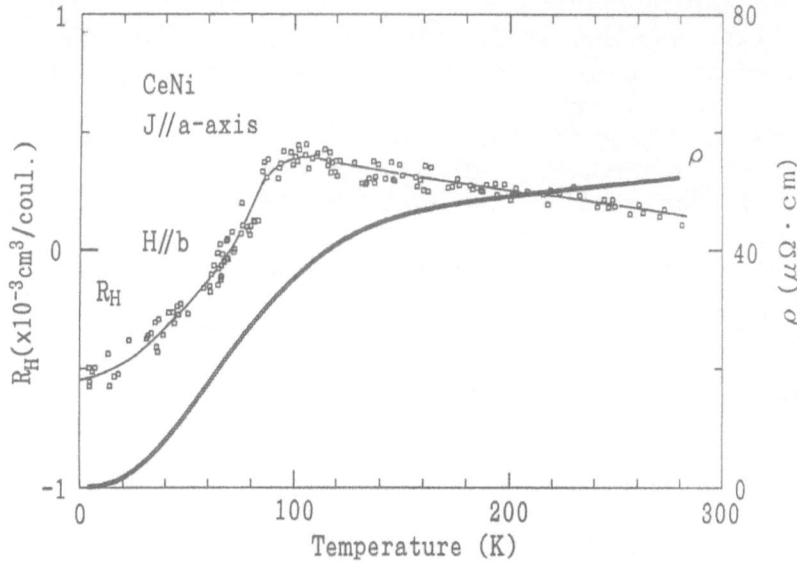

Fig. 7 Temperature dependence of R_H and ρ in CeNi.

susceptibility and Hall coefficient. The residual resistivity ρ_0 and the residual resistivity ratio ρ_{RT}/ρ_0 are estimated as $0.5\mu\Omega\cdot$cm and 250, respectively. The present sample indicates a high-quality single crystal. The magnetic susceptibility for field along the [001] direction increases with decreasing temperature, following the Curie-Weiss behavior. On the other hand, the magnetic susceptibility for field along the [110] direction is almost temperature-independent. This anisotropic behavior is reflected in the Hall coefficient at temperatures higher than 150K.

As the temperature is decreased, both Hall coefficients of $CeRu_2Si_2$ become large in magnitude, as shown in Figs. 5 and 6. The Hall coefficient makes a maximum around 10-20K and decreases steeply with decreasing temperature. We discuss the low-temperature behavior in the next section.

In CeNi the characteristic feature is the same as in $CeRu_2Si_2$, indicating a maximum around 100K, as shown in Fig. 7.

Anomalous Hall Coefficient at Low Temperatures

Recently Kohno and Yamada[6] have proposed the theoretical formula of the anomalous Hall coefficient, which is applied to the heavy fermion compounds at low temperatures. They have taken into account the skew scattering through the spin-orbit coupling of the f-orbitals. The formula is proportional to the square of the electrical resistivity including the residual resistivity ρ;

$$R_H = R_0 + c\ \rho^2 \qquad (4)$$

where c is independent of temperature and is proportional to the magnetic susceptibility. We have checked whether or not this formula is applied to the anomalous Hall coefficient in the heavy electron system.

Figures 8 and 9 show the ρ^2 dependence of the Hall coefficient in $CeRu_2Si_2$, $CeCu_6$ and $CeNi$. The experimental data are fitted well by the straight lines. The temperatures indicated by arrows mean the deviation from the straight lines. These values are lower than the corresponding Kondo temperatures.

We have also applied this relation to the well-known spin fluctuators of UPt_3 and UAl_2.[7] The straight lines approximately hold at low temperatures.

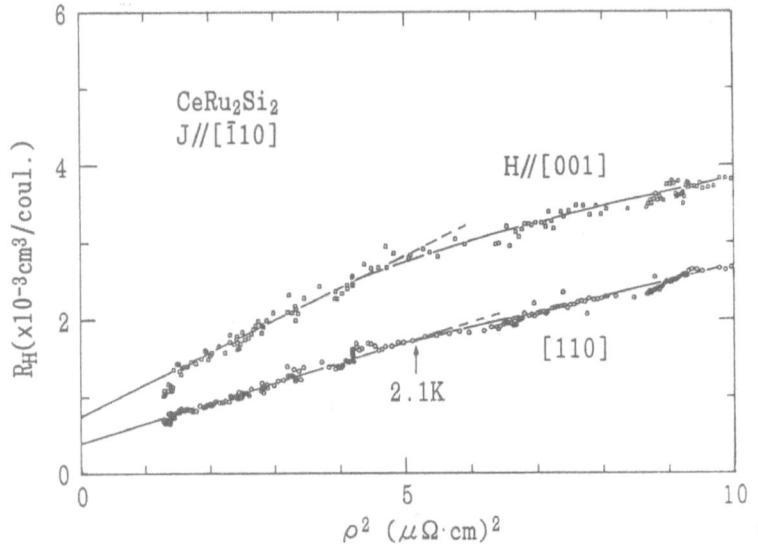

Fig. 8 ρ^2 dependence of R_H in $CeRu_2Si_2$.

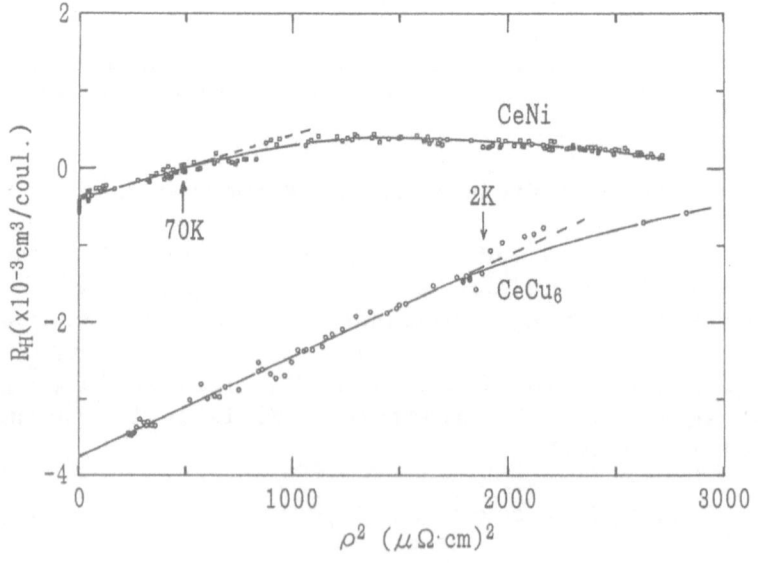

Fig. 9 ρ^2 dependence of R_H in $CeCu_6$ and $CeNi$.

ACKNOWLEDGEMENTS

We are grateful to Prof. K. Yamada for fruitful discussions. This work was supported by Grant-in-Aid for the Scientific Research from the Ministry of Education, Science and Culture and also by University of Tsukuba Project Research.

REFERENCES

1. Y. Ōnuki, T. Goto and T. Kasuya, in Mat. Sci. and Tech. vol. 3A, chap. 7, ed. K. H. J. Buschow, VCH, Weinheim (1991).
2. P. Coleman, P. W. Anderson and T. V. Ramakrishnan, Phys. Rev. Lett. 55:414(1985).
3. A. Fert, A. Håmzic and P. M. Levy, J. Mag. Mag. Mat. 63&64:353(1987).
4. Y. Ōnuki, I. Ukon, S. W. Yun, I. Umehara, K. Satoh, T. Fukuhara, H. Sato, S. Takayanagi, M. Shikama and A. Ochiai, J. Phys. Soc. Jpn. 61:293(1992).
5. Y. Ōnuki, T. Yamazaki, T. Omi, I. Ukon, A. Kobori and T. Komatsubara, J. Phys. Soc. Jpn. 58:2126(1989).
6. H. Kohno and K. Yamada, J. Mag. Mag. Mat. 90&91 :431(1990);K. Yamada, H. Kohno and S. Inagaki:preprint.
7. Y. Ōnuki, T. Yamazaki, I. Ukon, T. Komatsubara, A. Umezawa, W. K. Kwok, G. W. Crabtree and D. G. Hinks, J. Phys. Soc. Jpn. 58:2119(1989).

MAGNETORESISTANCE IN UTX COMPOUNDS

V. Sechovský,[1] L. Havela,[1] F.R. de Boer,[2]
H. Fujii[3], and T. Fujita [4]

[1]Department of Metal Physics, Charles University

  CS-121 16 Prague, Czechoslovakia

[2]Van der Waals - Zeeman Laboratory, University of Amsterdam

  NL-1018 XE Amsterdam, The Netherlands

[3]Faculty of Integrated Arts and Sciences

[4]Faculty of Science

  Hiroshima University, Hiroshima 730, Japan

INTRODUCTION

Transport properties of actinide intermetallic compounds are a subject of controversies originating in difficulties to understand mechanisms governing strongly correlated electron systems. Phenomenological similarities in the electrical resistivity behavior of some U and Ce compounds led to application of concepts like Kondo lattice for interpretation of both classes of materials. Adequacy of this model for many Ce-based compounds is well documented. In the case of U intermetallic compounds no direct experimental evidence proves the applicability of Kondo model with the condition $V_{k-f}^2/|E_F-E_f| \ll 1$ (V_{k-f} is the matrix element of the hybridization of $5f$ states with conduction electron states, $|E_F-E_f|$ is the energy separation of f states from the Fermi energy). Instead, $5f$ states are indicated states at E_F in most cases.

Experimental studies of transport properties in the context of magnetism in $5f$ electron systems were focused mostly on several uranium heavy-fermion compounds. Therefore, no systematic information connecting bulk transport properties with microscopic parameters is available up to now. Lack of experimental data is noticeable especially for magneto-transport. In this field we concentrate on materials, in which the the lower symmetry of magnetism can help to disentangle effects of different types of magnetic ordering. In particular, we study hexagonal compounds from the UTX series UNiAl, UNiGa and UPdIn, (ZrNiAl structure type), which display a strong uniaxial magnetic anisotropy.

An overview of ground state properties of UTX compounds[1] shows that a choice of transition (T) and non-transition (X) metal affects the degree of localization of uranium $5f$ states, which can be varied to a large extent while conserving the particular structure type. A reduced hybridization of $5f$ states (being pinned at E_F) with the $3d$ states of Ni or $4d$ states of Pd, which are located at considerably higher binding energies, is probably the main reason why the $5f$ magnetic moments in these three compounds are formed and become ordered at low temperatures.

MAGNETIC PROPERTIES

The type of ordering is essentially antiferromagnetic in all three cases below 19, 39 and 21 K, respectively[2-4]. The U magnetic moments are locked in the c-direction by huge magnetic anisotropy which is caused primarily by the participation of 5f electrons in bonding. The strong uniaxial magnetic anisotropy survives in the paramagnetic region[2,3,5]. Their collinear magnetic structures are built up by stacking basal-plane sheets of U magnetic moments. In UNiGa[6] and UPdIn,[7] the magnetic moments ($\mu_U \simeq 1.4$ and $1.5\ \mu_B$, respectively) are ferromagnetically ordered within these sheets whereas a long-periode modulation with a wave-vector $\vec{k} =$ (0.1, 0.1, 0.5) was observed in UNiAl[8].

UNiGa has the most complex magnetic phase diagram[6] of all three compounds. Below $T_N = 38.8$ K, we observe four AF phases separated by magnetic phase transitions at 37.5, 36.1 and 34.8 K. The ground-state phase is characterized by the stacking (+ + − + − −) of equal-moment ferromagnetic sheets, whereas structures modulated along the c-axis are observed in high temperature phases.

In case of UPdIn, two phases were found below T_N[7]. The ground state magnetic structure of UPdIn is characterized by a sequence (+ − + + −) along c-axis yielding a spontaneous magnetization of 0.3 μ_B (1/5 M_s). Above 8.5 K, the moments are sinusoidally modulated along the c-axis with the propagation vector $\mathbf{k} = (0, 0, 0.4)$.

ELECTRICAL RESISTIVITY

The electrical resistivity is anisotropic in all three compounds and exhibits anomalous temperature dependence (Figs 1 – 3). The high temperature parts (paramagnetic range) are nearly temperature independent with large absolute values of the resistivity (hundreds of $\mu\Omega$cm).

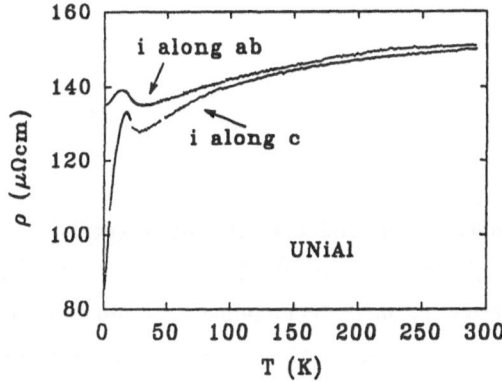

Figure 1. Temperature dependence of the electrical resistivity for $i \parallel c$ and $i \perp c$ in the UNiAl single crystal.

The resistivity of UNiAl[2] (Fig. 1) shows only a small anisotropy above 100 K ($\rho_{300K} \simeq 150\ \mu\Omega$cm). Upon decreasing temperature, the $\rho^{\parallel}(T)$ and $\rho^{\perp}(T)$ curves gradually deviate from one another. In contrast to UNiGa and UPdIn, lower resistivity values were measured for $i \parallel c$. Below 30 K, a minimum and a maximum are observed in both $\rho(T)$ curves at about 25 K (30 K) and 19 K (15 K), respectively. These features, connected with the magnetic ordering at 19 K, are more pronounced for $i \parallel c$. Moreover, at temperatures below the $\rho(T)$ maximum $\rho^{\parallel}$ decreases steeply with decreasing T to a low temperature value of 85 $\mu\Omega$cm ($\rho_0^{\parallel}/\rho_{300\ K} \simeq 0.57$) whereas 130 $\mu\Omega$cm ($\rho_0^{\perp}/\rho_{300\ K} \simeq 0.87$) is reached in the basal plane.

In UNiGa[11] (Fig. 2), a gradual upturn on the $\rho(T)$ curve for $i \parallel c$ is observed with lowering temperature. The upturn is suppressed by the magnetic field applied along c, and can be tentatively attributed to AF spin fluctuations with $\vec{q} \parallel c$, which are detected above T_N by magnetization measurements[3]. The two sharp peaks and the additional structure of the $\rho(T)$ curve between 30 and 40 K are connected with the four successive magnetic phase transitions. For $i \perp c$, the drop of the resistivity, which sets in at 39 K, reflects the the ferromagnetic ordering within the basal plane. At lower temperatures ($T < 20$ K), the resistivity follows a $\rho_0 + AT^2$ dependence. The ratio $\rho_0/\rho_{300K} \simeq 0.66$ for $i \parallel c$ is much larger than that for the current in the basal plane ($\simeq 0.21$).

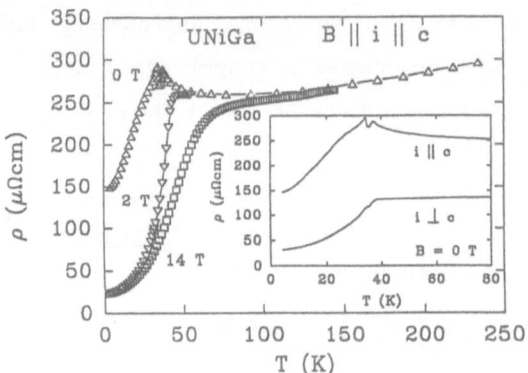

Figure 2. Temperature dependence of the electrical resistivity for $i \parallel c$ in the UNiGa single crystal in 0, 2 and 14 T. Insert: Low temperature section of the temperature dependence of the electrical resistivity for $i \parallel c$ and $i \perp c$. It is noted, that the intrinsic values of the resistivity for $i \parallel c$ can be much smaller than those presented. The measured resistivity was in some cases suddenly and irreversibly enhanced when passing magnetic phase transitions (no such effects were found for $i \perp c$). After the last measurement, a value of $\rho_{300K}^{\parallel} \cong 460$ $\mu\Omega$cm was recorded. The presented results for $i \parallel c$ have been normalized to the first-recorded resistivity value. But also this value can be enhanced substantially because the same crystal was used before for a number of low temperature magnetization measurements. These effects, are probably connected with large magnetostriction effects along the c-axis, which can produce microcracks in the sample.

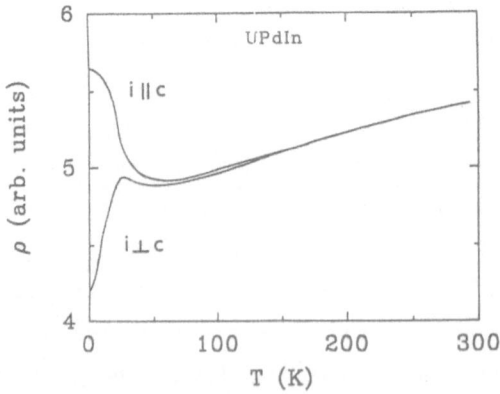

Figure 3. Temperature dependence of the electrical resistivity for $i \parallel c$ and $i \perp c$ in the UPdIn single crystal.

In UPdIn[5] (Fig. 3), the both $\rho(T)$ curves (for $i \parallel c$ and $i \perp c$) are almost identical in the high temperature range ($T > 50$ K). They pass a minimum around 50 K. Below this temperature, the resistance for i along c increases considerably and tends to saturation at 4.2 K. The resistance for i within the basal plane is, similar to UNiGa, more regular and resembles behaviour of ferromagnets. It decreases rapidly below 20 K.

The general tendency towards saturation observed for UTX compounds at high temperatures excludes application of any model yielding the flat $\rho(T)$ dependence as being due to mutual cancellation of a positive $d\rho/dT$ due to electron-phonon scattering and a negative "Kondo" $d\rho/dT$. It implies that the subtraction of resistivity of any background compound without f-component, as e.g. ThNiAl, is senseless. The Matthiessen rule is not obeyed here because of the mean free path approaching the interuranium spacing.

FIELD-INDUCED TRANSITIONS AND RESISTIVITY

By applying magnetic field along the c-axis, the metamagnetic transition to a ferromagnetically aligned phase is found in UNiAl (Fig. 4), UNiGa (Fig. 5), and UPdIn (Fig.6) at $\simeq$ 11, 0.8, and 16 T, respectively, all at $T = 4.2$ K. At higher temperatures, the ferromagnetic phase in UNiGa can be reached by a two step transition (Fig. 7) via a ferrimagnetic phase (+ + -). A double step transition is observed also in UPdIn at 4.2 K (Fig. 6). The first step corresponds to the transition from the (+ - + + -) phase to a phase with M = 1/3 M_s (presumably + + -) and the second to a full ferromagnetic-type alignment. The metamagnetic transitions are accompanied by dramatic reductions of the electrical resistivity (see Figs. 4 - 7).

The resistivity in UNiAl at 14 T (Fig. 4), is reduced by 70 % with respect to zero-field values for both current directions. Unlike UNiGa and UPdIn, where measurable resistance changes at 4.2 K are concentrated almost exclusively into the transition region, a noticeable decrease of ρ is found in UNiAl also below and above B_c (the critical field of the metamagnetic transition). This (similar to the γ-enhancement in field) can be related to moment-fluctuation effects, which strongly affect the magnetism in this compound. From the lack of saturation between 11.5 and 14 T one expects still a considerable decrease of the resistivity in higher fields, because the magnetization saturates only around 50 T. Above the metamagnetic transition in UNiGa (Fig. 3), the resistivity reduction amounts to $\simeq$ 0.1 and 0.4 $\rho(0\ T)$ for $i \parallel c$ and $i \perp c$, respective-

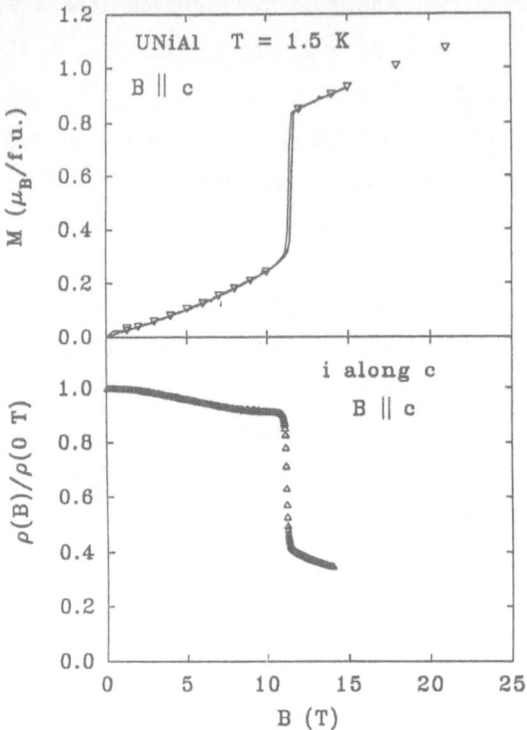

Figure 4. Magnetization curve (upper part) and magnetic-field dependence of the relative electrical resistivity for $i \parallel c$ (lower part) in the UNiAl single crystal at 1.5 K for $B \parallel c$.

ly. Despite the invariable γ-value of 40.8 mJ/mol K^2, values of the quadratic coefficient A are also strongly Both residual resistivity values in a field above the metamagnetic transition are only about 8% of the corresponding room-temperature values. The observed hysteresis effects reproduce well the hysteresis of magnetization.

The field dependence of the relative resistance in UPdIn, shown in Fig 6, displays two distinct drops, one around $B = 3$ T, the other at approximately 16 T. The low-field transition has a noticeable hysteresis and $\rho(B)$ in this region is rather time dependent in a constant field. This time dependence can be described as an exponential relaxation behaviour with relaxation times of the order of 100 ms. This observation is consistent with the relation between the width of the hysteresis loop and the field sweeping rate in magnetization measurements[5]. The high-field transition is practically without hysteresis. This was ascribed to a more complicated moment re-arrangement at the lower transition. The resistivity here is reduced only by a small portion, whereas the high-field transition depresses the resistivity far below its high-temperature values. Magnetic fields applied within the basal plane have no influence on the resistivity values in all three compounds, which proves that it is the scattering mechanism and not the electron-path variations which drive the huge magnetoresistance effects.

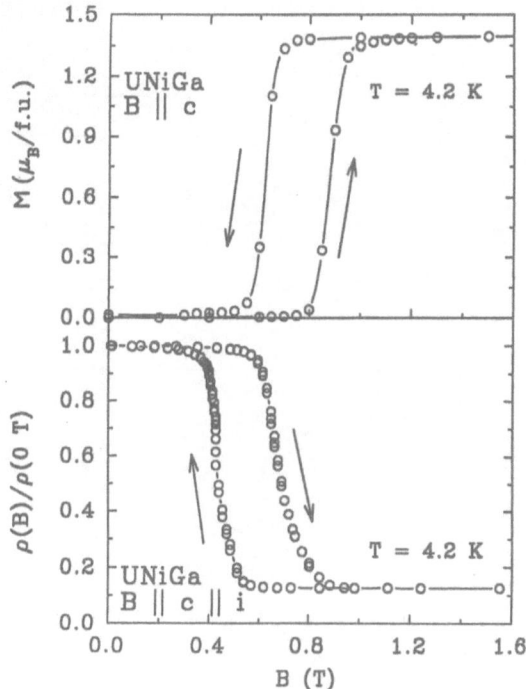

Figure 5. Magnetization curve (upper part) and magnetic-field dependence of the relative electrical resistivity for $i \parallel c$ (lower part) in the UNiGa single crystal at 4.2 K for $B \parallel c$. We note, that the different values of metamagnetic fields observed in the magnetization measurement and in the resistivity measurements for $i \parallel c$ and $i \perp c$ are due to different demagnetization factors of used samples. (also in Figure 6).

DISCUSSION AND CONCLUSIONS

All compounds under investigation display a significant anisotropy of ρ. Besides the general anisotropy due to the crystal-structure and Fermi-surface anisotropy, there is a significant anisotropy of ρ related to the anisotropy of $5f$-moment ordering. For transport of charge along sheets with ferromagnetically coupled moments, the $\rho(T)$ dependence mimics a behaviour of a ferromagnet. Relying on a local-moment picture, we can attribute the difference $\rho(300\ K) - \rho_0$ to the "spin-disorder scattering". The values of several hundred $\mu\Omega$cm point then to a very strong exchange interactions between $5f$ and conduction electrons (comparing to e.g $4f$-conduction electron interaction in lanthanides). Electron-phonon scattering cannot contribute substantially to $\rho(300\ K)$, as the $\rho(T)$ dependencies are flat at high T.

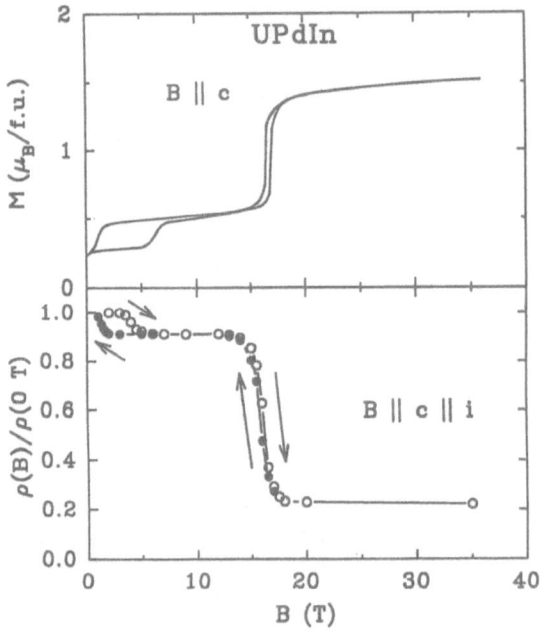

Figure 6. Magnetization curve (upper part) and magnetic-field dependence of the relative electrical resistivity for $i\ \|\ c$ (lower part) in the UPdIn single crystal at 4.2 K for $B\ \|\ c$.

The resistance for current sensing the AF coupling displays in all cases an upturn with decreasing T, which is pertinent to many antiferromagnets. The fact, that this upturn appears well above the ordering temperature, points to the presence of short-range AF correlations (AF spin fluctuations) on the scale of the electron mean free path.

The main issue remains a mechanism responsible for the huge drop of ρ at the metamagnetic transition which is preserved even in the low-temperature limit. One mechanism possible in non-local-moment systems, the change of the coupling strength between f and conduction electrons, is probably unimportant, because this would affect both the magnitude of magnetic moment themselves and their dynamics, reflected in variations of the γ-coefficient. But at least in UNiGa, where the drop of ρ is relatively largest, this was not observed. The invariability of γ makes questionable also a generally present impact of the Fermi surface reconstruction, which come into consideration due to the nesting of additional Brillouin zones because of additional periodicity in the AF state. Phenomenologically, the layered type of magnetism in UTX

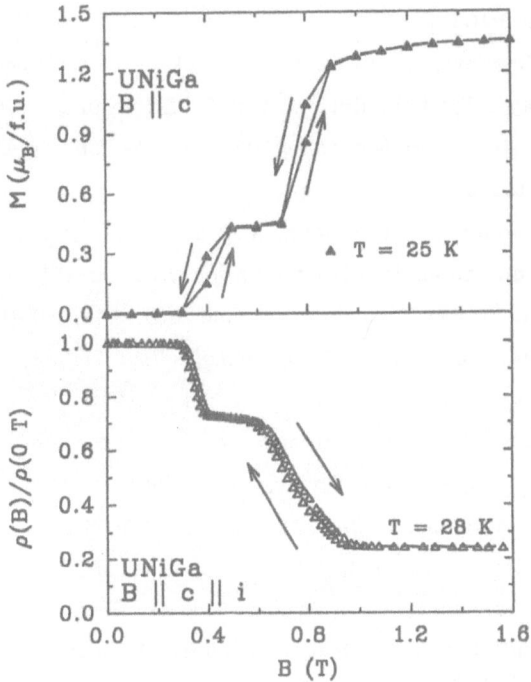

Figure 7. Magnetization curve (upper part) and magnetic-field dependence of the relative electrical resistivity for $i \parallel c$ (lower part) in the UNiGa single crystal at 25 and 28 K, respectively, for $B \parallel c$.

compounds is reminiscent of magnetic multilayers. One of explanations, used here to describe large magnetoresistance effects, is based on a picture of spin-up and spin-down split bands. The majority sub-band in one layer is a minority one (for one spin direction) in neighbouring layers for layers with AF coupling. Without spin-flip scattering, this reduces the density of states at E_F, which is available for charge transport in direction perpendicular to the layers.

Acknowledgements

Financial support of this work by a Grant-in-Aid for the International-Joint Research Program from the Ministry of Education, Science and Culture of Japan is gratefully acknowledged.

REFERENCES

1. V.Sechovsky, L.Havela, E.Brück, F.R.de Boer, and A.V.Andreev, 5f-ligand hybridization and magnetism in UTX compounds, *Physica B* 163:103 (1990)

2. L.Havela, V.Sechovsky, P.Nozar, E.Brück, F.R.de Boer, J.C.P.Klaasse, A.A.Menovsky, J.M.Fournier, M.Wulff, E.Sugiura, M.Ono, M.Date, and A.Yamagishi, Antiferromagnetic correlations in UNiAl, *Physica B*, 163:313 (1990)

3. L.Havela, V.Sechovsky, L.Jirman, F.R.de Boer, and E.Brück, Magnetic transitions in UNiGa, *J.Appl.Phys.* 93:4813 (1991)

4. E.Brück, F.R.de Boer, V.Sechovsky, and L.Havela, UPdIn - a new heavy-fermion compound, *Europhys.Lett.* 7:177 (1988)

5. H.Fujii, H.Kawanaka, M.Nagasawa, T.Takabatake, Y.Aoki, T.Suzuki, T.Kujita, E.Sugiura, K.Sugiyama, and M.Date, Anisotropic hybridization in a heavy-fermion compound with double magnetic transitions, *J.Magn.Magn.Mater.* 90&91:507 (1990)

6. P.Burlet, L.Jirman, V.Sechovsky, L.Havela, M.Diviš, Y.Kergadallan, J.C.Spirlet, J.Rebizant, E.Brück, F.R.de Boer, H.Nakotte, T.Suzuki, T.Fujita, and H.Maletta, Magnetic phase diagram of UNiGa, in: Proc. 22^{ièmes} Journées des Actinides, Méribel (1992), p. 125

7. H.Fujii, H.Kawanaka, T.Takabatake, E.Sugiura, K.Sugiyama, and M.Date, Magnetic and electrical properties in UCuSn, UPdIn and Th substituted UNiSn, *J.Magn.Magn.Mater.* 87:235 (1990)

8. J.M.Fournier and P.Burlet, Antiferromagnetic ordering in UNiAl, in: Proc. 21èmes Journées des Actinides, Montechoro (1991), p. 126

9. L.Jirman, V.Sechovsky, L.Havela, W.Ye, T.Takabatake, H.Fujii, T.Suzuki, T.Fujita, E.Brück, and F.R.de Boer, Magnetic and transport properties of UNiGa, *J.Magn.Magn.Mater.* 104-107:19 (1992)

10. H.Nakotte, E.Brück, F.R.de Boer, A.J.Riemersma, L.Havela, and V. Sechovský, Magnetoresistance and Metamagnetic Transitions in UPdIn, *Physica B* 179:269 (1992)

11. J.Mathon, Theory of magnetic multilayers. Exchange interactions and transport properties, *J.Magn.Magn.Mater.* 100:527 (1991)

ELECTRON TRANSPORT IN THE UTSn SERIES,

WHERE T = Ni, Pd, Pt, Cu and Au

R. Troć, B. Badurski and V.H. Tran

W. Trzebiatowski Institute of Low Temperature
and Structure Research, Polish Academy of Sciences
50-950 Wroclaw, Poland

1. INTRODUCTION

Studies of the magnetic and transport properties of the equiatomic ternary systems, UTM, in which U is combined with a transition metal T and an sp-element M have been a subject of great interest. This interest arises from the fact that these systems comprise a wide range of different ground states of uranium depending on the kind of T and M constituents. Such parameters like the occupancy of T-d shell, the size of M-element and the mutual coordination of T and M atoms with the uranium atoms (crystal structure) play an important role in the U5f-ligand hybridization[1]. As many authors have claimed, the latter is a dominating factor determining the observed magnetic and transport properties of these phases.

One interesting class of these systems are stannides, UTSn. In this work we have investigated these compounds with T = Ni, Pd and Pt or Cu and Au, which have mostly or complete filled d-states, respectively, what thus prevent or eliminate the U5f-Tnd hybridization. Also a large introduced atom into the compound as Sn diminishes greatly the effect of U5f-Msp hybridization. Therefore, the investigated ternary compounds behave mainly as the local-moment systems and their observed properties are predominantly governed by the geometry of U-surrounding, which is varied from compound to compound due to the appearance of different crystal structures in this series of compounds.

The magnetic and transport properties of UTSn compounds have intensively been discussed elsewhere[2,3] Here we have extended our studies by measuring except the electrical resistivity also the magnetoresistivity, $\Delta\rho/\rho$, in weak fields up to 1 T.

Previously, the magnetoresistivity for the cubic UNiSn and UPtSn phases as well as for the hexagonal UPdSn one has been measured by Palstra et al.[2]. We think that the transport properties are valuable tools in the study of these materials, especially near to any transition temperature, because the principal scattering mechanism here is believed to be associated mainly with the magnetic effects.

2. EXPERIMENTAL PROCEDURES AND RESULTS

For samples preparation and electrical resistivity measurements we refer to ref. 3. The samples for transport measurements were cut into bars with typical dimensions 1x1x5 mm[3]. Magnetoresistivity, $\Delta\rho/\rho$, was first measured at 4.2 K in applied fields up to 1 T and then keeping this field strength the temperature was varied from 4.2 K to that being usually much higher than the respective transition temperatures of a studied compound. The obtained results we present in Figs. 1-7 and discuss below in groups having the same crystal structure.

2.1 UNiSn

Recently we have found[3] that the UNiSn samples obtained just after melting (not annealed) reveal an orthorhombic structure of $CeCu_2$-type.

In contrast to the well known, low temperature (LT), MgAgAs-type phase of UNiSn[2], the orthorhombic phase behaves as a typical intermetallic actinide phase. The ρ-T curve of this phase is shown in Fig. 1. According to our previous results[3], the $\rho(T)$ function is strongly curvilinear up to about 200 K and then it saturates to a value of about 350 $\mu\Omega$cm. At low temperatures a T^2 dependence is marked. The temperature dependence of the derivative $d\rho/dT$ shows a sharp cusp at 12 K and some step-like change at about 20 K. $\Delta\rho/\rho$ at 4.2 K is negative and steeply decreases with rising temperature, showing a distinct kink at 20 K (see the inset of Fig. 1). At present, it is not completely clear whether this anomaly, seen in the vicinity of 20 K also in $\chi(T)$[3], is intrinsic or is caused by an impurity phase. As shown by Palstra et al.[2], the resistivity of the cubic UNiSn phase exhibits a semiconductor to metal transition at T_{SM} near T_N = 47 K. The semiconducting behaviour with an energy gap Δ/k = 800 K is ascribed to the particular band structure, relevant to the MgAgAs-type lattice. As recent studies[4] have shown, pressure strongly suppresses this energy gap. The thermoelectric power, Q, goes through an extremely sharp peak of 50μV/K at T_{SM}[5] The magnetoresistivity measured at temperatures up to 100 K and magnetic fields up to 7 T for this phase[2] is negative at all

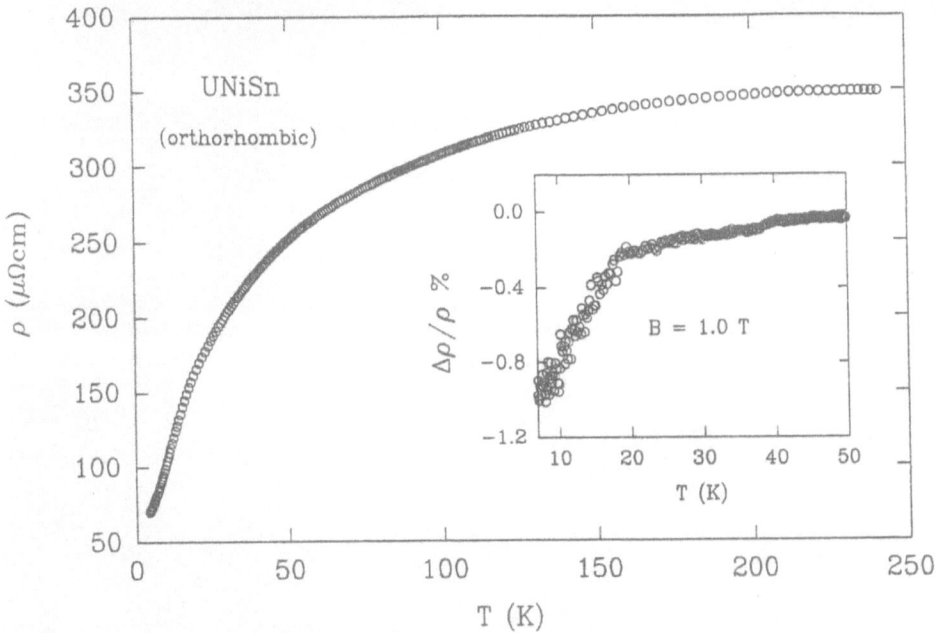

Figure 1. Temperature dependence of the electrical resistivity for the $CeCu_2$-type compound, UNiSn. Inset: the magnetoresistivity against the temperature at $B = 1$ T.

temperatures and varies quadratically with the magnetic field. Again, the largest effect in $\Delta\rho/\rho$ was found at temperatures near T_N, $\Delta\rho/\rho = -7.5$ % at 7 T comparing to -2% at 4.2 K and 5 T.

2.2 UPtSn

In our previous paper[3] we reported that except for the cubic (MgAgAs-type) UPtSn there exists also the hexagonal (Fe_2P-type) form, but for the off-stoichiometric sample, $UPt_{0.9}Sn_{1.1}$. The magnetic properties of these two forms appeared to be quite different. Whereas the susceptibility of the cubic phase increases largely below 35 K and then at about 20 K, it goes through a small maximum, the hexagonal phase is ferromagnetic below 28 K. Fig. 2 presents the $\rho(T)$ curve for the MgAgAs-type compound. As one can see from this figure, $\rho(T)$ behaves amazingly, i.e. the resistivity values are huge as those reported previously for this compound by Palstra et al.[2]. Despite that the shape of the $\rho(T)$ curve reflects the magnetic behaviour of this compound. As our results indicate (see Fig. 2), the resistivity increases with cooling down to about 35 K, passing through a flat maximum and then ρ below about 26 K decreases more rapidly. In addition, at $T = 7$ K a small minimum in $\rho(T)$ is observed (see the inset of Fig.2), an origin of which is not clear now.

The $\Delta\rho/\rho$ vs. T curve, taken at $B = 1$ T, is displayed in Fig. 3. The magnetoresistivity at low temperatures is positive, but at about 22 K changes its sign and at $T = 26$ K shows a sharp negative minimum, then at

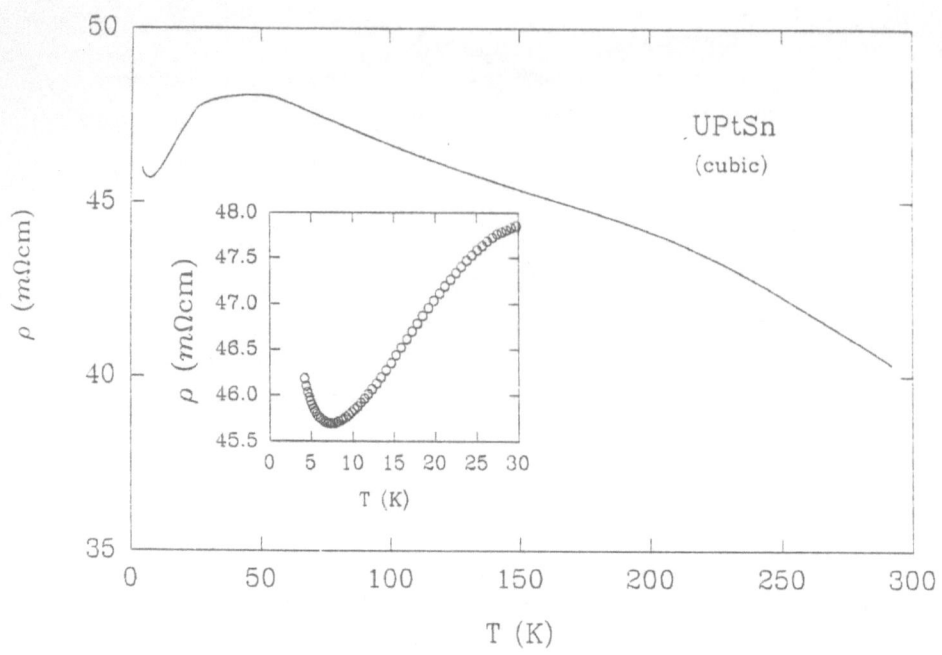

Figure 2. Temperature dependence of the electrical resistivity for the MgAgAs−type compound, UPtSn. Inset: low temperature behaviour of the resistivity

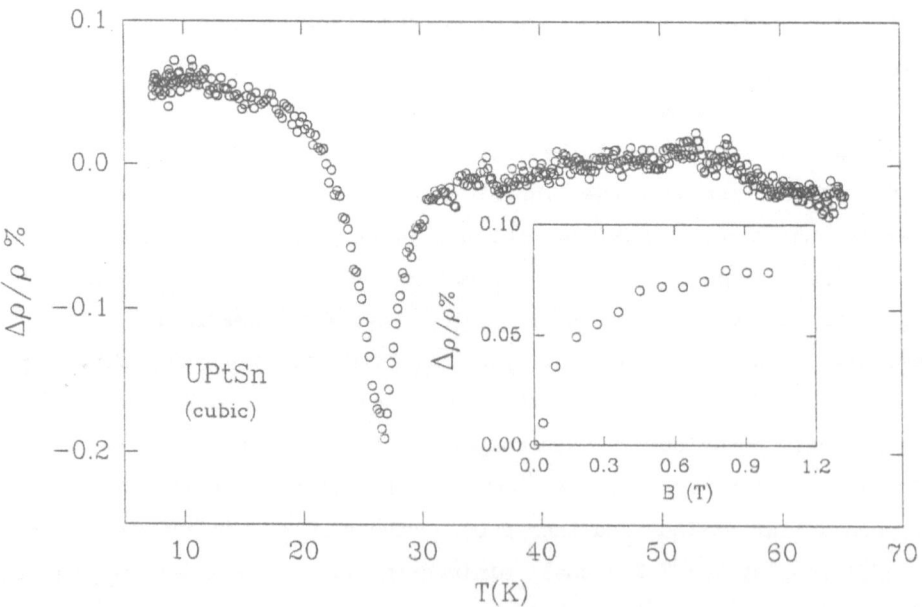

Figure 3. Temperature dependence of the magnetoresistivity for the MgAgAs−type compound, UPtSn Inset the magnetoresistivity at 4 2 K

35 K it becomes zero and finaly $\Delta\rho/\rho$ above about T = 55 K again slowly increases negatively. The behaviour described above exactly reflects the anomalies observed in the $\rho(T)$ curve. The similar temperature effect was observed by Palstra et al.[2] for their sample of UPtSn. The largest negative magnetoresistivity they also found around T = 30 K with a change in sign at 20 K. At 4.2 K and 5 T they reported $\Delta\rho/\rho$ = 0.5 %. The inset of Fig. 3 shows a positive increase in $\Delta\rho/\rho$ of our sample with applied magnetic field.

However, in contrast to the cubic UNiSn phase, having the same type of crystal structure, the thermoelectric power of cubic UPtSn, measured by Yamaguchi et al.[4], does not exhibit any notable structure and Q-values are close to zero over the temperature range, 4.2–300 K.

Fig. 4 displays the $\rho(T)$ curve of the hexagonal $UPt_{0.9}Sn_{1.1}$ form. Its behaviour is similar to that reported earlier[3]. There is a remarkable change in slope of $\rho(T)$ at 28 K, being typical for a ferromagnet. The field dependence of the magnetoresistivity at 4.2 K, shown in the upper inset of Fig. 4, is initially slightly negative and above B = 0.4 T becomes positive. The $\Delta\rho/\rho$ vs. T curve (the lower inset) shows a small jump at the ferromagnetic transition temperature, T_C = 28 K, and additionally one sees only slightly marked anomaly at about T = 24 K. The occurrence of two close lying features for $UPt_{0.9}Sn_{1.1}$ we have observed

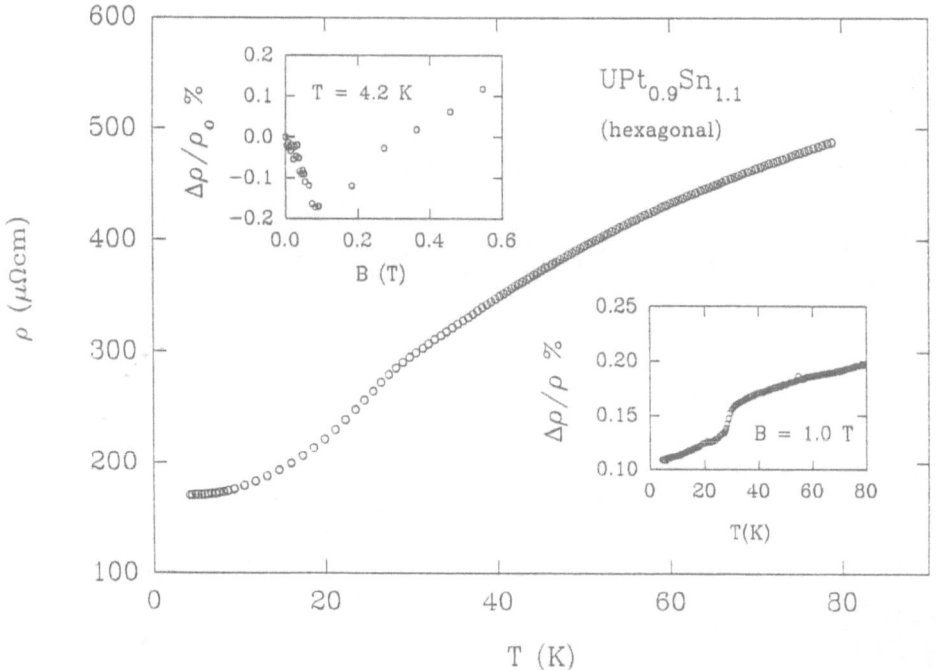

Figure 4. Temperature dependence of the electrical resistivity for the Fe_2P–type. compound, $UPt_{0.9}Sn_{1.1}$. Inset: the magnetic field (upper) and temperature (lower) dependences of the magnetoresistivity.

in our previous studies[3] in the temperature dependences of both the derivative $d\rho/dT$ and magnetization.

2.3 $CaIn_2$ - TYPE COMPOUNDS

In this hexagonal symmetry structure crystallizes a few UTM compounds. Among them also UPdSn, UCuSn and UAuSn. The characteristic feature of this structure is that the shortest U–U distances are not in the basal plane (d_{U-U} =4.6 – 4.7 Å), but along the c-axis (d_{U-U} = 1/2c = 3.60 – 3.65 Å). This leads to a linear chain of U-atoms spreading out along the c-axis, what should impact on the magnetic and transport properties of these systems.

a) UPdSn

Shown in Fig.5 is the ρ – T curve, taken at B = 0 and 1 T for newly obtained sample of UPdSn. Although an overall shape of this curve is similar to that reported earlier[2-4], the magnitudes of ρ are here considerably larger than those reported in previous studies. For example, the difference ρ_p - ρ_o, where ρ_p is an average value of the resistivities within the plateau (80 – 150) K and ρ_o is the residual resistivity, in our case is 4400 $\mu\Omega$cm, while for the samples studied previously this

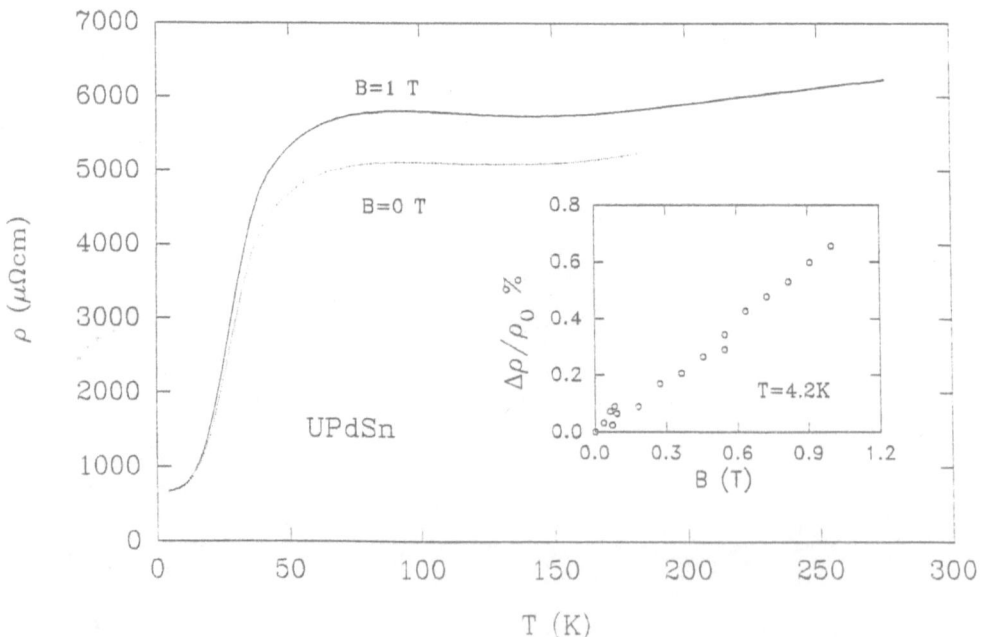

Figure 5. Temperature dependence of the electrical resistivity at B = 0 and 1 T for the $CaIn_2$-type compound UPdSn Inset the magnetic field change of the resistivity at 4 2 K

difference reaches the following values: 400[3], 800[2] and 2100[4] $\mu\Omega$cm. This large spread in these values strongly indicates a large-sample dependent effect taking place for this compound. Moreover, as Fig. 5 illustrates, ρ-values markedly increase when measured at B = 1 T. The temperature dependence of ρ is also strongly affected by pressure[4], especially above 100 K. It is also worth noting that the dρ/dT vs. T curve (not shown) has a fairly sharp maximum at T = 26 K and an inflection point at T = 41 K, which well correspond to the critical temperatures determined from the susceptibility[3,6,7] and neutron diffraction[7-9] measurements.

The inset of Fig. 5 shows the magnetoresistivity of UPdSn at 4.2 K. As seen, $\Delta\rho/\rho$ is positive and rises linearly with applied magnetic field. At B = 1 T it is equal to 0.7 %. In contrast to this result, Palstra et al.[3] have found at the same temperature, but at 5 T, that $\Delta\rho/\rho$ = -8%.

b) $UCu_{1+x}Sn_{1-x}$

Some basic properties of UCuSn have previously been studied by Fujii et al.[10] We have studied two samples, the stoichiometric UCuSn and non-stoichiometric $UCu_{1.1}Sn_{0.9}$ alloys. The electrical resistivity of both these materials behaves in similar manner, as was reported by Fujii et al.[10], i.e. ρ rises slightly with decreasing temperature down to 60 K. Below this temperature a characteristic bump in $\rho(T)$ with T_{max} = 25 K is observed, because the new periodicity generated by the antiferromagnetic order modifies the band structure below T_N = 60 K. The difference between these two materials is only seen in the magnitudes of ρ. Surprisingly, these for the non-stoichiometric phase are about 45 % lower in comparison to the stoichiometric one. In turn, the ρ-value at room temperature of our UCuSn sample is only half of that reported in Ref. 10.

Furthermore, the magnetic properties of UCuSn determined by us[11] seem to be more complex than those reported in Ref. 10. The cusp anomaly in $\chi(T)$ at T_N = 60 K, observed by Fujii et al.[10], exactly corresponds to what we have found for the non-stoichiometric sample. It appears that the magnetic susceptibility of our x = 0 sample increases suddenly below 77 K and goes through a broad maximum centered at T = 30 K. It is interesting to note that some sign of such a maximum is also seen for $UCu_{1.1}Sn_{0.9}$[11]. However, it is surprising that the difference in magnetic behaviour of our UCuSn sample is not reflected in its temperature dependence of ρ. At low temperatures the resistivity varies as $\rho(T) = a\ T^2$ with the coefficient a = 1.23 and 0.52 $\mu\Omega$cm/K^2 for the x = 0 and 0.1 samples, respectively.

The magnetoresistivity was measured only for the $UCu_{1.1}Sn_{0.9}$ sample.

At 4.2 K $\Delta\rho/\rho$ increases negatively as a B^2, reaching a value of -0.11% at B = 1 T.

c) *UAuSn*

The electrical properties of UAuSn have previously been studied by Palstra et al.[3] and recently by de Boer et al.[6]. The former authors present a flat $\rho(T)$ curve between 4.2 - 300 K, while the latter ones have observed an upturn in $\rho(T)$ developing below 80 K.

Our $\rho(T)$ data for UAuSn are depicted in Fig.6. One can see from this

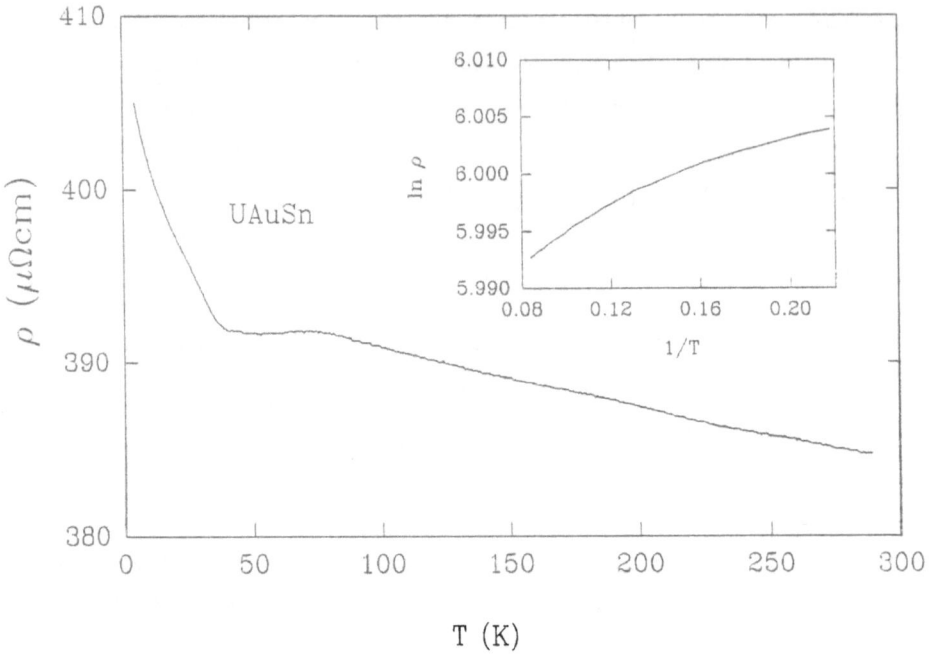

Figure 6. Temperature dependence of the electrical resistivity for the $CaIn_2$-type compound, UAuSn. Inset: the $\ln\rho$ vs. $1/T$ plot below 12 K.

figure that ρ monotonically increases if the temperature is decreased from room temperature to about 75 K, passes a plateau down to T_N = 35 K and below this temperature $\rho(T)$ rapidly increases exponentially, without any sign of saturation at 4.2 K. This exponential increase in $\rho(T)$ is interpreted as a result of the opening of the energy gap in the density of states. In the inset of this figure, the $\ln\rho$ vs. T^{-1} plot is shown. The energy gap estimated from this plot between 4.2 - 8 K is about 10 K. The mechanism of the gap formation in UAuSn is likely associated with a complex, long-range magnetic ordering, which finds its support in the well-resolved Mössbauer spectra measured on the same sample[12]. It is interesting to note that an analogical situation was found in UPdIn[13], for which ρ measured along the c-axis indicates that a pseudogap is formed on the part of Fermi surface in connection with the development of an incommensurate, sinusoidally modulated magnetic structure below T_N = 20

K^{14}. The possible randomness in the occupancy of the Au and Sn lattice sites may prevent the formation of long-range magnetic order in some samples of UAuSn, as suggested in Ref.6, and hence, we think, no distinct change in the resistivity near the critical point can be observed.

The $\Delta\rho/\rho$ vs. T plot around the Néel temperature of UAuSn is given in Fig.7. Despite the extremely smallness of the effect, one can easy

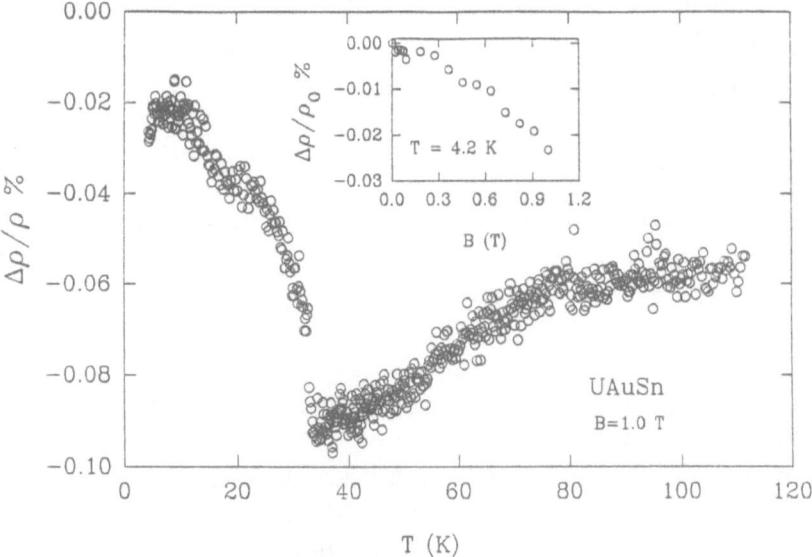

distinguish from this plot the transition temperature, $T_N = 35$ K, where $\Delta\rho/\rho$ achieves a fairly sharp negative minimum.

Like for the Cu-based sample, the magnetoresistivity of UAuSn at 4.2 K is negative following a B^2 law, but the effect is about an order of magnitude smaller.

SUMMARY

We present results of the temperature dependences of the electrical resistivity measured in zero and magnetic fields up to 1 T for the UTSn phases with T from the end of the trasition-metal series. These phases crystallize in different lattices and exhibit various magnetic transitions at low temperatures. We have found that the resistivity behaviour for most of the studied compounds is strongly sample-dependent. It results from the fact that most properties of the UTSn phases are influenced considerably by small changes in the chemical compositions. It also implies that some of UTSn compounds occur in different crystal structure forms.

Especially the strange results obtained for the cubic phase of UPtSn in comparision to previous data require further studies of this compound. It is also interesting to note that the temperature dependence of the magnetoresistivity in weak magnetic fields may yield usefull information on the magnetic transitions taking place in these phases. Our such data obtained for UAuSn clearly indicate that this compound is antiferromagnetically ordered below T_N = 35 K, but probably with a complex magnetic structure. The latter is inferred from the low temperature electrical resistivity behaviour of this compound, which indicates the opening of the energy gap of about 10 K on the Fermi surface. Single crystal measurements should clear up these findings.

ACKNOWLEDGEMENTS

The work was supported by the research program of the KBN Grant Nr. 202969101.

REFERENCES

1. D.D. Koelling, B.D. Dunlap, and G.W. Crabtree, *Phys. Rev.* B31:4966 (1985).

2. T.T.M Palstra, G.J. Nieuwenhuys, R.F.M. Vlastuin, J van der Berg, J.A. Mydosh, and K.H.J. Buschow, *J. Magn. Magn. Mat.*, 67:331 (1987).

3. V.H. Tran and R. Troć, *J. Magn. Magn. Mat.*, 102:74 (1991).

4. M. Kurisu, H. Kawanaka, T. Takabatake, and H. Fujii, *J. Phys. Soc. Jap.*, 60:3792 (1991).

5. Y. Yamaguchi, J. Sakurai, F. Teshima, H. Kawanaka, T. Takabatake, and H. Fujii, *J. Phys: Cond. Matter* 2:5715 (1990).

6. F.R. de Boer, E. Brück, H. Nakotte, A.V. Andreev, V. Sechovsky, L. Havela, P. Nozar, C.J.M. Denissen, K.H.J. Buschow, B. Vaziri, M. Meissner, H. Maletta, and P. Rogl, *Physica B* 176:275 (1992).

7. R.A. Robinson, A.C. Lawson, K.H.J. Buschow, F.R. de Boer, V. Sechovsky, and R.B. von Dreele, *J. Magn. Magn. Mat.*, 98:147 (1991).

8. R.A. Robinson, A.C. Lawson, J.W. Lynn, and K.H.J. Buschow, *Phys. Rev.* B45:2939 (1992).

9. R. Troć, V.H. Tran, J.Rossat-Mignod, M. Bonnet, R. Kmiec, A. Szytuła, M. Kolenda, R. Kruk, K. Łątka, and T. Tomala, *J. Magn. Magn. Mat.*, (submitted).

10. H. Fujii, H. Kawanaka, T. Takabatake, E. Sugiura, K. Sugiyama, and M. Date, *J. Mag. Magn. Mat.*, 87:235 (1990).

11. V.H. Tran and R. Troć, *Physica B* (in press).

12. R. Kruk et. al. (to be published).

13. H. Fujii, H. Kawanake, M. Nagasawa, T. Takabatake, Y. Aoki, T. Suzuki, T. Fujii, E. Sugiura, R. Sugiyama and M. Date, *J. Magn. Magn. Mat.*, 90&91: 507 (1990).

14. E. Suigiura, K. Sugiyama, H. Kawanaka, T. Takabatake, H. Fujii, and M. Date, *J. Magn. Magn. Mat.*, 90&91:65 (1990).

THERMAL CONDUCTIVITY OF Ce AND Yb
BASED KONDO COMPOUNDS

E. Bauer

Institut für Experimentalphysik
Technische Universität Wien
A - 1040 Wien, Austria

INTRODUCTION

It is well known that the electronic thermal conductivity of simple metals traces the behaviour of the electrical conductivity of the same material. The resemblance between both transport quantities is theoretically proven by the so called Wiedemann - Franz law. However, there are some constraints since, for example, small angle scattering processes do not contribute to the electrical resistivity. Moreover, the lattice thermal conductivity cannot be neglected in certain cases. In spite of such discrepancies, each information on details of the electrical resistivity allows a better understanding of the much more complex behaviour of the temperature dependent thermal conductivity.

There are a lot of fundamental papers considering both the Kondo effect and crystal field splitting and their impact on the temperature dependent electrical resistivity $\rho(T)$ and on the temperature dependent thermopower $S(T)$.[1,2,3,4]

• The magnetic contribution to the electrical resistivity $\rho_{mag}(T)$ due to both effects is characterized by logarithmic ranges for temperatures which are much larger or smaller than the temperature of a certain crystal field level Δ_i/k_B. Around Δ_i/k_B a maximum in $\rho_{mag}(T)$ occurs. This dependence has been concluded from solutions of the Coqblin - Schrieffer model in a perturbation type calculation.[1] Usually, this distinct behaviour serves to state pronounced Kondo scattering processes in presence of crystal field splitting. As an archetypal example for this behaviour, $CeAl_2$ is mentioned.[5] Kondo type interactions in this compound yield to a negative logarithmic behaviour at low and high temperatures and a maximum around 80 to 90 K, marking the Γ_8 quartet above the Γ_7 ground state.

However, as the separation of the different CF levels becomes closer, just an integral effect can be observed, i.e. a negative logarithmic range in the crystal field ground state and in the full multiplet of the considered ion. Moreover, if the high temperature Kondo temperature T_K^H (i.e. the Kondo temperature in the full multiplet of the Hund ground state) becomes comparable with the overall crystal field splitting Δ_{CF}, the deduced behaviour of $\rho_{mag}(T)$ may be degenerated in the sense that just one maximum in $\rho_{mag}(T)$ appears. Above which, an extended range with an (-lnT) behaviour occurs.[6] As representatives for this case we mention CeCu$_6$,[7] CeAl$_3$,[8] CeCu$_4$Al [9] or YbCu$_4$Ag.[10]

• Extraordinary large thermopower values, which exceed those of normal metals by one ore two orders of magnitude are observed for Kondo compounds. The reason of which is an asymmetric relaxation time τ_k with respect to the energy around the Fermi level.[3] If crystal field splitting in the compounds is negligible, Kondo interaction processes yield to an universal behaviour of $S(T)$, scaled by just a single temperature, the Kondo temperature T_K. Additionally, a maximum appears around $T = T_K$. However, as crystal field splitting or other interaction mechanisms like the RKKY interaction become important, the universal behaviour of $S(T)$ is lost.[2,11] If the degeneracy of the ground state is lifted by the crystal field, an extremum in $S(T)$ at a certain fraction of Δ_{CF} should occur ($S^{max} = (1/3 - 1/6)\Delta_{CF}$).[2] Very recently it has been shown that the maximum in $S(T)$ observed for Ce - or the minimum for Yb systems can be related also with the value of the high temperature Kondo temperature T_K^H.[12] However, it is noticed that the value of $(1/3 - 1/6)\Delta_{CF}$ is usually of the same magnitude as the value of T_K^H.

In the following sections we will show that also thermal conductivity measurements on Ce and Yb based compounds can be used to study the Kondo effect, crystal field splitting or the contribution due to RKKY interaction. For this, a short section with basic theoretical concepts, followed by a number of experimental results on Ce and Yb based Kondo compounds will be presented. To deduce from the total measured thermal conductivity the contribution due to the interaction of conduction electrons with the almost localized magnetic moments, a comparison with equivalent nonmagnetic compounds is necessary. Besides it is necessary to consider the lattice contribution to the total thermal conductivity.

THEORETICAL CONCEPTS

As demonstrated in many textbooks,[13,14] the temperature dependent thermal conductivity $\lambda(T)$ may be described using the linearized Boltzmann equation. In the usual way the thermal conductivity can be written as

$$\lambda = \frac{1}{T}\left[K_2 - \frac{(K_1)^2}{K_0}\right] \tag{1}$$

where the integrals K_n are given by

$$K_n = \frac{k_F^3}{3\pi^2 m}\int \epsilon_k^n \cdot \tau(\epsilon_k)\left(-\frac{df_k}{d\epsilon}\right)d\epsilon_k \tag{2}$$

with m the electron mass, k_F the Fermi wave vector and ϵ_k the electron energy. Once, the relaxation time $\tau(\epsilon_k)$ of a certain scattering event is determined, the thermal conductivity λ can be calculated. Following the treatment of Bhattacharjee and Coqblin,[15] the thermal conductivity, originated from Kondo scattering processes in presence of crystal field splitting can be obtained using the third order perturbation theory in the framework of the Coqblin - Schrieffer model. The associated relaxation time, including second and third order terms is given by

$$\frac{1}{\tau_k} = \frac{mkv_0c}{\pi\hbar^3(2j+1)}(R_K + S_K) \tag{3}$$

k is the wave vector, v_0 the sample volume and c is the concentration of impurity moments with total angular momentum j. R_K and S_K are mainly functions of J^2 and J^3, respectively. The exchange integral J is calculated applying the Schrieffer Wolff transformation to the hybridization matrix element of the Anderson Hamiltonian.[16] Equation (1), (2) and (3) then yields[15]

$$\lambda = \frac{\hbar^3 k_F^2(2j+1)}{3\pi m^2 v_0 c}\frac{1}{T}(W_2 - W_3) \tag{4}$$

with

$$W_2 = \int \epsilon_k^2 \left(-\frac{\partial f_k}{\partial \epsilon_k}\right)\frac{d\epsilon_k}{R_k} \tag{5}$$

and

$$W_3 = \int \epsilon_k^2 \left(-\frac{\partial f_k}{\partial \epsilon_k}\right)\frac{S_k}{R_k^2}d\epsilon_k \tag{6}$$

Here, $S_K \ll R_K$ has been assumed. Using the f = 1/2 approximation, the electronic thermal conductivity $\lambda_{e,mag}(T)$, confined by scattering processes of conduction electrons with almost localized magnetic moments is calculated. The principal result, expressed in terms of the electronic thermal resistivity $(1/\lambda_{e,mag} \equiv W_{e,mag})$ times temperature traces closely the well known characteristics of the electrical resistivity of Kondo compounds in presence of crystal field splitting.[17]

$$W_{e,mag} \cdot T = \frac{R}{L_0}\left(\mathcal{V}^2 + \frac{(\alpha^2-1)}{(2j+1)\alpha}J^2\right) + \frac{2R}{L_0}n(E_F)J^3 \cdot \frac{(\alpha^2-1)}{2j+1}\ln\left(\frac{k_BT}{D}\right) \tag{7}$$

The customary notation[1,15] has been used. $\alpha = 2j+1$ is the effective degeneracy of the 4f state, $R = (3m\pi v_0 c)/(e^2 h k_F^2)$ and $L_0 = (\pi^2 k_B^2)/(3e^2)$ is the Sommerfeld value of the Lorenz number. Equation (7) shows that $W_{e,mag} \cdot T$ is logarithmic for temperatures which are much larger or smaller than the temperature of a certain crystal field level Δ_i/k_B.

RESULTS

In this section thermal conductivity results obtained for some Ce and Yb based Kondo compounds are summarized. The temperature dependence of the thermal conductivity $\lambda(T)$ for such compounds is illustrated in figure 1. $\lambda(T)$ of the presented compounds exhibits no clear cut anomalies, which can be immediately associated with the Kondo effect, crystal field splitting or magnetic ordering. The reason for the undramatic temperature variation of the thermal conductivity is that the anomalous behaviour, originated from the Kondo interaction of conduction electrons with almost localized magnetic moments is masked by contributions caused from scattering processes of conduction electrons with static imperfections and from the interaction with lattice vibrations. Additionally, the phonon system is able to carry heat in condensed matter, which can exceed that of the electronic system. It is therefore necessary to clear the total measured quantity from the mentioned contributions to figure out the behaviour expected in the scope of equation (7).

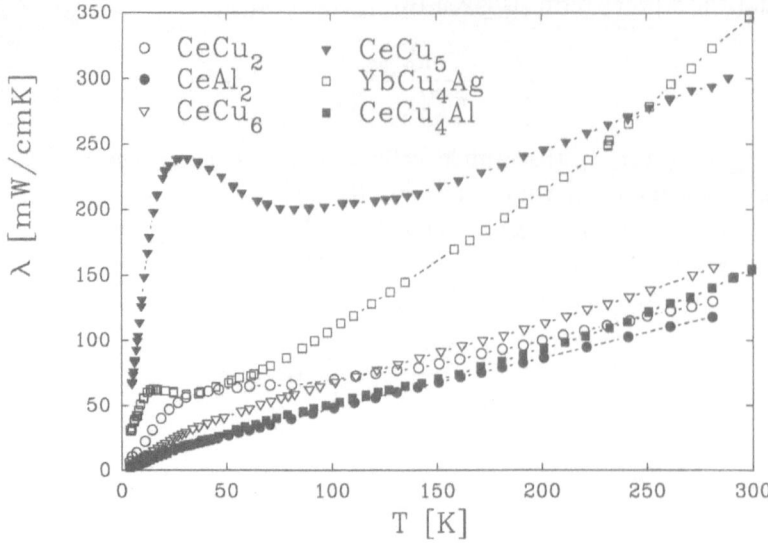

Figure 1. Temperature dependent thermal conductivity λ of various Ce - and Yb - based compounds

The usual way to deduce the magnetic contribution from the measured data is a comparison with isostructural nonmagnetic compounds. These latter systems should account for the electron - phonon interaction and the interaction of conduction electrons with static imperfections of the sample. Both contributions are assumed to be equal in equivalent magnetic and nonmagnetic compounds. For such a comparison, we show $\lambda(T)$ for a variety of La - and Lu - based samples (figure 2), equivalent to those of figure 1. Some of the compounds presented exhibit at low temperature a maximum in $\lambda(T)$ which usually characterize metals and compounds with low scattering rates on static imperfections. The appearance of this type of maximum in nonmagnetic compounds has the same meaning as a small value of the residual resistivity in $\rho(T)$. Compared to the compounds depicted in figure 1, $\lambda(T)$ of the La - and the Lu - based samples is much larger. This is explained from additional scattering processes of the conduction electrons with the magnetic moments of Ce^{3+} and Yb^{3+} ions. Such interaction mechanisms increase the thermal resistivity, consequently, the total thermal conductivity is lowered.

DISCUSSION

Usually, the thermal conductivity λ of metallic systems consists of a sum of an electronic, λ_e, and a lattice contribution, λ_l.

According to Matthiessen's rule, the electronic thermal resistivity W_e can be expressed by

$$1/\lambda_e \equiv W_e = W_{e,o} + W_{e,ph} + W_{e,mag} \tag{8}$$

where (e,o), (e,ph) and (e,mag) are the contributions owing to scattering processes of electrons with static imperfections, with thermally excited phonons and magnetic moments, respectively. While $W_{e,0} \propto 1/T$ in the whole temperature range, the contribution $W_{e,ph}$ follows from the classical Wilson law. The dependence of $W_{e,mag}$ is accounted for by equation (7). An expression similar to equation (8) can be derived for the lattice thermal conductivity λ_l. Makinson[18] has pointed out that there are mainly two processes which limit the phonon mean free path:

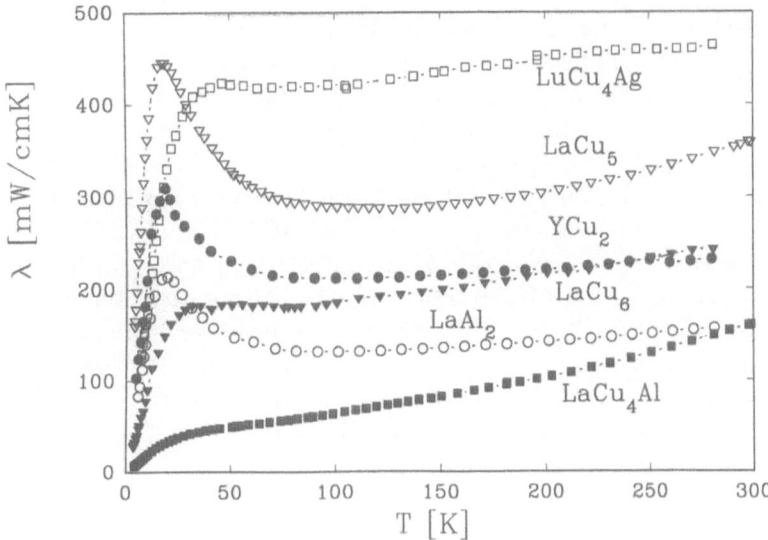

Figure 2. Temperature dependent thermal conductivity λ of various La - and Lu - based compounds

1. Phonon - phonon Umklapp processes become important at high temperatures, resulting in a strong decrease of λ_l ($\lambda_{l,ph} \propto 1/T$).

2. At low temperatures and metallic systems, the interaction of lattice vibrations with the conduction electron system is the most effective scattering process for phonons ($\lambda_{l,e} \propto T^2$).

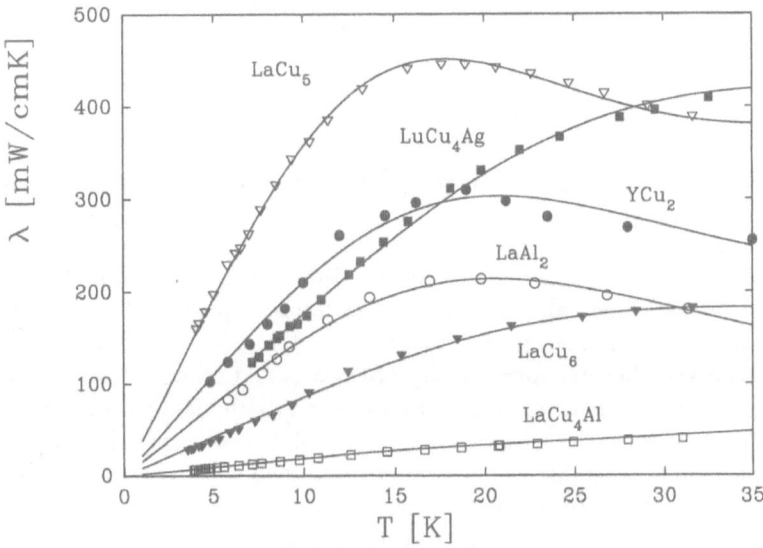

Figure 3. Low temperature thermal conductivity λ of various La - and Lu - based compounds The full lines are least squares fits according equation (9)

Taking into account these interaction processes and their particular temperature dependencies, it is, in principle, possible to analyze the total thermal conductivity. Considerable simple conditions[19] can be found for nonmagnetic compounds at low temperatures

$$\lambda = \lambda_e + \lambda_l = \frac{1}{W_{e,0} + W_{e,ph}} + \lambda_{l,e} = \frac{1}{\alpha/T + \beta T^2} + \delta T^2 \qquad (9)$$

Figure 3 displays $\lambda(T)$ for some nonmagnetic La - and Lu - compounds in the low temperature range. The full lines in this figure are least squares fits according equation (9). These fits reveal satisfactorily agreement with the data and give therefore quantitative information concerning λ_l of a certain nonmagnetic compound, at least at low temperatures. The compounds shown in figure 3 are characterized by a lattice contribution to the total thermal conductivity, which does not exceed a few percent around 10 K. The only compound which behaves different is LaCu$_4$Al. $\lambda(T)$ of this compound is much smaller than the thermal conductivity of the other compounds (cf. figure 2). The reason of which are enhanced scattering processes, arising from atomic disorder owing to substitutions of Cu by Al in LaCu$_5$. The data presented in figure 3 (except LaCu$_4$Al) are therefore characterized by a thermal conductivity which is mainly determined by the electron system. The lattice contribution, at least at low temperatures, is negligible and will therefore be omitted in the further discussion.

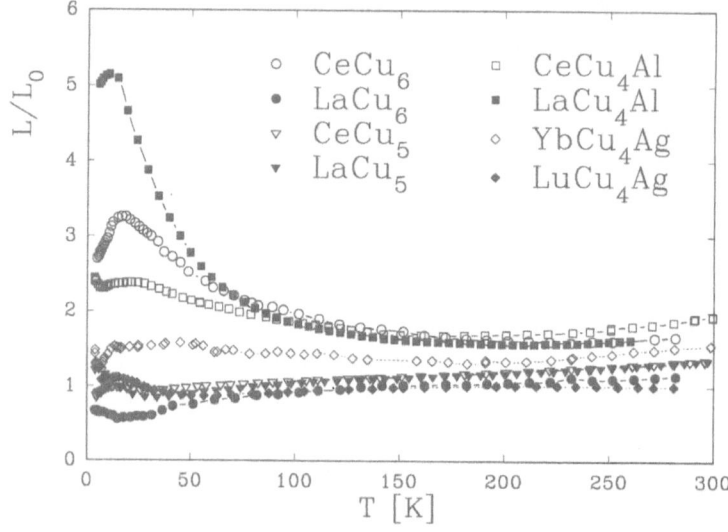

Figure 4. Temperature dependent Lorenznumber L, normalized to the Sommerfeldvalue L_0 for various La-, Ce-, Yb- and Lu compounds.

Information concerning the lattice contribution to the total thermal conductivity of compounds can be gained also from the temperature dependent Lorenz number $L(T)$. Theoretically, $L(T)$ follows from the Wiedemann - Franz law which relates the electrical resistivity and the thermal conductivity ($L = \lambda \cdot \rho / T$). Since this law accounts just for electronic contributions, a contribution of the lattice yields to deviations from the Sommerfeld value L_0 of the Lorenz number. Simple free electron theory shows that L_0 is determined from constants only, amounting to $2.45 \cdot 10^{-8} W\Omega K^{-2}$. As can be seen from figure 4, the normalized temperature dependent Lorenz numbers for most of the compounds (including also CeAl$_2$ and CeCu$_2$) do not deviate strongly from L_0, therefore the lattice thermal conductivity can be considered as small in both the magnetic and the nonmagnetic compounds. However, CeCu$_6$, CeCu$_4$Al and LaCu$_4$Al do not belong to that group of materials where $\lambda_l(T)$ seems to be negligible, since L/L_0 of the latter compounds exceed the Sommerfeld value by up to 500 %. Because of the fact that the latter group contains magnetic and nonmagnetic compounds, the reason for the unusual large deviations of L from L_0 cannot be found from magnetic interactions, i.e. the Kondo effect in the Ce compounds. Theoretically, it has been shown[15] that in spite of the Kondo effect in metallic samples, the Lorenz number L remains relatively close to its Sommerfeld value in the whole temperature domain. The conclusion which therefore can be drawn for this group is that the lattice thermal conductivity is, at least, of the same order of magnitude as the electronic contribution to the thermal conductivity. This infers that $\lambda_l(T)$ cannot be neglected immediately. A considerable large lattice contribution to $\lambda(T)$ for $LaCu_4Al$ has also been deduced from the analysis of the low temperature thermal conductivity.[20]

In the remaining part of this chapter the evaluation of the magnetic thermal resistivity for typical examples is demonstrated and results will be discussed in the scope of equation (7). The most reliable and simplest analysis can be performed for those compounds where the lattice thermal conductivity is of minor importance.

CeAl₂ and CeCu₂. The total thermal conductivity $\lambda(T)$ for $CeAl_2$ and $CeCu_2$ has been shown in figure 1. The associated Lorenz numbers for both compounds prove the dominance of the electronic thermal conductivity. In spite of this, the $\lambda(T)$ data do not reveal unusual features originated from the Kondo effect in presence of crystal field splitting. To emphasize these effects, we have chosen a T/λ vs T representation (figure 5)

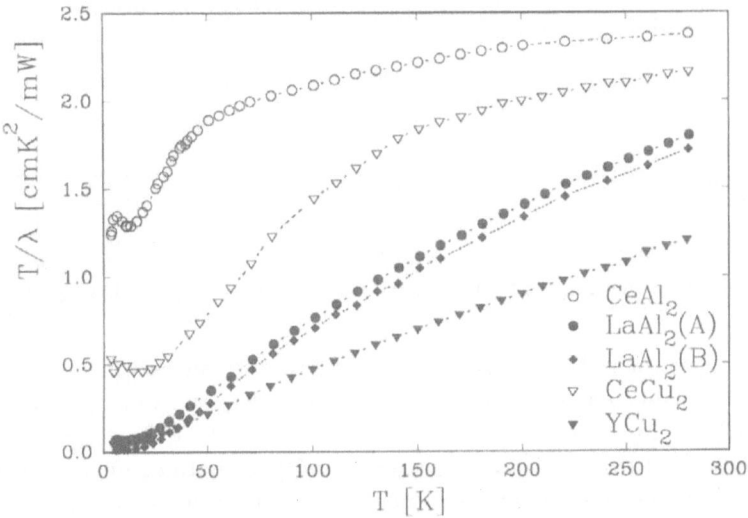

Figure 5. Temperature dependent thermal conductivity λ for $CeAl_2$ and $CeCu_2$ plotted as T/λ vs T

which follows from the application of the Wiedemann - Franz law $(T/\lambda(T) \propto \rho(T))$. For the purpose of comparison, the equivalent nonmagnetic compounds (two different heat treated $LaAl_2$ samples and YCu_2) are added in the same representation. $T/\lambda(T)$ for $CeAl_2$ and $CeCu_2$ is characterized by a Kondo - like minimum at low temperatures, whereas at higher temperatures a pronounced saturation tendency is obtained. The overall behaviour for both compounds resembles just those characteristics known from $\rho(T)$ measurements on these samples. The behaviour of the nonmagnetic isostructural compounds $LaAl_2$ and YCu_2 is reminiscent of the temperature dependent electrical resistivity of simple metallic compounds. Deviations of $T/\lambda(T)$ from the behaviour expected from resistivity measurements are usually originated from a non negligible lattice thermal conductivity. It is therefore noticed that this type of plot additionally informs concerning whether or not the lattice contribution is of importance.

Under the assumption that the lattice contribution to the total thermal conductivity is negligible, i.e. $\lambda(T) \approx \lambda_e(T)$, the magnetic contribution to the thermal resistivity, $W_{e,mag}$ can be found from a comparison of the equivalent magnetic (m) and nonmagnetic (nm) compounds.[20]

$$1/\lambda^m - 1/\lambda^{nm} \equiv W^m - W^{nm} = \Delta W \approx W_{e,mag} \tag{10}$$

The magnetic thermal resistivity $\Delta W \approx W_{e,mag}$ for CeCu$_2$ and CeAl$_2$ is plotted as $\Delta W \cdot T$ vs ln$T$ in figure 6. The temperature dependent behaviour of $\Delta W \cdot T$ for both

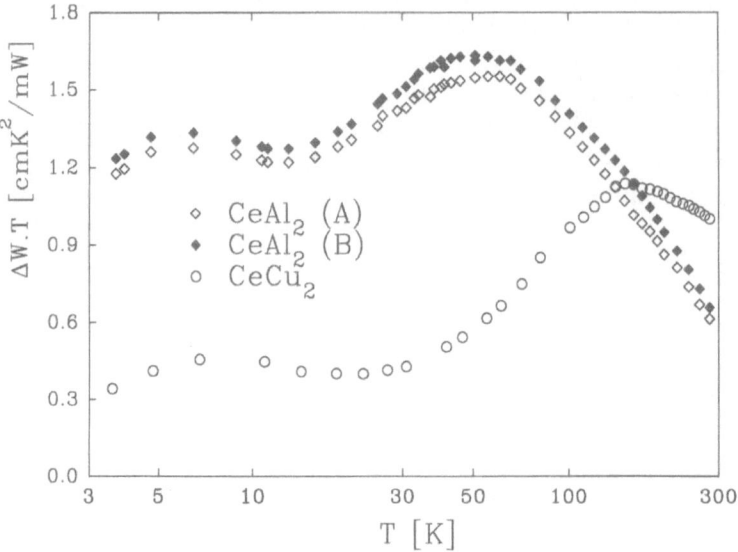

Figure 6. Magnetic contribution to the thermal resistivity for CeAl$_2$ and CeCu$_2$ plotted as $\Delta W \cdot T$ vs lnT. .

compounds is characterized by two nearly logarithmic ranges which are separated by a maximum roughly around 75 K and 140 K for CeAl$_2$ and CeCu$_2$, respectively. The plots of $\Delta W \cdot T$ are in fact almost identical with those of the respective ρ_{mag} data. The dependence of the magnetic thermal resistivity for both compounds is in excellent agreement with the theoretical predictions, following from equation (7): Firstly, the logarithmic slope of $\Delta W \cdot T$ is much larger at high than at low temperatures. This is associated with the ratio Q of the effective degeneracy of the full j $= 5/2$ multiplet and of the crystal field ground of the Ce ion. Since both compounds are known to exhibit a doublet as ground state, Q should amount to about 11.[1] Secondly, the maxima at 75 K and 140 K correspond roughly to the overall crystal field splitting. From inelastic neutron scattering experiments it is known that the quartet state Γ_8 of CeAl$_2$ is situated about 8 meV above the Γ_7 ground state.[21] For the orthorhombic compound CeCu$_2$, inelastic neutron scattering experiments by Loewenhaupt et al.[22] reveal crystal field doublets 9 and 23 meV above the ground state. Differences in the magnitude of the overall crystal field splitting Δ_{CF} deduced from neutron scattering experiments and the maximum in $\Delta W \cdot T$ are thought to be originated from the subtraction procedure (equation (10)) applied to the thermal conductivity data. This subtraction accounts essentially for the phonon contribution in the magnetic compound. However, it is well known that for example La - based compounds like LaAl$_2$ exhibit a strong electron - phonon coupling which frequently lead to a superconducting instability. Such an enhanced coupling of the conduction electrons with lattice vibrations results in a much steeper slope for the quantity $T/\lambda(T)$ (compare figure 5). Consequently, the position of the maximum in $\Delta W \cdot T$ for the magnetic compounds sensitively follows from the choice of the appropriate phonon contribution $W_{e,ph}(T)$.

The next group for which the magnetic contribution to the total thermal conductivity is discussed, includes compounds, characterized by $T_K^H \approx \Delta_{CF}$. As representative, YbCu$_4$Ag is chosen.

YbCu$_4$Ag. The magnetic contribution to the electrical resistivity for this compound exhibits a rather extended range with a (-lnT) dependence and a maximum around 70 K. Below this maximum, $\rho_{mag}(T)$ steadily decreases, approaching a T^2 behaviour in the low temperature region. This latter feature is associated with a Fermi liquid state. Most of the physical properties of YbCu$_4$Ag are explained in the scope of the Kondo lattice model, assuming a characteristic temperature T_0 of about 150 K.[10] The cubic symmetry of the crystal structure gives rise to a lifting of the eightfold degenerate ground state into doublets (Γ_6 and Γ_7) and a quartet (Γ_8). Inelastic neutron scattering experiments do not lead to an undoubted crystal field level scheme. However, it has been concluded that the overall crystal field splitting Δ_{CF}/k_B most likely amounts to 180 K,[23] which is of the same magnitude as T_0.

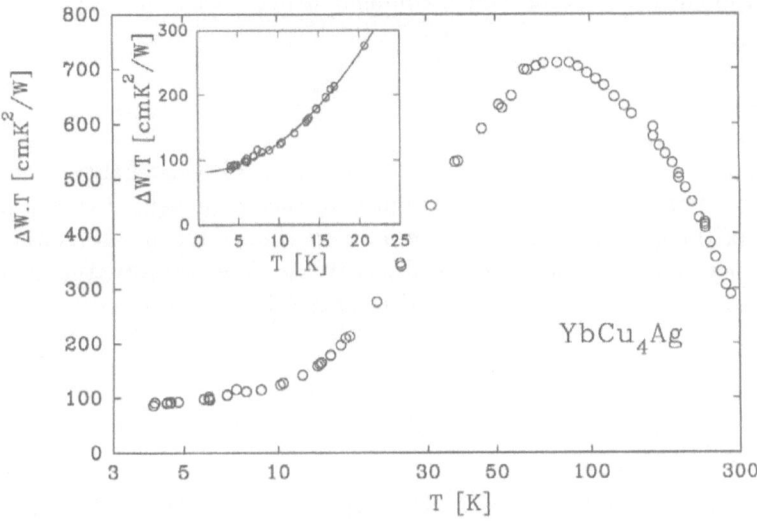

Figure 7. Magnetic contribution to the thermal resistivity for YbCu$_4$Ag plotted as $\Delta W \cdot T$ vs lnT. The inset shows $\Delta W \cdot T$ at low temperatures on a linear scale. The full line is a least squares fit according to equation (11).

Both the small electrical resistivity and the Lorenz number of the order of L_0 allows to neglect the lattice thermal conductivity for analysing $\lambda(T)$ of YbCu$_4$Ag. Figure 7 displays the magnetic thermal resistivity of YbCu$_4$Ag plotted as $\Delta W \cdot T$ vs lnT. ΔW has been evaluated according equation (10) and using LuCu$_4$Ag as nonmagnetic reference compound. The features deduced for this quantity resemble closely the behaviour of $\rho_{mag}(T)$; i.e. the maximum around 70 K and a (-lnT) range above which. The overall behaviour of $\Delta W \cdot T$ is clearly explained in the framework of the Kondo effect and crystal field splitting. The logarithmic behaviour observed at high temperatures is characteristic of the Kondo effect in the full j = 7/2 multiplet and can be accounted for by equation (7) with $\alpha = 2j + 1$. The proximity of the Kondo temperature T_K and the crystal field splitting Δ_{CF} yields to a different behaviour of $\Delta W \cdot T$ vs lnT compared with that of CeAl$_2$ or CeCu$_2$. The inset of figure 7 shows $\Delta W \cdot T$ vs T below 25 K. It is well known, that a characteristic sign of a Kondo lattice compound is the Fermi liquid state at low temperatures. This causes for example a T^2 dependence of the electrical resistivity, while for the magnetic contribution to the thermal resistivity also a power law can be expected. Thus, for the low temperature range, the thermal resistivity is accounted for by

$$T \cdot W_e = T \cdot W_{e,0} + T \cdot W_{e,mag} = \alpha + A \cdot T^n \tag{11}$$

A least squares fit (full line in the inset) indicates that the exponent n in equation (11) is close to 2 (n = 2.03). This exponent demonstrates also the close relation of the thermal and the electrical resistivity, which, of course, can be read off from the Wiedemann - Franz law.

The magnetic contribution for compounds with a non negligible lattice contribution, e.g. $CeCu_6$ or $CeCu_4Al$ has been evaluated recently.[24,20] Phonon - phonon Umklapp processes, which limit the phonon mean free path reduce the lattice thermal conductivity rapidly at elevated temperatures. It is therefore supposed that λ_l vanishes with increasing temperatures like $1/T$, yielding to the possibility that $\Delta W \cdot T$ can be evaluated in the usual way (equation (10)). Indeed, both compounds $CeCu_6$ and $CeCu_4Al$ are characterized by a logarithmic behaviour of $\Delta W \cdot T$ at higher temperatures, satisfying the predictions of equation (7) for Kondo compounds in presence of crystal field splitting.

Summary

Various examples of well known Kondo compounds have been analysed for their magnetic contribution to the thermal resistivity. In each case, ranges with a logarithmic behaviour of the quantity $\Delta W \cdot T$ have been obtained, tracing closely the behaviour of the temperature dependent electrical resistivity. Theoretically, this particular behaviour follows from the combined influence of the Kondo effect and crystal field splitting.

Acknowledgements

Parts of the work have been supported by the "Austrian Science Foundation" under project no. P7608-PHY. I am indebted to the "Österreichische Forschungsgemeinschaft" for financial support.

References

1 D Cornut and B Coqblin, Influence of the crystalline field on the Kondo effect of alloys and compounds with cerium impurities, *Phys Rev* B5 4541 (1972)

2 A K Bhattacharjee and B Coqblin, Thermoelectric power of compounds with cerium influence of the crystalline field on the Kondo effect, *Phys Rev* B13 3441 (1976)

3 S Maekawa, S Kashiba, S Takahashi and M Tachiki, Kondo effect versus crystal field, *in* "Theory of Heavy Fermions and Valence Fluctuations", T Kasuya and T Saso eds, Springer Series in Solid State Sciences, Vol 62, Berlin (1987)

4 S Maekawa, S Kashiba, M Tachiki and S Takahashi, Thermopower in Ce Kondo systems, *J Phys Soc Japan* 55 3194 (1986)

5 E Bauer, E Gratz, W Mikovits, H Sassik and H Kirchmayr, Transport phenomena in $CeAl_2$, *J Magn Magn Mat* 29 192 (1982)

6 A Guessous, *PhD Thesis*, University of Grenoble (1987)

7 Y Onuki, Y Shimizu and T Komatsubara, Magnetic properties of a new Kondo lattice intermetallic compound $CeCu_6$, *J Phys Soc Japan* 53 1210 (1984)

8 D Jaccard, R Cibin and J Sierro, Resistivity and magnetoresistance of $CeAl_3$ single crystals, *Helv Phys Acta* 61 530 (1988)

9 E Bauer, Anomalous properties of Ce-Cu and Yb-Cu based compounds, *Adv Phys* 40 417 (1991)

10 C Rossel, K N Yang, M B Maple, Z Fisk, E Zirngiebl and J D Thompson, Strong electronic correlations in a new class of Yb-based compounds $YbXCu_4$ (X = Ag, Au, Pd), *Phys Rev* B35 1914 (1987)

11 K Fischer, Thermoelectric power of heavy - fermion compounds, *Z Phys* B76 315 (1989)

12 D Jaccard, A Basset, J Sierro and J Pierre, Thermopower of $Ce_x Y_{1-x} InCu_2$ and $CeInCu_y Ag_{2-y}$, *J Low Temp Phys* 80 285 (1990)

13 A H Wilson "The Theory of Metals", second edition, Cambridge University Press, London (1958)

14 J M Ziman "Electrons and Phonons" 4th edition, Clarendon, Oxford, (1960)

15 A K Bhattacharjee and B Coqblin, Thermal conductivity of cerium compounds *Phys Rev* B38 338 (1988)

16 P W Anderson, Localized magnetic states in metals, *Phys Rev* 124 41 (1961)

17 E Bauer, E Gratz, G Hutflesz, A K Bhattacharjee and B Coqblin, Thermal conductivity of Ce-based Kondo compounds, *J Magn Magn Mat* 108 159 (1992)

18 R E B Makinson, Thermal conductivity of metals, *Proc Camb Phil Soc* 34 474 (1938)

19 P G Klemens, Theory of thermal conductivity of solids, *in* "Thermal Conductivity", vol 1, R P Tye ed , Academic Press, London (1969)

20 E Bauer, E Gratz, G Hutflesz and H Muller H, Thermal conductivity in $Ce(Cu, Al)_5$ compounds, *J Phys Cond Mat* 3 7641 (1991)

21 P Fulde and M Loewenhaupt, Magnetic excitations in crystal - field split 4f systems, *Adv Phys* 34 589 (1985)

22 M Loewenhaupt, M Prager, E Gratz and B Frick, Magnetic excitations in $CeCu_2$, *J Magn Magn Mat* 76-77 415 (1988)

23 A Severing, A P Murani, J D Thomson, Z Fisk and C K Loong, Neutron scattering experiments on $YbXCu_4$ and $ErXCu_4$ (X = Au, Pd and Ag), *Phys Rev* B41 1739 (1990)

24 E Bauer, E Gratz and Y Peysson, Thermal conductivity of $(Ce_x La_{1-x})Cu_6$ compounds, *J Magn Magn Mat* 63-64 303 (1987)

HIGH–PRESSURE RESISTIVITY AND LATTICE PARAMETERS OF CeRu$_2$Si$_2$

P. Haen[1], J.-M. Laurant[1], K. Payer[2], and J.-M. Mignot[3]

[1]CRTBT, CNRS, BP 166, 38042 Grenoble-Cédex 9, France
[2]Inst. Exp. Phys., Techn. Universität, Wiedener Hauptstr. 8-10, 1040 Wien, Austria
[3]Laboratoire Léon Brillouin, CEA-CNRS, CE Saclay, 91191 Gif sur Yvette, France

INTRODUCTION

The atomic volume is an important parameter for investigating the ground state of heavy fermion systems. This has been shown, for instance, by an extensive series of pressure and alloying experiments on the tetragonal compound CeRu$_2$Si$_2$.[1-5] This material is known to be non-superconducting at ambient pressure, at least down to $T = 20$ mK.[1,6] The absence of long-range order, also taken for granted until recently, is now questioned on the basis of muon-spin relaxation results.[7] However, the magnetic moments involved are extremely small ($\approx 10^{-3}$ μ_B), and this is probably why no evidence for a magnetic transition is found in the low-temperature resistivity. The dominant contribution to the latter quantity is thus of the usual Fermi-liquid type, with a quadratic, AT^2, term holding below $\approx 0.7 - 1$ K at ambient pressure.[1,5,6,8]

The effect of hydrostatic pressure on the electrical resistivity was first investigated by Thompson et al.[4] up to 16.8 kbar. Later on, measurements to lower temperatures ($T_{min} \approx$ 25 mK) were reported,[5] allowing a reliable estimate of the (*temperature*) electronic Grüneisen parameter $\Gamma_T \approx 200$. Most remarkably, the *field* Grüneisen parameter, deduced from the magnetoresistance or the magnetization under pressure,[1,5] was found to have almost exactly the same value, $\Gamma_H \approx 200$, emphasizing the role of low-energy quasiparticle excitations in the metamagnetic process. Some questions, however, remained unsolved, such as the possibility that a superconducting state might occur at higher pressures, or the occurrence of a crossover into a less-heavy, lower-Γ, Fermi liquid state as the unit-cell volume becomes further reduced at higher applied pressures. New experiments were thus undertaken to extend the experimental range of pressure to $P_{max} \approx 100$ kbar. In this review, we will present the results and compare them in detail to those of refs. 4 and 5. We will show that the three sets of data are entirely consistent and allow the pressure dependence of the Grüneisen parameter Γ to be determined. For this purpose, we will make use of the compressibility data obtained in a separate x-ray diffraction experiment up to 70 kbar, which we performed on CeRu$_2$Si$_2$ material from the same source. A brief account of these results has already been given in ref. 9.

EXPERIMENTAL CONDITIONS

The present high-pressure resistivity experiments were carried out using a Bridgmann anvil device described previously.[10] In the present study, we used tungsten carbide anvils, which provide a larger high-pressure volume (2 mm in diameter) for the sample assembly. Quasi-hydrostatic pressure conditions were ensured by a soft solid pressure medium (steatite). All pressure changes were performed at room temperature (RT). The pressure was determined from the superconducting transition temperature T_c of a strip of Pb foil located close to the sample, according to the $T_c(P)$ calibration of ref. 11, and its homogeneity could be estimated from the width of the transition ($\approx 5\%$ in this series of measurements). The resistances of both the specimen and the Pb manometer were measured in a true four-wire geometry, using a high-sensitivity $a.c.$ device. The CeRu$_2$Si$_2$ specimen (tetragonal, $I4/mmm$ structure) was prepared by cleaving the starting single-crystal parallel to the (0 0 1) plane, then polishing it down to the desired thickness ($e \approx 200$ μm). As in our previous experiments[1,5,6] the electrical current was applied within the basal plane. The contacts consisted of Pt wires pressed onto the samples. In a few cases during the present experiment, one of them eventually became unstable upon cooling down, which explains why some of the experimental $R(T)$ curves remained incomplete. However, the Pb manometer was not affected and P could still be determined from its critical temperature. Finally, no good measurements could be performed below 23 kbar (including $P = 0$) upon release of pressure, due to cracks forming within the sample and/or loss of contacts.

The x-ray experiments under pressure were performed in the diffractometer shortly described in ref 12. Fine (< 0.5 μm) CeRu$_2$Si$_2$ powder prepared by crushing a small fragment of single-crystal material was mixed up with NaCl which served both as a pressure medium and a manometer. As the absorption coefficients of Ce and Ru are both very high, the proportion of CeRu$_2$Si$_2$ was only of about 15 %. The mixture was introduced in the hole ($\varnothing \approx 0.3$ mm) of an stainless steel gasket with thickness $e \approx 0.3$ mm, and squeezed between two single-crystal diamond anvils. The load is applied by a stainless steel press using a gas piston multiplier. The measurements are performed in the transmission mode, with the x-ray beam (molybdenum K$_\alpha$) from a rotating-anod generator passing through the anvils. The diffraction pattern are recorded by a gas (krypton-ethan) position sensitive detector.

ROOM TEMPERATURE RESISTANCE AND CALIBRATION
TO ABSOLUTE VALUES

Moderate pressure data for the variation of electrical resistivity of CeRu$_2$Si$_2$ at RT are available from refs. 4 and 5. Both measurements indicate a significant increase in $\rho(P)$. However, the absolute values of the resistivity are appreciably larger in ref. 4, which most likely reflects the difficulty in accurately determining the geometrical factor of a small, irregular-shaped, platelet. Since we are quite confident in the value $\rho(295\ \mathrm{K}) = 85$ μΩcm at ambient pressure, which was obtained as the average of several measurements on different specimens, we chose to use it for all subsequent normalizations. More precisely, since samples from different batches do not necessarily have the same residual resistivity ρ_0, it is preferable to normalize the *temperature dependent part*, $\Delta\rho \equiv \rho - \rho_0$, of the resistivity to a constant value, namely $\Delta\rho(295\ K) = 83\ \mu\Omega cm$. If this normalization is applied to the data of ref. 4, an excellent agreement is obtained with our own results, not only at RT as a function of pressure (fig. 1), but also regarding the T-dependences at fixed pressure to be discussed in the following.

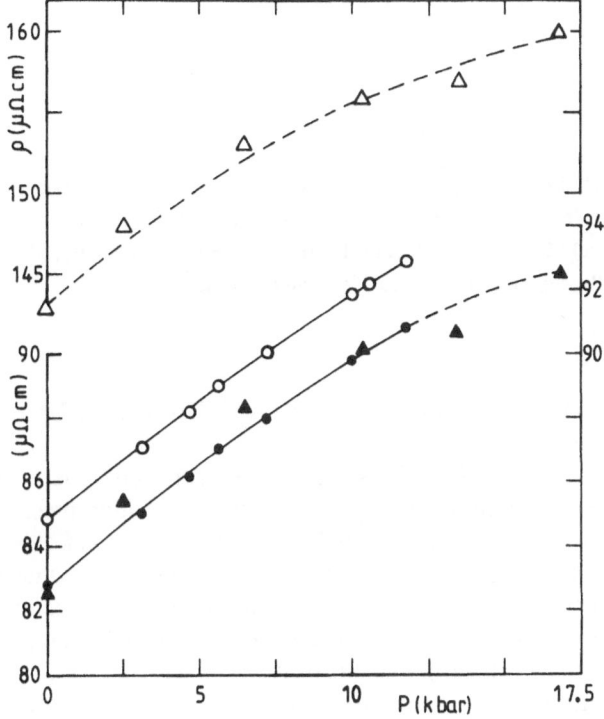

Figure 1. : Pressure dependence of the electrical resistivity of CeRu$_2$Si$_2$ at 295 K, ($\triangle$) : ref. 4, (o) : ref. 5, and of $\rho-\rho_0$: (●) : ref. 5, (▲) : ref. 4 after normalization of the data of ref. 5.

The same procedure can then be applied to the Bridgman anvil results, for which it is difficult to estimate the sample dimensions precisely. Fig. 2 shows the resistance of CeRu$_2$Si$_2$ at RT plotted as a function of pressure. The numbers represent the order in which the pressures were successively applied (7 values ranging from 23 to 105 kbar). One notes immediately that the pressure dependence has reversed sign with respect to the preceding regime, suggesting that a maximum occurs in $R(P)$ at about 20 kbar. Unfortunately, as noted in the previous section, the pressure could not be reduced sufficiently in the present experiments to provide an overlap with the previous data. Nevertheless, a downward curvature is clearly observed in the data on both sides of the maximum, and a smooth interpolation can be obtained by normalizing the high pressure values to $\Delta\rho$ (295 K) = 93 $\mu\Omega$cm for P = 23 kbar. The resulting variation of $\Delta\rho(P)$ at RT is shown in fig. 3. The main features of the curve are unambiguous, even though the shape of the maximum at 20 kbar may depend slightly on the normalization adopted. Above this maximum, $\Delta\rho$ decreases strongly between 30 and 50 kbar, and finally tends towards a low value, comparable to that of LaRu$_2$Si$_2$, at the highest pressures.

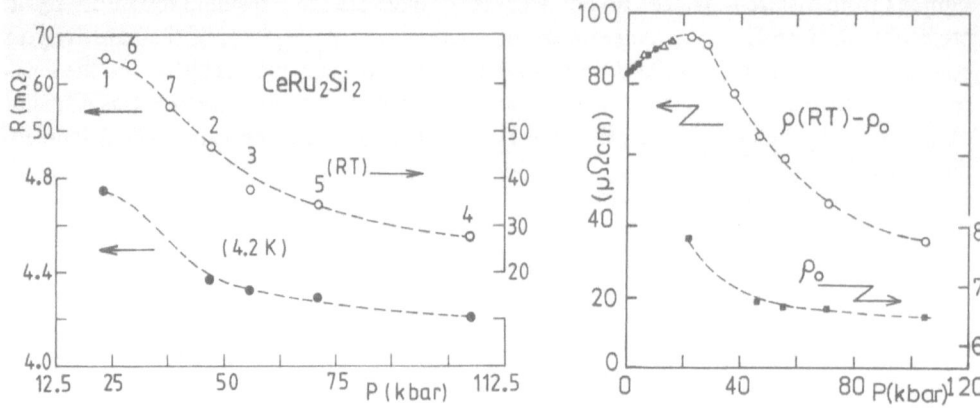

Figure 2. (Left Frame) : Pressure dependence of the electrical resistance of CeRu$_2$Si$_2$ at room temperature (open circles) and 4.2 K (closed circles) measured in Bridgmann anvils (this work).

Figure 3. (Right Frame) : Pressure variation of the room temperature resistivity, (ρ-ρ_0), and of the residual resistivity, ρ_0, of CeRu$_2$Si$_2$; ● : ref 5; △ : ref. 4; o and ■ : present work.

LOW-TEMPERATURE RESISTIVITY AT FIXED PRESSURES

Fig. 4 shows a selection of $\Delta\rho(T)$ isobars for pressures comprised between 0 and 105 kbar. The plot includes data from refs. 4 and 5, as well as our high-pressure results. In order to obtain consistent values, we have i) normalized the different sets of data as explained in the preceding section, and ii) subtracted out the residual resistivities ρ_0, which are found to differ substantially from one case to another. In ref. 5, ρ_0 is pressure independent (≈ 2 $\mu\Omega$cm) up to 10.6 kbar; in ref. 4 it also appears to be constant, with a comparable value ($\rho_0 \approx 2.7$ $\mu\Omega$cm), up to 16.8 kbar; at higher pressures ($P \geq 23$ kbar), on the other hand, ρ_0 is larger by almost a factor of 4, and decreases as P increases (fig. 3). It should be kept in mind, however, that the latter variation may be partly due to pressure-induced sample deformations. At such high pressures, the residual resistivity is already obtained for $T \approx 4.2$ K. In fact, $\rho(T)$ is even found to increase slightly on further cooling down to 1.3 K.

From the curves plotted in fig. 4, it can be concluded that the three sets of experimental data are quite consistent with one another, and that our normalization procedure is basically correct. Let us first remark that no evidence for superconductivity is found in the present measurements. As already noted in refs. 4 and 5, pressure qualitatively affects the shape of the $\rho(T)$ curves. The shoulder observed near 25 K at ambient pressure shifts rapidly to higher temperatures at very low pressures, then merges into the large upturn of $\rho(T)$ occurring at intermediate temperatures. It has been suggested[4,5,6] that this anomaly is related to the Kondo, or spin-fluctuation, temperature of this compound. The pronounced negative curvature in $\rho(T)$ observed above ≈ 200 K at ambient pressure has been ascribed to crystal field effects,[5,6] in agreement with the splitting estimated from the specific heat[2] (220 K) or the thermal expansion[13] (280 K). This curvature actually becomes a slight maximum if one subtracts out the resistivity of the reference compound LaRu$_2$Si$_2$ (see refs. 5, 8, and 14). It is still visible for $P = 29$ kbar, but has practically disappeared at 38 kbar. It can be inferred that the spin-fluctuation temperature is close to RT for this range of pressures. At even higher pressures, the curves show an upward curvature similar to that observed in mixed-valence compounds and, finally, the 105 kbar isobar is essentially that for a normal metal, although the ρ-values remain larger than in LaRu$_2$Si$_2$ (i // a, taken from ref. 8).

The gradual transformation of CeRu$_2$Si$_2$ first into an intermediate valence system, then into a normal metal, by the application of pressure is also reflected in the functional dependence of $\rho(T)$ at low temperature[5]. In the logarithmic plot of $(\rho - \rho_0)$ as a function of T (fig. 5) one notes the gradual depression of the Fermi-liquid AT^2 term and its extension to higher temperatures (the phonon scattering correction is unessential in this range of pressures,

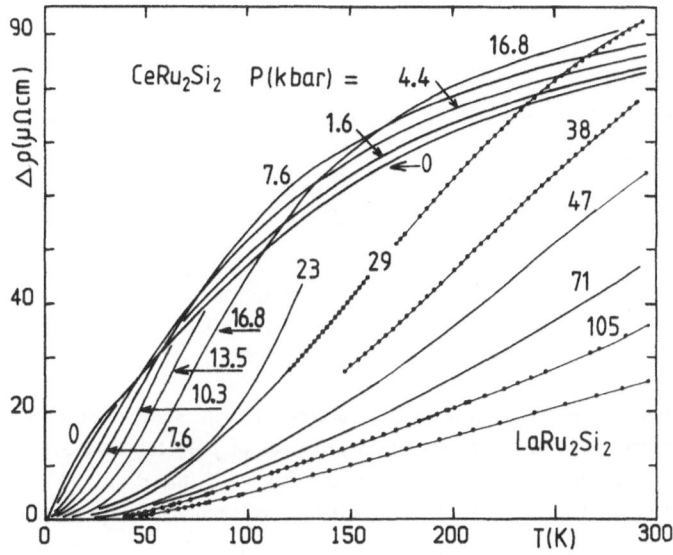

Figure 4. : Variation of $(\rho - \rho_0)$ as a function of temperature for CeRu$_2$Si$_2$ at different pressures : this study in Bridgmann anvils and selected curves from ref. 4 (P = 10.3, 13.5, 16.8 kbar) and from ref. 5 (P = 0, 1.6, 4.4, 7.6 kbar); data from ref. 8 for LaRu$_2$Si$_2$ ($i // a$) are plotted for comparison.

as can be seen from the LaRu$_2$Si$_2$ data). However, above 47 kbar, it becomes difficult to define this quadratic term because the error bars resulting from the uncertainty in ρ_0 become quite large. Furthermore, additional contributions with larger T–exponents ($n = 3$–5) appear above $\approx$ 20 K. Finally, for the 105 kbar curve, the AT^2 term, if any, is almost identical to that of LaRu$_2$Si$_2$ (below $\approx$ 10-15 K), and thus more likely to arise from a Baber scattering mechanism. (A more relevant reference would probably be YRu$_2$Si$_2$, whose volume is comparable to that of CeRu$_2$Si$_2$ at 105 kbar).

From the results in fig. 4 the coefficient A of the AT^2 term can be determined with a good precision between 0 and 29 kbar. As P increases, the error bars becomes larger, especially after correcting A for the residual T^2 contribution of LaRu$_2$Si$_2$. Following refs. 4 and 5, we can introduce a characteristic temperature T^* by writing the quadratic term as $AT^2 = \alpha \left(T/T^*(P)\right)^2$. The variation of $A^{-1/2} \propto T^*$ is plotted in fig. 6 as a function of pressure. The slope decreases markedly above 15–20 kbar, then seems to saturate at the highest pressures.

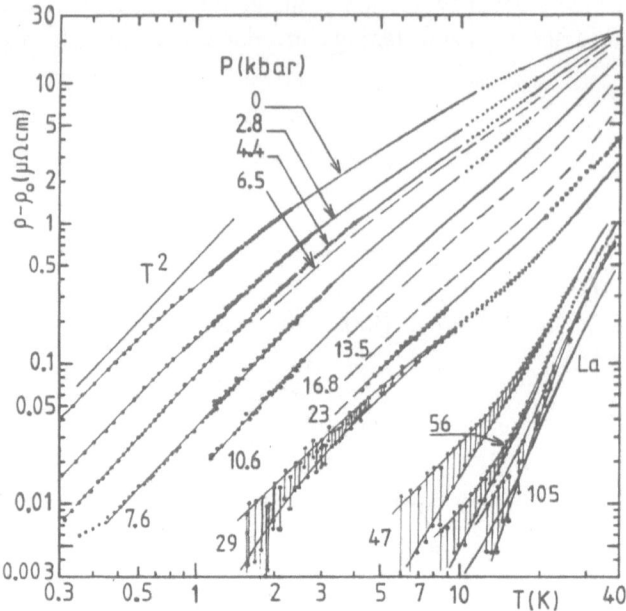

Figure 5. : Log-log plot of $\rho-\rho_0$ in $CeRu_2Si_2$ as a function of temperature for different pressures showing the change in the exponents of the power-law dependence of $\rho(T)$ for $0 \leq P \leq 10.6$ kbar : ref. 5, for $P = 13.5$ and 16.8 kbar : ref. 4, and for $P \geq 23$ kbar : this work (the $P = 10.3$ kbar curve of ref. 4 is superimposed to the $P = 10.6$ kbar curve of ref. 5). Continuous line noted La : $\rho-\rho_0$ plot for $LaRu_2Si_2$.

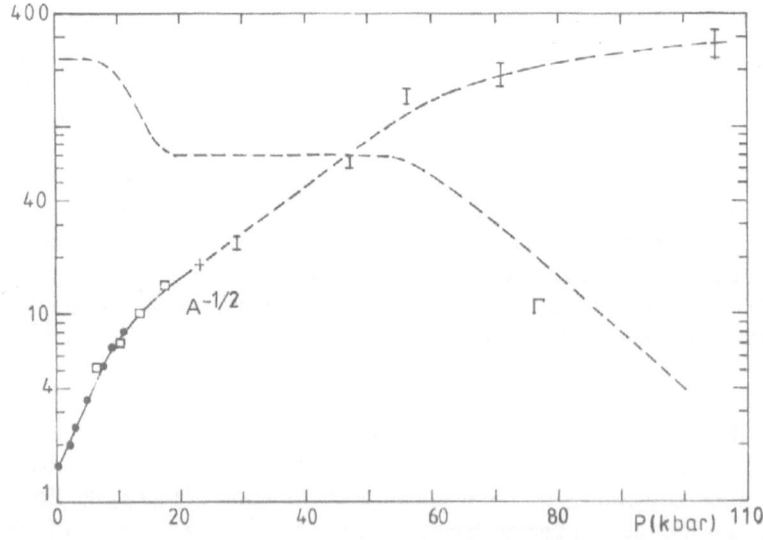

Figure 6. : Pressure dependence of $A^{-1/2}$ (A is coefficient the AT^2 resistivity, expressed in $p\Omega cm/K^2$) and of the Grüneisen coefficient Γ of $CeRu_2Si_2$ (in log scale) : ● : ref 5; △ : ref. 4; I : present work.

Fig. 7 shows the relative variation of the lattice parameters a and c and of the unit cell volume $(\Delta V)/V = 2(\Delta a)/a + (\Delta c)/c$ deduced from the x-ray experiments. $(\Delta a)/a$ and $(\Delta c)/c$ were determined from the positions of the (2 0 0) and (0 0 4) reflections, respectively. A larger effect is seen on the c-parameter of the tetragonal structure, as compared to the basal plane parameter a. Yet, the scatter in the data points is also larger, the reason being that the (2 0 0) reflection is well resolved, whereas the (0 0 4) one is less intense. Within the precision of the present data, the volume compressibility remains surprisingly constant over the entire range of pressure, possibly indicating some incipient softening associated with the anomalous electronic properties of $CeRu_2Si_2$. A detailed comparison with the reference compound $LaRu_2Si_2$ would be useful to clarify this point. Measurements with improved accuracy and to higher pressures would probably reveal an upward concavity in the curves, as in, e.g., $CeAl_2$.[12,15] The volume compressibility deduced from the ΔV data is plotted at the bottom of fig. 7. Its average value is $\kappa = 0.82 \pm 0.05$ (Mbar)$^{-1}$. The low pressure points seem to be erroneous, perhaps because the first x-ray spectra used as reference were not really measured at zero pressure as expected, but with a small pressure generated in the process of closing the cell.

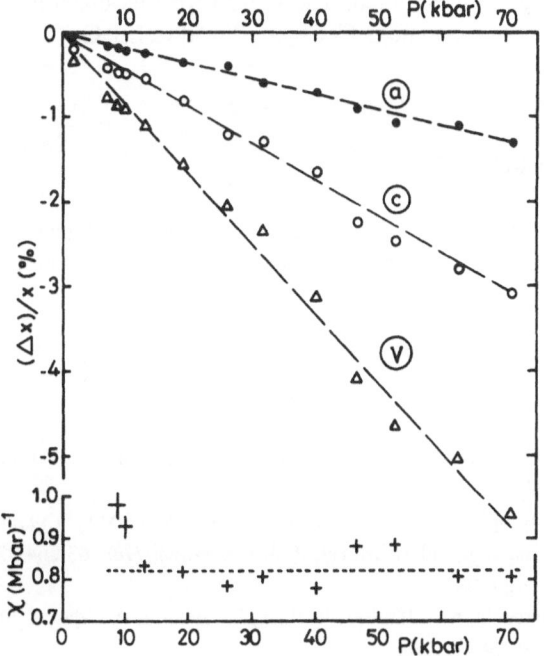

Figure 7. : Pressure dependence of the room temperature lattice parameters a and c, and of the specific volume of $CeRu_2Si_2$ up to 70 kbar ; lower curve : volume compressibility.

Using the above value of the volume compressibility, the pressure dependence of the electronic Grüneisen coefficient Γ of $CeRu_2Si_2$ can be determined from that of $A^{-1/2}$. Both quantities are plotted together in fig. 6. At low pressures, Γ first remains constant up to ≈ 8 kbar, with a very high value of about 200, then decreases markedly to less than 5 at the maximum pressure. The existence of a plateau between about 20 and 50 kbar is an interesting feature which might characterize an intermediate valence regime in $CeRu_2Si_2$. However, this observation must certainly be taken with caution in view of the various sources of uncertainty affecting the determination of A, and hence of Γ.

CONCLUSION

This study provides a comprehensive view of the effect of pressure on the electrical resistivity of $CeRu_2Si_2$ based on a critical comparison of earlier data, as well as a novel experimental work extending the pressure range investigated previously to more than 100 kbar. As pressure increases, a gradual crossover is observed from an initially heavy fermion regime to a mixed-valence one, then towards a state in which the resistivity behaves essentially like that of a normal metal. Whereas similar effect have already been reported for other heavy fermion materials, special care has been taken in the present experiments to quantitatively define the changes occuring in the low-temperature Fermi liquid regime. This analysis results in the dramatic decrease found for the electronic Grüneisen parameter, which is a direct consequence of the quenching of the heavy fermion state. The compression curves determined for both latice parameters a and c are surprisingly linear, possibly indicating some anomalous softening associated with the instability of the $4f$ shell in this material. Finally, it must be noted that no evidence for a superconducting transition was detected in the pressure and temperature ranges investigated. The reason for this difference between the present compound and its Cu-based analogs $CeCu_2Si_2$[10] and $CeCu_2Ge_2$[16] is an unsolved and most intriguing question.

ACKNOWLEDGEMENTS

We wish to thank Drs. J. Flouquet, F. Lapierre, D. Jaccard, C. Ayache for their stimulating interest to this work, and Dr. J.D. Thompson for allowing us to use the data of ref. 4.

REFERENCES

1. J.-M. Mignot, J. Flouquet, P. Haen, F. Lapierre, L. Puech, and J. Voiron, *J. Magn. Magn. Mat.* 76&77:97 (1988).

2. M.J. Besnus, J.P. Kappler, P. Lehmann, and A. Meyer, *Solid State Commun.* 55:779 (1985) ; M.J. Besnus, P. Lehmann, and A. Meyer, *J. Magn. Magn. Mat.* 63&64:323 (1987) ; P. Lehmann, thesis, Strasbourg (1987).

3. R.A. Fisher, C. Marcenat, N.E. Phillips, P. Haen, F. Lapierre, P. Lejay, J. Flouquet, and J. Voiron, *J. Phys. France* 49:1555 (1988).

4. J.D. Thompson, J.O. Willis, C. Godart, D.E. Mac Laughauglin, and L.C. Gupta, *J. Magn. Magn. Mat.* 47&48:281 (1985).

5. J.-M. Mignot, A. Ponchet, P. Haen, F. Lapierre, and J. Flouquet, *Phys. Rev. B* 40:10917 (1989).

6. P. Haen, J. Flouquet, F. Lapierre, P. Lejay, and G. Remenyi, *J. Low Temp. Phys.* 67:391 (1987).

7. A. Amato, R. Feyerherm, J. Flouquet, F.N. Gygax, P. Lejay, A. Schenk, and U. Zimmermann, SCES'92, Sendai, Sep. 7-11, 1992, to be published in *Physica B*.

8. F. Lapierre and P. Haen, *J. Magn. Magn. Mat.* 108:167 (1992).

9. K. Payer, P. Haen, J.-M. Laurant, J.-M. Mignot, and J. Flouquet, SCES'92, Sendai, Sep. 7-11, 1992, to be published in *Physica B*.

10. B. Bellarbi, A. Benoit, D. Jaccard, J.-M. Mignot, and H.F. Braun, *Phys. Rev. B* 30:779 (1984).

11. A. Eichler and J. Wittig, *Z. Angew. Phys.* 25:319 (1968).

12. B. Barbara, J. Beille, B. Cheaito, J.-M.Laurant, M.F. Rossignol, A. Waintal, and S. Zemirli, *J. Physique France* 48:635 (1987).

13. A. Lacerda, A. de Visser, P. Haen, P. Lejay, and J. Flouquet, *Phys. Rev. B* 40:8759 (1989).

14. J.D. Thompson, J.O. Willis, C. Godart, D.E. Mac Laughauglin and L.C. Gupta, *Solid State Commun.* 568:169 (1985).

15. I. Vedel, A. M. Redon, J.-M. Mignot, and J.M. Léger, *J. Phys. F (Met. Phys.)* 17:849 (1987).

16. D. Jaccard, K. Behnia, and J. Sierro, *Phys. Lett. A.* 163:475 (1992).

COLLAPSE OF THE HEAVY FERMION STATE
UNDER HIGH PRESSURE

Tomoko Kagayama and Gendo Oomi

Department of Physics, Faculty of General Education
Kumamoto University, Kumamoto 860
Japan

INTRODUCTION

There has been a lot of investigation about the electronic and magnetic properties of heavy fermion (HF) systems containing certain lanthanide or actinide elements, because these compounds give useful information for studying the role of strong electron correlations in metals.[1,2] The compounds in the HF systems are characterized by a extremely large coefficient γ of the linear term in the electronic specific heat, a large value of the coefficient of T^2–term in the electrical resistivity $\rho(T)$ at low temperature, a log T term in the $\rho(T)$ at high temperature and so forth.

Basically these anomalous properties are dominated by a complicated interplay of two effects, the (concentrated) Kondo effect and the effect of crystalline electric field (CEF). Main features of the effects are that the former gives rise to the logarithmic temperature dependence and a peak in the $\rho(T)$ curve and the latter shows a Schottky type anomaly in the specific heat $C(T)$[3] and the thermal expansion coefficient $\alpha(T)$.[4]

It is well known that the electronic states of HF systems are strongly dependent on the change in pressure or volume.[5] This fact indicates the large change of the characteristic temperature of the systems, the so–called Kondo temperature T_K, by an application of pressure. Usually the HF system has low T_K of the order of several degree Kelvin but the compounds of intermediate valence state (IVS) shows a relatively high T_K of ~100 K. So we expect an interesting transition (or a crossover) in the electronic state from the HF state to IVS by changing the T_K or by changing the pressure. Recently we have reported such pressure–induced crossover for several Ce[6,7] and U[8] compounds, and found a systematics for the value of $JN(0)$ at ambient pressure, where J is the $s-f$ exchange interaction and $N(0)$ the density of state at Fermi level.

In the present paper we summarize the recent experimental results of electrical resistivity and lattice constant at high pressure for the Ce–compounds. The results are discussed on the recent theoretical works and analyzed on the basis of simple phenomenological theory.

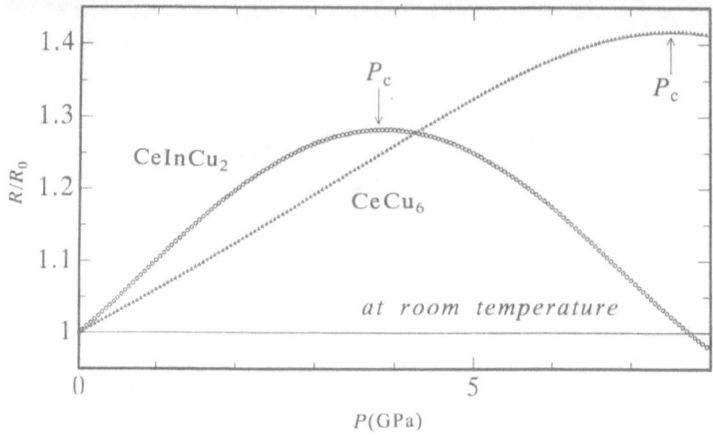

Figure 1. Pressure dependence of the relative change of the electrical resistance R/R_0 at room temperature for $CeInCu_2$ and $CeCu_6$.

SURVEY OF EXPERIMENTAL RESULTS OF ELECTRICAL RESISTANCE AND LATTICE CONSTANTS AT HIGH PRESSURE FOR Ce–BASED HEAVY FERMION COMPOUNDS, $CeInCu_2$ and $CeCu_6$

Here we review the recent experimental results of Ce–based heavy fermions at high pressure mainly focused on the typical two HF compounds, $CeInCu_2$ and $CeCu_6$. $CeCu_6$ crystallizes in the orthorhombic crystal structure with a large γ value of 1.5 J/mol K^2 and a strong T^2 dependence at low temperature below 1 K.[9] Cubic Heusler type compound $CeInCu_2$ is also characterized as a HF material having $\gamma \sim 1.2$ J/mol K^2 at 1 K and also a strong T^2 dependence below 2.5 K.[10,11] The electrical resistance R/R_0 at room temperature is shown in Fig. 1 as a function of pressure P (GPa), where R and R_0 are the electrical resistance at a pressure P and ambient pressure, respectively. The R increases with increasing pressure until it shows a maximum at P_c (3.8 GPa for $CeInCu_2$ and 7.5 GPa for $CeCu_6$) and then begins to decrease. The maximum in the $R(P)$ curve stems from Kondo effect, in which the characteristic temperature T_K (~4 K at ambient pressure for $CeInCu_2$ and $CeCu_6$) increases with increasing pressure. As will be shown later, the maximum in $R(P)$ corresponds to the crossover from the HF to IVS induced by pressure.

In order to examine whether the crossover is accompanied with a volume discontinuity or not, we observed the pressure dependence of volume of $CeInCu_2$[12] and $CeCu_6$[13] at room temperature. Figure 2 shows the volume V/V_0 as a function of pressure, where V and V_0 are the volume at high and ambient pressure, respectively. There is no discontinuity in the compression curves within experimental error and further no crystal structure change up to 14 GPa. Considering the fact, it is concluded that the pressure–induced crossover in $CeInCu_2$ and $CeCu_6$ occurs continuously without any discontinuous change in volume.

The temperature dependence of electrical resistivity $\rho(T)$ of $CeInCu_2$ at various pressures up to 8 GPa is shown in Fig. 3.[14] The values of ρ at high pressure were corrected by taking into account the change of geometrical factor l/S of the specimen at high pressure, which was obtained by the data of the thermal expansion[4] and of the compressibility.[12] At ambient pressure, ρ increases gradually with decreasing temperature, reaches a maximum at 27 K and then decreases by further cooling. This behavior is similar to those of typical HF compounds. The temperature of resistivity-maximum T_{max} is found to increase with increasing pressure and the maximum in $\rho(T)$ curves tends to be smeared at high pressure. The change in the overall behavior in the $\rho(T)$ curve in Fig. 3 implies a crossover from HF to IVS at high pressure, i.e., a collapse of HF state at high pressure.

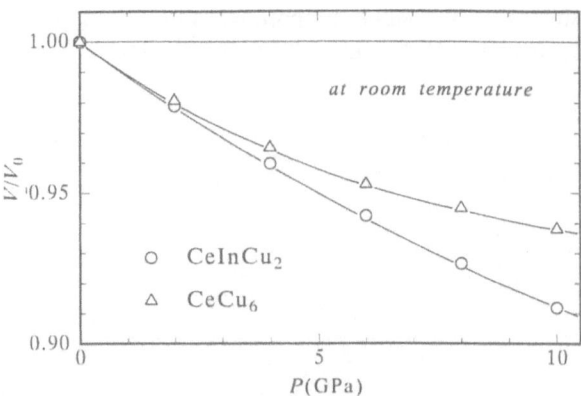

Figure 2. The relative change in the volume V/V_0 of CeInCu$_2$ and CeCu$_6$ as a function of pressure at room temperature.

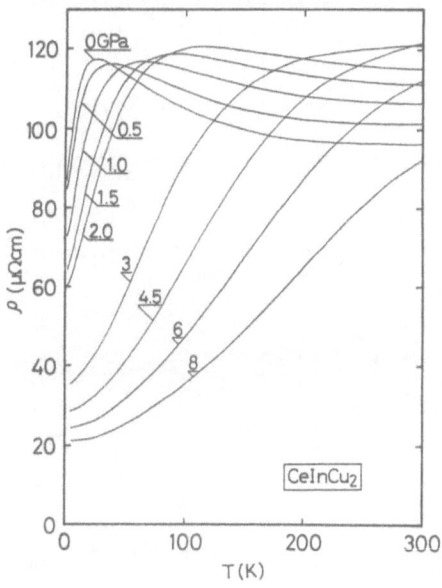

Figure 3. Electrical resistivity ρ of CeInCu$_2$ at high pressures as a function of temperature.

The results for CeCu$_6$ are shown in Fig. 4. The overall behavior of $\rho(T)$ is almost similar to that of CeInCu$_2$. Crossover in the electronic states of CeCu$_6$ also occurs at high pressure. This result is in qualitative agreement with the previous one.[15]

Figure 5 shows the T_{max} as a function of pressure for CeInCu$_2$ and CeCu$_6$. T_{max}'s are found to increase with increasing pressure. The increasing rate of T_{max} of CeInCu$_2$ is larger than that of CeCu$_6$. Since T_{max} is roughly proportional to the Kondo temperature T_K, this result indicates that T_K is enhanced by an application of pressure.

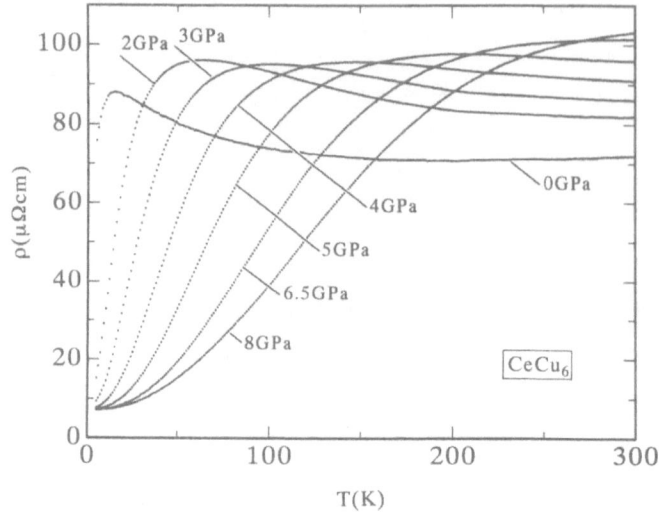

Figure 4. Temperature dependence of the electrical resistivity ρ of CeCu$_6$ at various pressures.

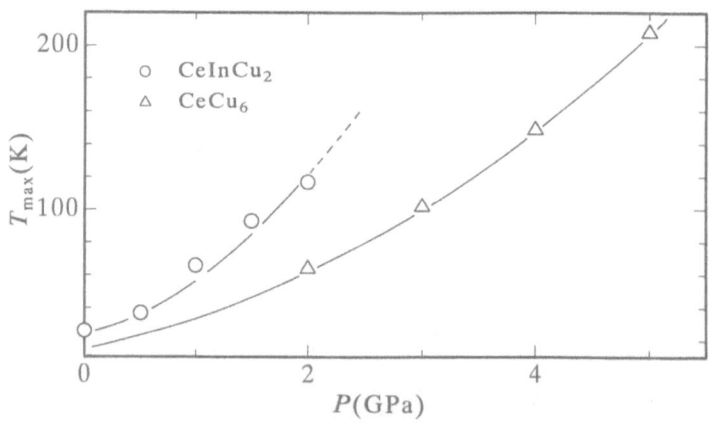

Figure 5. The temperature showing the maximum in the $\rho(T)$ curve, T_{max}, for CeInCu$_2$ and CeCu$_6$ as a function of pressure.

In order to examine the Fermi liquid behavior in the $\rho(T)$ curve at low temperature, we plotted in Figs. 6 and 7 the electrical resistivity $\rho(T)-\rho_0$ as a function of T^2 for CeInCu$_2$ and CeCu$_6$ at various pressures, where ρ_0 is the residual resistivity. It is seen that $\rho(T)$ shows the T^2-dependence and also the temperature range having T^2-behavior expands with increasing pressure. The coefficient A of T^2-term in logarithmic scale, which was estimated from the plot in Figs. 6 and 7, is shown as a function of pressure in Fig. 8 for CeInCu$_2$ and

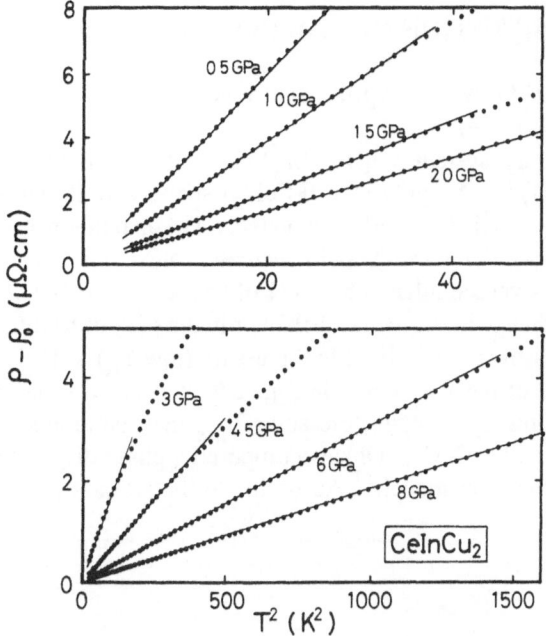

Figure 6. $\rho(T)-\rho_0$ as a function of T^2 for CeInCu$_2$ at various pressures.

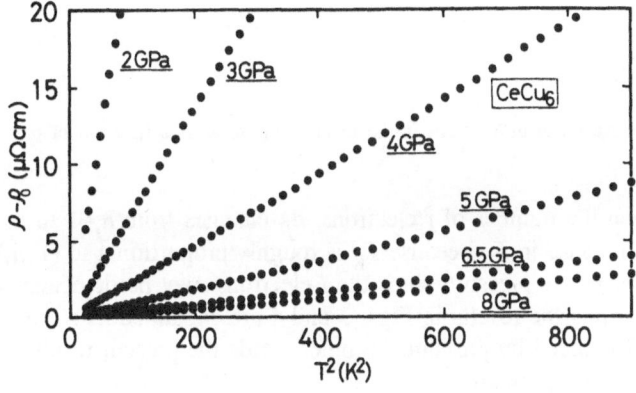

Figure 7. $\rho(T)-\rho_0$ as a function of T^2 for CeCu$_6$ at various pressures.

CeCu$_6$. A decreases largely with increasing pressure below ~4 GPa and then shows a small pressure variation above ~5 GPa. The closed relation between A and T_{max} will be discussed in detail in the following section.

SIMPLE INTERPRETATION OF THE RESULTS AT HIGH PRESSURE

Crossover from HF to IVS by Applying Pressure

The pseudobinary system Ce(In$_{1-x}$Sn$_x$)$_3$ is well known to show a crossover from HF(x=0) to IVS(x=1).[16] The $\rho(T)$ at x=0(CeIn$_3$) shows a well−defined maximum around 50 K, which is characteristic of HF compounds. The maximum becomes less prominent with increasing x. At x=1(CeSn$_3$), $\rho(T)$ increases only monotonously with temperature (T<300 K), which is very similar to the $\rho(T)$ of CeInCu$_2$ at 8 GPa. Taking these facts into consideration, the change in the overall behavior in the $\rho(T)$ of CeInCu$_2$ observed in Figs. 3 and 4 implies a crossover from HF at low pressure (low T_K) to IVS at high pressure (high T_K). The maximum of the R–P curve in Fig. 1 is considered to be due to a shift of T_{max} to higher temperatures by applying pressure: T_{max} may be around 300 K at 3.8 GPa for CeInCu$_2$ and 7.5 GPa for CeCu$_6$. Ohkawa proposed a phase diagram of HF systems on the basis of periodic Anderson model.[17] According to that, a crossover from HF state to IVS

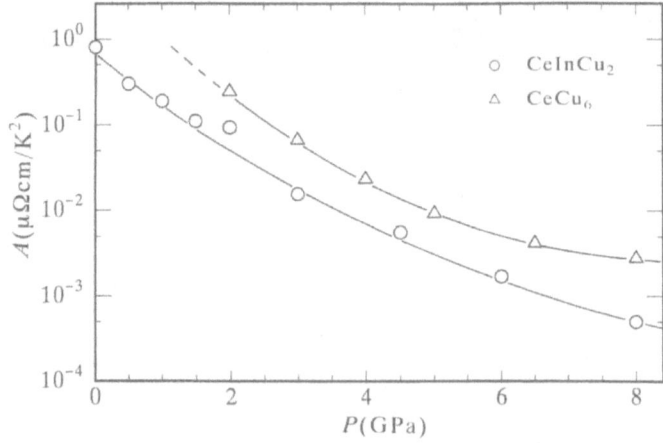

Figure 8. The coefficient A of T^2−term in the logarithmic scale as a function of pressure for CeInCu$_2$ and CeCu$_6$.

takes place when the number of f electrons, n_f, changes from $n_f \approx 1$ to n_f<1. A decrease in n_f indicates an increase in T_K because T_K is roughly proportional to $(1-n_f)/n_f$. In the present case IVS is induced by pressure since the f electrons may be delocalized at high pressure to increase T_K. Thus the results in Figs. 3 and 4 are explained as a crossover from the HF state to the IVS induced by pressure. In other words the present results show a collapse of HF state by applying pressure.

Volume Dependence of the Resistivity–Maximum Temperature T_{max} and the Coefficients of T^2-term A

In this section we attempt to analyze quantitatively the present data to elucidate the characteristics of the electronic structure of HF compounds. According to the theory of Yoshimori and Kasai (YK),[18] A is proportional to T_K^{-2} and T_K is approximately proportional to T_{max}. Then we have the following relation,

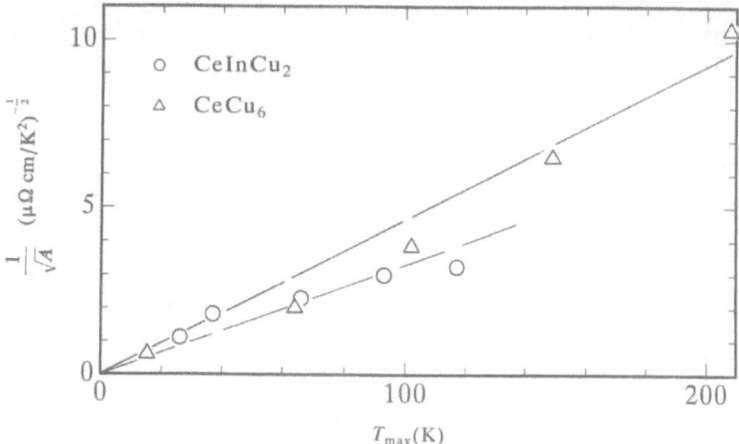

Figure 9. $1/\sqrt{A}$ plotted against T_{max} for CeInCu$_2$ and CeCu$_6$. The solid lines show the linear dependence.

$$T_{max} \propto \frac{1}{\sqrt{A}} \, .$$ (1)

Figure 9 shows the plot of $1/\sqrt{A}$ against T_{max} for both CeInCu$_2$ and CeCu$_6$, in which $1/\sqrt{A}$ is found to increase linearly with T_{max} as is shown by a solid line. This fact indicates that the relation (1) is valid and the large decrease in the magnitude of A or the increase of T_{max} at high pressure in Figs. 5 and 8 is due to the increase of T_K by an application of pressure.

The Kondo temperature T_K is described as

$$T_K \propto \exp\left[-\frac{1}{|JN(0)|} \right],$$ (2)

where J is the exchange interaction between the conduction electron and the localized $4f$ spin and $N(0)$ the density of states at the Fermi level. The increase of T_{max} with pressure indicates the increase of $|JN(0)|$. In other words, $|JN(0)|$ is one of the most fundamental parameters because T_K is mainly dominated by its change under the application of external forces such as pressure or magnetic field. The large change in the magnitude of $|JN(0)|$ gives rise to the large change in A, T_{max} or T_K.

Here we assume the following volume dependence of $|JN(0)|$,[19]

$$|JN(0)| = |JN(0)|_0 \exp\left[-q\frac{V-V_0}{V_0} \right],$$ (3)

where $|JN(0)|_0$ is the value of $|JN(0)|$ at ambient pressure and q is the dimensionless constant. q may be defined as a Grüneisen parameter of $|JN(0)|$,

$$q = -\frac{\partial \ln|JN(0)|}{\partial \ln V}\bigg|_{V=V_0},$$ (4)

and usually has a value between 6 and 8. From eqs. (2) and (6) the Grüneisen parameter for T_K is written,

$$-\frac{\partial \ln T_K}{\partial \ln V}\bigg|_{V=V_0} = \frac{q}{|JN(0)|_0} . \tag{5}$$

Further we obtain the following relations,

$$\ln \frac{T_{\max}(P)}{T_{\max}(0)} = -\frac{1}{2} \ln \frac{A(P)}{A(0)} = -\frac{q}{|JN(0)|_0} \frac{V_0-V}{V_0} . \tag{6}$$

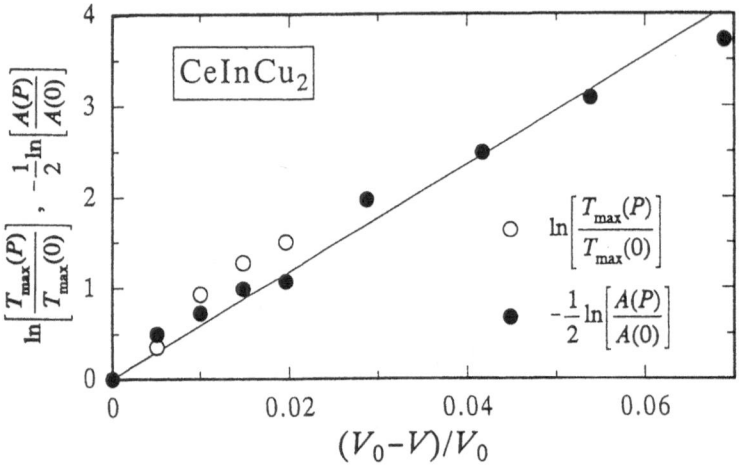

Figure 10. The values of $\ln[T_{\max}(P)/T_{\max}(0)]$ and $-\frac{1}{2}\ln[A(P)/A(0)]$ as a function of volume change $(V_0-V)/V_0$ for CeInCu$_2$. The solid line is obtained from the least square fitting.

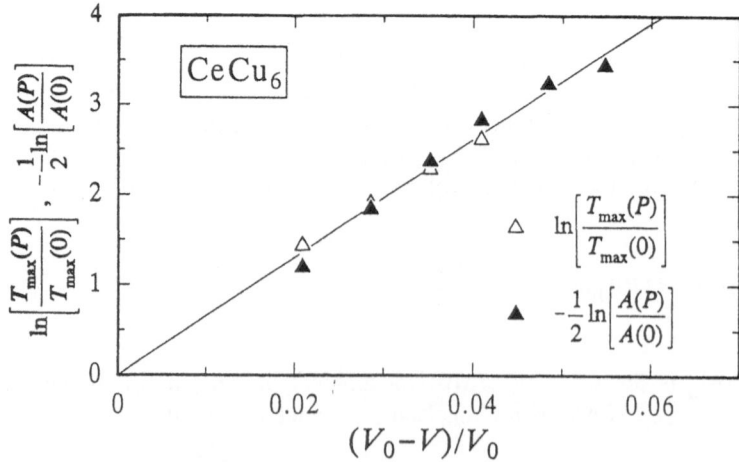

Figure 11. The values of $\ln[T_{\max}(P)/T_{\max}(0)]$ and $-\frac{1}{2}\ln[A(P)/A(0)]$ as a function of volume change $(V_0-V)/V_0$ for CeCu$_6$. The solid line is obtained from the least square fitting.

Figures 10 and 11 show the plot of $\ln[T_{max}(P)/T_{max}(0)]$ and $-\frac{1}{2}[\ln A(P)/A(0)]$ as a function of $(V_0-V)/V_0$. The linear relationship is found in the plots as is shown by solid lines, which were obtained by a least square fitting. From the result in Figs. 10 and 11, we estimated the values of $q/|JN(0)|_0$ to be 59 and 65 for $CeInCu_2$ and $CeCu_6$, respectively. These values are larger than that of the IVS compound $CePd_3$ of 9 by about an order of magnitude. This confirms that the electronic state of HF compounds is very sensitive to a change in volume or pressure. Since the value of q was obtained to be 6 for $CeCu_6$,[21] $|JN(0)|_0$ of $CeCu_6$ is 0.09. Assuming the same value of $q(=6)$ for $CeInCu_2$, we obtained the value of $|JN(0)|_0$ to be 0.10.

It was reported that the values of $|JN(0)|_0$ are 0.125, 0.13 and 0.115 for CeTe, CeBi and CeSb, respectively.[22] These values are the same order of magnitude as those obtained above. On the other hand the value of 0.5 was estimated for the intermediate valence α–Ce.[23] Thus the value of $|JN(0)|_0$ of HF compounds may be around 0.1.

SUMMARY

From the measurements of the electrical resistivity and lattice constants of $CeCu_6$ and $CeInCu_2$ at high pressure we obtained the following results:
1) a crossover from HF (low T_K) to IVS (high T_K), i.e., a collapse of HF is induced by an application of pressure,
2) the crossover occurs without any discontinuous change in volume,
3) the coefficient of T^2–term, A, decreases rapidly with pressure, and
4) the values of $|JN(0)|_0$ are estimated to be about 0.1 for $CeInCu_2$ and $CeCu_6$.

Acknowledgments

The authors would like to express their sincere thanks to Prof. Y.Ōnuki and Prof. T.Komatsubara for supplying them single crystalline $CeInCu_2$ and $CeCu_6$. They also greatly acknowledge Prof. N.Mōri, Dr. H.Takahashi and Dr. Y.Uwatoko for their encouragement and assistance throughout the present experiment. They are deeply indebted to Prof. B.R.Cooper, Dr. J.D.Thompson and Prof. K.Yamada for their stimulating discussion. This work was supported partly by Grant–in–Aid for Scientific Research from the Ministry of Education, Science and Culture and the 1991 special fund of Kumamoto City.

REFERENCES

1. G.R.Stewart, *Rev.Mod.Phys.* 56:755(1984).
2. N.B.Brandt and V.V.Moshchalkov, *Adv.in Phys.* 33:373(1984).
3. M.Kato, K.Satoh, Y.Maeno, Y.Aoki, T.Fujita, Y.Ōnuki and T.Komatsubara, *J.Phys.Soc.Jpn.* 56:3661(1987).
4. G.Oomi, T.Kagayama, Y.Ōnuki and T.Komatsubara, *Physica B* 163:557(1990).
5. J.D.Thompson, Frontiers in solid state sciences, to be published.
6. T.Kagayama, G.Oomi, H.Takahashi, N.Mōri, Y.Ōnuki and T.Komatsubara, *Phys.Rev.B* 44:7690(1991).
7. G.Oomi, T.Kagayama, H.Takahashi, N.Mōri, Y.Ōnuki and T.Komatsubara, *J.Alloys and Compounds* (1992) to be published.
8. K.Iki, G.Oomi, Y.Uwatoko, H.Takahashi, N.Mōri, Y.Ōnuki and T.Komatsubara, *J.Alloys and Compounds* 181:71(1992).
9. Y.Ōnuki, Y.Shimizu and T.Komatsubara, *J.Phys.Soc.Jpn.* 54:304(1985).
10. Y.Ōnuki, T.Yamazaki, A.Kobori, T.Omi, T.Komatsubara, S.Takayanagi, H.Kato and N.Wada: *J.Phys.Soc.Jpn.* 56:4251(1987).
11. T.Kagayama, G.Oomi, R.Yagi, Y.Iye, Y.Ōnuki and T.Komatsubara, *J.Phys.Soc.Jpn.* 61:2632(1992).

12. T.Kagayama, K.Suenega, G.Oomi, Y.Ōnuki and T.Komatsubara, *J.Magn.&Magn.Mater.* 90&91:449(1990).

13. G.Oomi, A.Shibata, Y.Ōnuki and T.Komatsubara, *J.Phys.Soc.Jpn.* 57:152(1988).

14. T.Kagayama, G.Oomi, H.Takahashi, N.Mōri, Y.Ōnuki and T.Komatsubara, *J.Magn.&Magn.Mater.* 108:(1992)103.

15. S.Yomo, L.Gao, R.L.Meng, P.H.Hor, C.W.Chu and J.Susaki, *J.Magn.&Magn.Mater.* 76&77:257(1988).

16. R.A.Elenbaas, C.J.Schinkel and C.J.M.Deudekom, *J.Magn.&Magn.Mater.* 15–18:979(1980).

17. F.J.Ohkawa, *J.Magn.&Magn.Mater.* 52:217(1985).

18. A.Yoshimori and H.Kasai, *J.Magn.&Magn.Mater.* 31–34:475(1983).

19. M.Lavagna, C.Lacroix and M.Cyrot, *J.Phys.F* 13:1007(1983).

20. T.Kagayama, G.Oomi, H.Takahashi, N.Mōri, Y.Ōnuki and T.Komatsubara, *J.Alloys and Compounds* 181:185(1992).

21. A.Shibata, G.Oomi, Y.Ōnuki and T.Komatsubara, *J.Phys.Soc.Jpn.* 55:2086(1988).

22. B.R.Cooper, private communication.

23. J.W.Allen and R.M.Martin, *Phys.Rev.Lett.* 49:1106(1982).

SEVERAL ASPECTS ON THERMOPOWER
OF Ce COMPOUNDS

J. Sakurai

Department of Physics
Toyama University
Gofuku, Toyama 930, Japan

ABSTRACT

The thermopower (TEP) of Ce compounds shows a variety of anomalies. First, the sign of the TEP is considered in relation to the impurity Kondo resonance peak at the Fermi energy on the basis of Mott's equation of the TEP. Next, a relation between the TEP and the electronic specific heat developped by Fischer and by others on the basis of the impurity Anderson model are verified, by referring to available data, to approximately hold for the heavy Fermion compounds and for the intermediate compounds for the temperature range $T > T_{coh}$, T_{coh} being the temperature of an onset of a coherence effect. Finally, the TEP behaviour of the compounds with different types of magnetic correlations is shown, and the occurrence of a negative peak of the TEP for the antiferromagnetic Kondo compounds is explained in the spirit of Mott's equation.

INTRODUCTION

In Ce compounds, a coexistence and a competition between incoherent or coherent Kondo interaction and RKKY interaction give rise to interesting and different types of anomalies in physical properties. The thermopower (TEP) is one of transport properties which is sensitive to the details of the band structure and the scattering spectrum of conduction electrons. Therefore, TEP measurements on variety of Ce compounds are expected to be fruitful for understanding details of electron interactions.

As a matter of fact, a large number of the TEP data on Ce compounds has been accumulated until now, and Mott's expression of the TEP succeeded to give simple and clear physical meanings on these data as exemplified in the next section. However, discussions of TEP data often seem to be of a qualitative nature. Although the TEP can reflect very sensitively details of the structures of electron interactions, an absolute value of the TEP alone is not as simple to give a physical meaning as for other quantities, the specific heat, for example.

However, a simple relation between the TEP and the electronic specific heat at the low temperature limit was rather recently proposed by Kawakami et al[1], Houghton et al[2] and by Fischer[3] on the single impurity Anderson model. According to this relation, meaning of a TEP value can be as simple and fundamental as a specific heat value. In the subsequent section, we try to verify this relation by experimental data of the TEP and the specific heat of on Ce compounds. In the lack of enough numbers of available data on dilute Ce systems, the heavy Fermion compounds and the intermediate valence compounds of Ce were also considered.

The largest numbers of Ce compounds may be probably classified as the antiferromagnetic Kondo compounds. The TEP curves of them are very often characterized with an existence of a deep and negative peak at a temperature several times higher than the Neel temperatures T_N, and thus the occurrence of the peak is sometimes considered to be an evidence of onset of the magnetic correlation. However, theoretical and quantitative explanations for the peak have not be fully achieved so far. In the subsequent section, we try to show several types of TEP data with regard to different natures of the magnetic correlations.

MOTT'S EQUATION OF THERMOPOWER

Mott's equation[4] for the thermopower S is often referred to in discussions of TEP data because of simple and clear physical insights it provides. The equation is written as follows,

$$S = -\frac{\pi^2 k^2}{3 \mid e \mid} T \left(\frac{\partial \ln D}{\partial \varepsilon} + \frac{\partial \ln \tau}{\partial \varepsilon} \right)_{\varepsilon = \varepsilon_F} \tag{1}$$

where, D and τ are the state density and the relaxation time of conduction electrons, respectively, and the energy derivative is at the Fermi energy ε_F. Jaccard et al[5] explained the sign of the TEP of Ce compounds on the base of this equation.

For Ce compounds for $T > T_K$, a narrow resonance with its state density $N(\varepsilon)$ at Fermi energy is considered to take place with its energy centered at ε_f above ε_F, and hence with the derivative of $N(\varepsilon)$ being positive at $\varepsilon = \varepsilon_F$. The state is incoherent and plays a role of scatterer for conduction electrons. Since τ is inversely proportional to $N(\varepsilon)$, the sign of S is expected to be positive from eq. (1) in accordance with the experimental data for Ce compounds of the heavy Fermion state or of the intermediate valence state in the temperature range $T > T_K$.

While for these Ce compounds at sufficiently low temperature $T < T_{coh}$, T_{coh} being the temperature of the onset of formation of coherent Kondo lattice, $4f$ electrons of Ce construct a coherent Kondo-lattice resonance with the state density $N(\varepsilon)$ and they participate to conduction. Hence, we introduce the $N(\varepsilon)$ in the first term D in eq. 1), and the sign of S becomes opposite to the sign of the derivative of $N(\varepsilon)$ at $\varepsilon = \varepsilon_F$. Experimentally, the sign of S for the heavy Fermion compounds, $CeCu_6$[6] for example, and for the intermediate valence compounds, $CeNi$[7] and $CeSn_3$[8] for examples, is positive down to low temperatures. Therefore, the formation of the Kondo lattice seems to introduce a dip in $D(\varepsilon)$ so that its derivative becomes negative at $\varepsilon = \varepsilon_F$.

THERMOPOWER AND SPECIFIC HEAT

Mott's equation succeeds to explain satisfactorily some of qualitative features of the thermopower. However, one can seldom discuss a numerical value of TEP on the base of this equation. The equation expressed by a derivative form seems to be more suited to bring the structure of the TEP to light rather than its actual value.

In respect to relation between numerical value of the TEP and other quantities, the pioneering concept by Lord Kelvin[9] must be cited at first; on a simple electron gas without scattering, the thermopower was shown to be equal to the entropy flow carried by electrons in a temperature gradient. For an actual discussion, however, the model of a simple electron gas must be modified and other competing mechanisms have to be taken into account as well.

Recently, TEP theories[1-3] of the heavy Fermion compounds were developed on the single impurity Anderson model. Here, we refer to Fischer's paper. The model leads to the Kondo term of order J^3V and the resonance term of order J^2V, J and V being the exchange interaction and the potential interaction, respectively. The Kondo term, a relation between the TEP and the constant γ of the specific heat similarly to the Kelvin's concept, is positive and is given as follows,

$$S = \frac{2\pi}{|e|N}\gamma T \cot(\pi n_f/N) \qquad (2)$$

where N is the degeneracy of $4f$-impurity, and n_f, the occupation number of a $4f$-impurity, is near to 1 for Ce compounds. On the other hand, the resonance term in the presence of a magnetic correlation between Kondo impurities gives rise to a negative contribution.

We tried to compare eq. (2) with experimental data. However, data on Ce dilute alloys have not been accumulated much. Therefore, we include in our comparison data on the heavy Fermion compounds in appropriate temperature range $T > T_{coh}$, where each $4f$ electron is expected to behave independently and incoherently as an impurity Kondo ion. Some of the heavy Fermion compounds, $CeAl_3$[10] and $CeCu_2Si_2$[5] for examples, have a negative peak of the TEP at low temperature. They were excluded from the present comparison. They might have weak magnetic correlations.

In Fig. 1, the TEP divided by the temperature S/T for several Ce compounds is plotted against the specific heat constant. We immediately see a strong correlation between both the quantities. The line in Fig. 1 is after eq. (2) with $N = 6$ for the ground state of spin-orbit coupling and with the assumption of $n_f = 1$.

Only a single data of a dilute Ce alloy in Fig. 1 is for $Ce_{0.1}La_{0.9}Cu_6$[6] at 2 K, where data on both the quantities are available. Huge enhancements of both S/T and γ take place below this temperature[11]. Thus, further studies are waited for.

As the heavy Fermion compounds, $CeCu_6$[6,11,12] $CePtIn$[13,14], $CeInCu_2$[15], $CeCu_4Al$[16], $CeCu_3Al_2$[17], $CeCu_4Ga$[18] and $CeCu_3Ga_2$[19] are referred to. For these compounds, both S/T and γ show huge enhancements at low temperatures. Thus, the curves with temperature as a hidden parameter are plotted in Fig. 1. The temperature range of each curves as written in Fig. 1 is from 2 K to 10 K, for example, where data of both the quantities are available. The condition $T \geq T_{coh}$ for an impurity scattering is considered to be satisfied since T_{coh} for the heavy Fermion compounds is at most a few degrees.

The curves for $CeCu_6$, $CePtIn$, $CeInCu_2$, $CeCu_3Ga_2$ and $CeCu_3Al_2$ are noticed not to be far from the theoretical line. While, the curves for $CeCu_4Al$ and $CeCu_4Ga$ are away from the line. However, if they are multiplied by an appropriate factor, they could be brought near to the line. Apart from the numerical factor, all curves resembles each other qualitatively.

Thus, S/T and γ are closely related for the dilute Ce compounds and for the heavy Fermion compounds for $T \geq T_{coh}$, and eq. (2) crudely represents the relation between the two quantities. However, the numerical agreement between experimental data and eq. (2) is rather poor. There may be many reasons for this. In addition to eq. (2), there are other contributions to the TEP. The temperature range of application of eq.

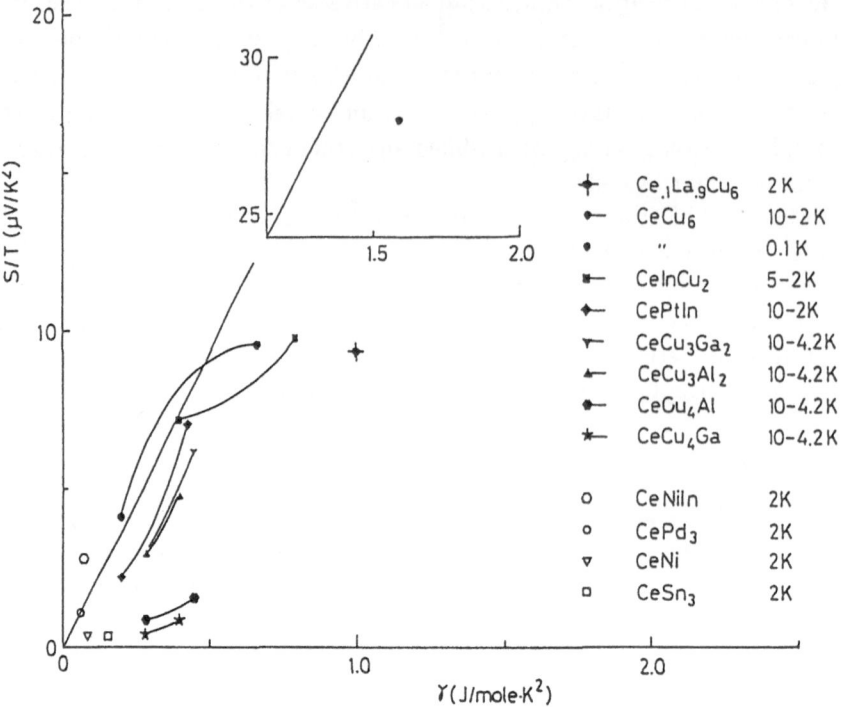

Fig. 1 The thermopower divided by the temperature S/T for some of the heavy Fermion compounds and the intermediate valence compounds is plotted as a function of the constant of the electronic specific heat γ.

(2) is limited and must be very carefully chosen. Moreover, eq. (2) does not consider the anisotropy of TEP for samples with low crystal-symmetries. From an experimental point of view, the value of TEP at low temperature is often sample-dependent, and difficulties of obtaining reproducible data must be stressed. When we consider all these, we are led to conclusion that eq. (2) is quite satisfactory as a first approximation for the relation between S/T and γ for an impurity Kondo systems of Ce compounds and alloys.

For $CeCu_6$, the values of S/T and γ are well established down to milli Kelvin region[11,12]. The curve for the compound in Fig. 1 is down to 2 K. With the decrease of temperature below 2 K down to the region $T \geq T_{coh}$, both S/T and γ become larger, and at the same time, the deviation from the theoretical line becomes larger. Only below 0.1 K, both S/T and γ get saturated and attain to constant values with the accomplishment of the heavy Fermion state. A point for this low temperature limit is marked in Fig. 1. This point turns out not to be far from the theoretical line.

We also refer to several intermediate valence compounds, namely, $CeSn_3$[8,20], $CeNi$[7,21], $CePd_3$[22,23] and $CeNiIn$[13,24]. The values of γ for these compounds are constant at low temperature range, say below 20 K, and are around 0.1 $J/mole \cdot K^2$ due to their high values of T_K. Correspondingly, the S curves of these compounds smoothly extrapolate to zero as T approaches to zero at the same temperature range. The points for the compounds in Fig. 1 are at 2 K or at 10 K and are supposed to be at the low temperature limit. The theoretical line passes among the scattering of the points, and it seems to serve again as a crude approximation for the relation between S/T and γ for the compounds.

THERMOPOWER, ANTIFERRO- OR FERROMAGNETIC CORRELATION AND MAGNETIC ORDER

Although the implication of Mott's equation is clear, further steps into TEP theories on magnetic Kondo compounds turn out to be very complicated and hard to discuss with a simple physical meaning. Fischer[3] discribed that the negative peak of thermopower originates from the resonance term in the presence of magnetic correlations. Matho[25] discussed the importance of the antiferromagnetic correlation to give rise to the negative peak of thermopower.

Experimentally, the negative peak of thermopower for antiferromagnetic Kondo compounds of Ce occurs at a temperature several times higher than T_N, and thus, the thermopower sensitively detects a precursory short-range antiferromagnetic correlation. However, we note that a magnetic correlation is not always of a short-range and antiferromagnetic. Thus, the question arise, 1) how the onset of a long-range antiferromagnetic correlation at T_N is reflected on the S curve? and 2) how the S curve looks like if the magnetic correlation in question is ferromagnetic? In the rest of this paper, we try to answer these questions by referring to available data on the TEP.

As a typical example of antiferromagnetic Kondo compounds of Ce, we refer to CePdSn[26]. The magnetic resistivity $\rho_m = \rho_{CePdSn} - \rho_{LaPdSn}$ and the thermopower S are shown in Fig. 2 a) and b), respectively. Two steps of lagarithmic increase of ρ_m as

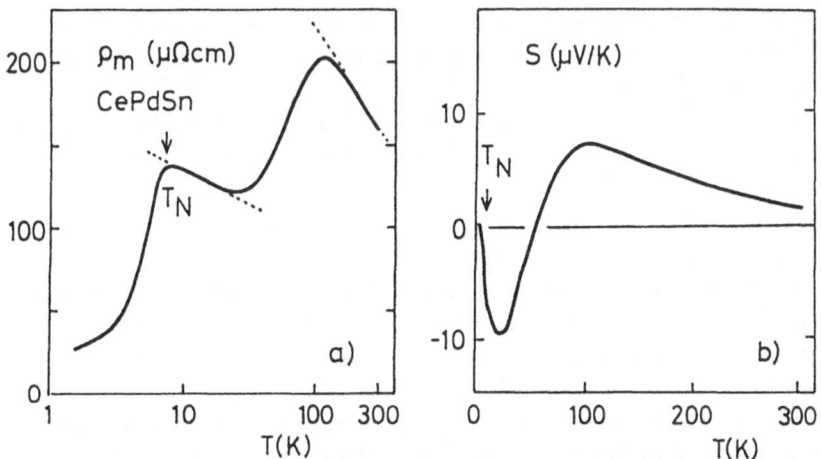

Fig. 2 a) The magnetic resistivity ρ_m and b) the thermopower S of CePdSn plotted as a function of temperature. The two steps of Kondo resistivity are marked by the dotted lines. The arrows indicate the Neel temperature T_N.

marked by the dotted lines in Fig. 2 a) are due to the Kondo effect of the six-fold degeneracy of Ce $4f$ electron for $T > \Delta$, Δ being the energy separation of the crystalline electric field (CEF) levels, and to the Kondo effect of the two-fold degeneracy of the CEF ground level for $T < \Delta$. The huge peaks of S at about 100 K in Fig. 2 b) corresponds to the high-temperature peak of ρ_m in Fig. 2 a). The temperature of the peak is relevant to the separation of the CEF splitting, and the amplitude of the peak to the Kondo interaction according to Bhattacharjee et al[27]. The negative peak of S, sensing precursory short-range antiferromagnetic correlation, is at 25 K much higher

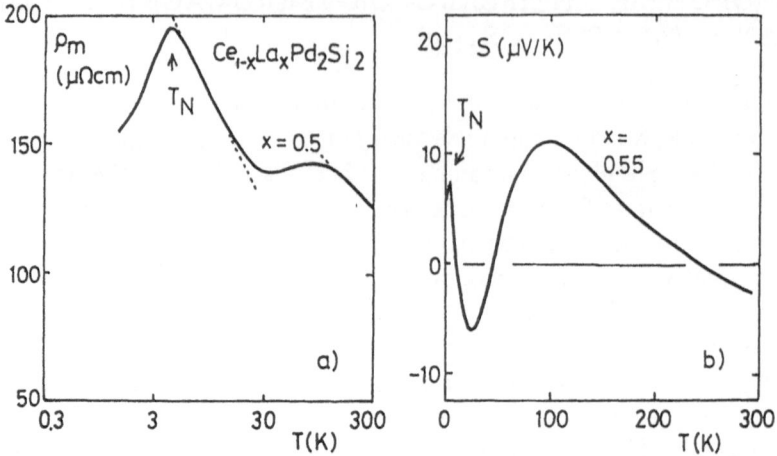

Fig. 3 a) The magnetic resistivity ρ_m for $Ce_{1-x}La_xPd_2Si_2$ with $x = 0.5$ and b) the thermopower S for $Ce_{1-x}La_xPd_2Si_2$ with $x = 0.55$. The two steps of the Kondo resistivity are marked by the dotted lines. The arrows indicate the Neel temperature T_N.

than T_N of 7.0 K. Below T_N, ρ_m manifests a rapid decrease as seen in Fig. 2 a), while, the S curve shows only a faint change in its derivative at T_N. Similar behaviour of S was observed for other antiferromagnetic Kondo compounds, $CePtSn$[28], $CePb_3$[29], $CeCu_2$[30] and $CeAl_2$[5] for examples, for which the thermopower shows only a faint or almost no singularity at T_N.

However, there are some cases where the TEP shows indeed a sharp peak with a singularity at T_N. We refer to $Ce_{1-x}La_xPd_2Si_2$, for example, a pseudo-binary system of an antiferromagnetic Kondo compound $CePd_2Si_2$ with its non-magnetic La isomorphic compound $LaPd_2Si_2$. In Figs. 3 a) and b), the ρ_m curve[31] and the TEP curve[32] for $Ce_{1-x}La_xPd_2Si_2$ are shown, respectively. Two steps of lagarithmic increase of ρ_m and a steep decrease below T_N on one hand in Fig. 3 a), and a huge peak of S at 100 K and a negative peak at 25 K on the other hand in Fig. 3 b) are similar to those for CePdSn. A notable difference is that the S curve for $Ce_{1-x}La_xPd_2Si_2$ continues to increase below 20 K and makes a prominently sharp peak at T_N. In addition, we notice that the maximum value of ρ_m at T_N is much higher than another peak at 100 K as seen in Fig. 3 a), making a contrast to the ρ_m curve for CePdSn. Judging from nearly parallel slope of the two dotted lines for ρ_m in Fig. 3 a), the Kondo interaction for $T < \Delta$ is somehow very much enhanced in comparison to that for $T > \Delta$. A ρ_m curve with a notable Kondo increase probably shows a large decrease below T_N as well, giving rise to a sharp peak at T_N. This in turn may be related to the existence of the sharp peak in the S curve at T_N. Thus, in some cases an antiferromagnetic long-range order at T_N can give rise to a sharp peak in the S curve.

Our next question is whether a negative peak of S exists for the ferromagnetic Kondo compounds of Ce. Unfortunately, only a few compounds are known in this category. Among them, we refer to $CeNi_{1-x}Pt_x$. The ρ_m curve[33] and the S curve[34] for $CeNi_{1-x}Pt_x$ with $x = 0.5$ and 1.0 are shown in Figs. 4 a) and b), respectively. The Kondo interaction in CePt is much weaker than that in $CeNi_{0.5}Pt_{0.5}$, as seen from the difference of the slopes of the dotted lines for the Kondo resistivity in Fig. 4 a). Thus, the S curve for CePt alone seems somewhat to be of a normal metal. However, it

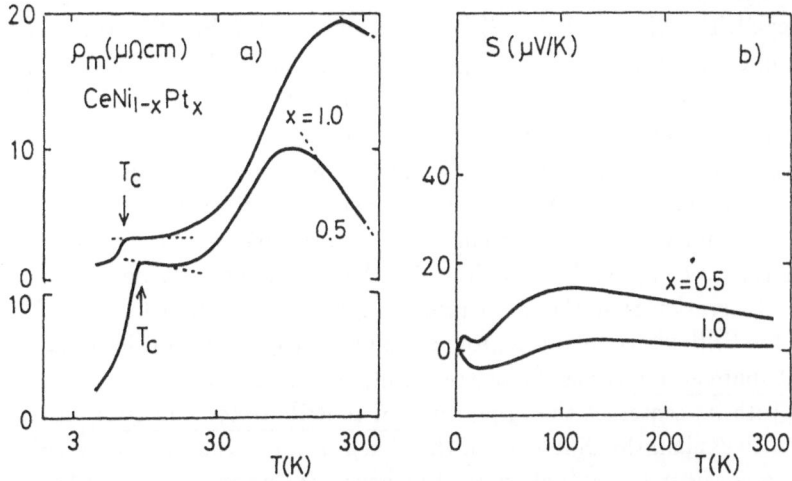

Fig. 4 a) The magnetic resistivity ρ_m and b) the thermopower S for $CeNi_{1-x}Pt_x$ with $x = 0.5$ and 1.0. The two steps of the Kondo resistivity are marked by the dotted lines. The arrows indicate the Curie temperature T_C.

retained the common features to the S curves for $CeNi_{0.5}Pt_{0.5}$ proving its Kondo character. Thus, on the S curves of both the two samples, we observe an existence of a broad negative peak at about 25 K, although its amplitute is not large. The Curie temperature T_C for the two samples are around 10 K. Therefore, the negative peak of S is probably the indication of the precursory short-range ferromagnetic correlation. On the other hand, nothing appears for the S curves at T_C. A small peak appears for the S curve for $CeNi_{0.5}Pt_{0.5}$ at 7 K. We do not know its origin.

The final case is a ferromagnetic Kondo compound where the S curve show a

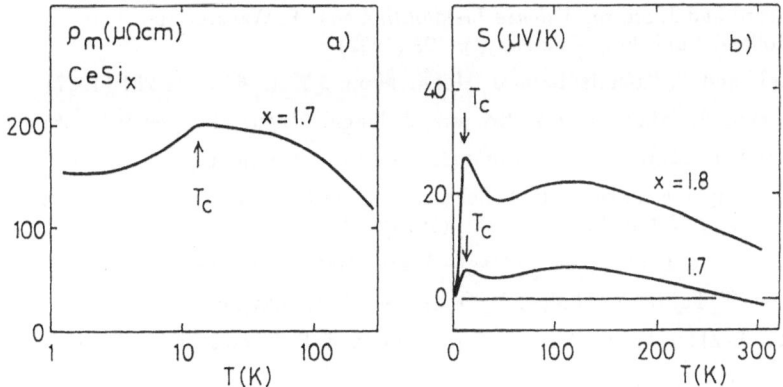

Fig. 5 a) The magnetic resistivity ρ_m for $CeSi_x$ with $x = 1.7$ and the thermopower S for $CeSi_x$ with $x = 1.7$ and 1.8. The arrows indicate the Curie temperature T_C.

strong peak with a singularity at T_C. The ρ_m curve for CeSi$_{1.7}$[35] and the S curves for CeSi$_{1.7}$ and CeSi$_{1.8}$[36] are shown in Fig. 5 a) and b), respectively. The ρ_m curve do not show well resolved two steps of the logarithmic Kondo increase observed in Figs. 2 a) - 4 a). We do not know the reason for this. On the other hand, the S curves in Fig. 5 b) continue to increase notably below 40 K and they manifests a sudden fall at $T_C = 10K$, giving rise to an eminent peak at T_C in contrast to the S curve in Fig. 4 b).

From a comparison of the strength of the negative peak of S for the antiferromagnetic Kondo compounds with those for the ferromagnetic Kondo compounds, we make an observation that the short-range magnetic correlation for $T > T_N$, with its center in the Brillouin zone at the propagation vector τ_m of the antiferromagnetic structure, contribute effectively as the scatter, while the short-range magnetic correlation for $T > T_C$, with its center at the origin, do not contribute too effectively as the scatter. Therefore, to explain the difference of the S behaviour in the spirit of mott's equation, we have to consider the actual shape and hence an inhomogeneity of the Fermi surface in the Brillouin zone. Although the amplitude of each component of short-range antiferromagnetic correlation is small, the integration over an effective area of the scatterers on the Fermi surface becomes just as important as the care for of a long-ranged magnetic correlation, and gives rise to the sharp negative peak of the TEP.

In conclusion, we have seen several types of the S curves associated with the different types of magnetic correlations. The S curves of the antiferro- or ferromagnetic Kondo compounds of Ce are seen to manifest sometimes a strong peak at T_N or at T_C reflecting an onset of long-range magnetic correlation. The negative peak of S seems to exist for the antiferro- as well as for ferromagnetic Kondo compounds, although it may be much more eminent for the former compounds.

Hearty thanks are due to Prof. K. Yamada for his very helpful discussions.

REFERENCES

1) N. Kawakami and A. Okiji, Jpn. J. Appl. Phys. 26, Suppl. 26-3, 499 (1987)

2) A. Houghton, N. Read and H. Won, Phys. Rev. B 35, 5123 (1987)

3) K. H. Fischer, Z. Phys. B 76, 315 (1989)

4) see for example, J. M. Ziman, *Electrons and Phonons*, Oxford University Press (1960)

5) D. Jaccard and J. Sierro, *Valence Instabilities*, eds. P. Wachter and H. Boppart, north Holland Publishing Company, p 409 (1982)

6) Y. Onuki and T. Komatsubara, J. Magn. Magn . Mat., 63 & 64 281 (1987)

7) T. Ohyama, J. Sakurai and Y. Komura, J. Magn. Magn. Mat., 63 & 64 581 (1987)

8) J. Sakurai, K. Murata and Y. Komura, Solid State Commun., 49 287 (1984)

9) see for example, F. J. Blatt, P. A. Schroeder, C. L. Foiles and D. Greig, *Thermoelectric Power of Metals*, Plenum Press, (1976) p 164

10) D. Jaccard and J. Flouquet, J. Magn. Magn. Mat., 47 & 48 45 (1985)

11) K. Satoh, T. Fujita, Y. Maeno, Y. Onuki and T. Komatsubara,

12) H. Sato, J. Zhao, W. P. Pratt, Jr., Y. Onuki and T. Komatsubara, Phys. Rev. B 36 8841 (1987)

13) H. Fujii, KY. Uwatoko, M. Akayama, K. Satoh, Y. Maeno, T. Fujita, J. Sakurai, H. Kamimura and T. Okamoto, Jpn. J. Appl. Phys. 26 Suppl. 26-3 549 (1987)

14) T. Fujita, K. Satoh, Y. Maeno, Y. Uwatoko and H. Fujii, J. Magn. Magn. Mat. 76 & 77 133 (1988)

15) S. Takayanagi, S. B. Woods, N. Wada, T. Watanabe, Y. Onuki, A. Kobori, T. Komatsubara, M. Imai and H. Asano, J. Magn. Magn. Mat., 76 & 77 281 (1988)

16) E. Bauer, E. Gratz and N. Plillmayr, Solid State Commun., 62 271 (1987)

17) E. Bauer, N. Pillmayr, E. Gratz, D. Gignoux and D. Schmitt J. Magn. Magn. Mat. 67 L 143 (1987)

18) E. Bauer, N. Pillmayr. E. Gratz, D. Gignoux, D. Schmitt, K. Winzer and J. Kohlmann, J. Magn. Magn. Mat., 71 311 (1988)

19) E. Bauer N. Pillmayr, E. Gratz, D. Gignoux and D. Schmitt, Physica letters A 124 445 (1987)

20) K. A. Gschneidner, Jr., S. K. Dhar, R. J. Stierman and T. -W. E. Tsang, J. Magn. Magn. Mat., 47 & 48 51 (1985)

21) Y. Ishikawa, K. Mori, T. Mizushima, A. Fujii, H. Takeda and K. Sato, J. Magn. Magn. Mat., 70 385 (1987)

22) H. Sthioul, D. Jaccard and J. Sierro, *Valence Instabilities*, P. Wachter and H. Boppart eds. North Holland Publishing Company p 443 (1982)

23) N. Pillmayr, unpublished

24) H. Fujii, T. Inoue, Y. Andoh, T. Takabatake, K. Satoh, Y. Maeno, T. Jujita, T. Sakurai and Y. Yamaguchi, Phys. Rev. B 39 6840 (1989)

25) K. Matho and M. T. Beal-Monod, J. Phys. F 4 848 (1974)

26) J. Sakurai, Y. Yamaguchi, K. Mibu and T. Shinjo, J. Magn. Magn. Mat., 84 157 (1990)

27) A. K. Bhattacharjee and B. Coqblin, Phys. Rev. B 13 3441 (1976)

28) J. Sakurai, R. Kawamura, T. Taniguchi, S. Nishigori, S. Ikeda, H. Goshima, T. Suzuki and T. Fujita, J. Magn. Magn. Mat., 104-107 1415 (1992)

29) J. Sakurai, H. Kamimura and Y. Komura, J. Magn. Magn. Mat., 76 & 77 287 (1988)

30) E. Gratz, E. Bauer, B. Barbara, S. Zemirli, F. Steglich, C. D. Bredl and W. Lieke, J. Phys. F 15 1975 (1985)

31) M. J. Besnus, A. Braghta and A. Meyer, Z. Phys. B 83 207 (1991)

32) Y. Bands, J. Sakurai and E. V. Sampathkumaran, Submitted to J. Magn. Magn. Mat.,

33) D. Gignoux and J. C. Gomez-Sal, Phys. Rev. B 30 3967 (1984)

34) J. Sakurai and M. Horie, unpublished

35) N. Sato, H. Mori, T. Satoh, T. Miura and H. Takei, J. Phys. Soc. Jpn., 57 1384 (1988)

36) J. Sakurai and Y. Murashita, Physica Letters A 150 113 (1990)

TRANSPORT AND THERMODYNAMIC PROPERTIES OF UNi$_2$Al$_3$ and UPd$_2$Al$_3$

C. Geibel, A. Böhm, C.D. Bredl, R. Caspary, A. Grauel, A. Hiess, C. Schank, F. Steglich, G. Weber

Institut für Festkörperphysik, TH Darmstadt, W-6100 Darmstadt, Germany

Abstract

We present and discuss the transport and thermodynamic properties of the new heavy-fermion superconductors UNi$_2$Al$_3$ and UPd$_2$Al$_3$. Analysis of the normal-state properties of UPd$_2$Al$_3$ indicates pronounced crystal-field effects and suggests a 5f^2 configuration with a lowest-lying singlet. From the shifts of the maxima in the resistivity and susceptibility to higher temperatures when replacing Pd by Ni we deduce an increase by a factor of 3 in the crystal field splitting Δ. The energy scale for the screening of the 5f electrons appears to scale with Δ, leading to the observed reduction of the ordered moment on going from the Pd- to the Ni-compound. The specific heat in the superconducting state of UPd$_2$Al$_3$ shows many similarities to the specific heat of other heavy fermion superconductors. The temperature dependence of the thermodynamic critical field determined from the specific heat is close to the result of the BCS-theory.

INTRODUCTION

Since the discovery of the large electronic specific heat in CeAl$_3$ /1/ and of superconductivity in CeCu$_2$Si$_2$ /2/, heavy fermion (HF) compounds have been the subject of intensive experimental and theoretical research. It is now commonly accepted, that the HF-state in Cerium-based compounds can be well described within the frame of the Kondo-lattice theory /3/. In contrast the situation for Uranium-based HF-compounds is still not clear. The difficulties begin with the unknown number of electrons in the f-shell: Uranium can adopt a U^{3+}, U^{4+} or U^{5+}-ionic state corresponding to a 5f^3, 5f^2 and 5f^1 configuration, respectively. With very few exception, it is generally not possible to determine the ionic state directly. Further, the spatial extent of the f-electrons in comparison to the valence electrons is larger for U than for Ce. Thus, hybridization between 5f electrons and valence electrons from ligand atoms can be much stronger. The importance of 5f-band-structure effects for the description of the HF-state in U compounds is still an open question. At last, the pairing mechanism and the pairing state in both U- and Ce-based HF-superconductors is also controversially discussed.

The discovery of the two new HF superconductors UNi$_2$Al$_3$ and UPd$_2$Al$_3$ in 1991 /4/5/

has opened a new opportunity to investigate these problems. Transport, susceptibility and specific-heat measurements indicate in UNi_2Al_3 and UPd_2Al_3 antiferromagnetic (AF) ordering at $T_N=4.5K$ and $T_N=14.5K$, respectively, and a transition into the superconducting state at $T_c=1.1K$ and $T_c=2.0K$, respectively /6/. Lateron, neutron diffraction and μ^+SR measurements revealed quite a large ordered moment for HF superconductors, approximately $0.1\mu_B$ in UNi_2Al_3 /7/ and $0.85\mu_B$ in UPd_2Al_3 /8/, and proved the coexistence of superconductivity and magnetic order below T_c in both compounds. Here, we present a status report on the transport and thermodynamic properties of these two compounds. These results give strong evidence for a tetravalent Uranium state ($5f^2$ configuration) and a singlet crystal-field (CF) ground state. They further suggest that the superconducting state can be described quite well with classical BCS-type models.

CRYSTAL FIELD AND KONDO EFFECT

Stimulated by the pronounced maximum in the susceptibility of UPd_2Al_3 and its resemblence to the susceptibility of $PrNi_5$ which crystallizes in the related $CaCu_5$ structure, already in the first paper on UPd_2Al_3 we speculated about a possible $5f^2$ configuration (see Ref. 8 in /5/). Lateron, we found that the specific heat of the 5f-electrons in UNi_2Al_3 could be well fitted by a Bethe-Ansatz for a single ion Kondo effect and a doublet-singlet CF transition, implying a $5f^2$ configuration /9/. This led us to perform a crystal-field analysis of the 5f-derived specific heat and the susceptibility of UPd_2Al_3 /10/. A more detailed description of the experiments and the calculation is given in /11/. In the following discussion, we can safely ignore the $5f^1$ configuration, because its small free-ion moment $\mu_{eff}=2.54\mu_B$ is significantly lower than the experimental value of $3.4\mu_B$ obtained at high temperature on a polycrystalline sample, which in turn corresponds well to the free-ion moment for both the $5f^2$ ($\mu_{eff}=3.58\mu_B$) and $5f^3$ ($\mu_{eff}=3.62\mu_B$) configuration. The discrimination between configurations $5f^2$ and $5f^3$ is more difficult and it is easily shown that fits to the specific heat alone are not sufficient for an unequivocal assignment.

The specific heat of the 5f electrons as obtained by subtracting the specific heat of $ThPd_2Al_3$ from that of UPd_2Al_3 shows the AF transition at 14.5K and a broad peak at 55K which in the following will be addressed as a Schottky-type anomaly (Fig. 1). The dashed curve corresponds to a fit with the five doublets of a $5f^3$ configuration. The entropy connected with the excess specific heat found experimentally at low temperatures amounts to nearly Rln2, as expected for a lowest-lying doublet state. The fit, therefore, must be considered quite good. However, Fig. 1 also contains a fit with the three singlets and the three doublets (two singlets lowest) of the $5f^2$ configuration, which we will more fully discuss later. If we assume that part of the entropy connected with the Schottky anomaly is transferred to the onset of magnetic order by the exchange interaction, this fit can also be considered quite good.

Obviously additional information is needed for an unequivocal assignment and for this we turn to the magnetic susceptibility. The latter is very anisotropic for UPd_2Al_3, with large values and a pronounced temperature dependence for the field applied in the basal plane ($\chi_\perp$) and small values and a weak temperature dependence for the field applied along the c-axis ($\chi_\parallel$) (Fig. 2) /12/. In a magnetic field, the Hamiltonian reads

$$H = H_{CF} + g\mu_B \ J \cdot H \tag{1}$$

$$H_{CF} = B_2^0 O_2^0 + B_4^0 O_4^0 + B_6^0 O_6^0 + B_2^6 O_6^6 \qquad (2)$$

where g and J are the Landé factor and the total angular momentum of the free ion, respectively, and O_n the Stevens equivalent operators and B_n the CF parameters. In comparing the susceptibilities χ_{CF} derived from (1) and (2) with experiment, exchange

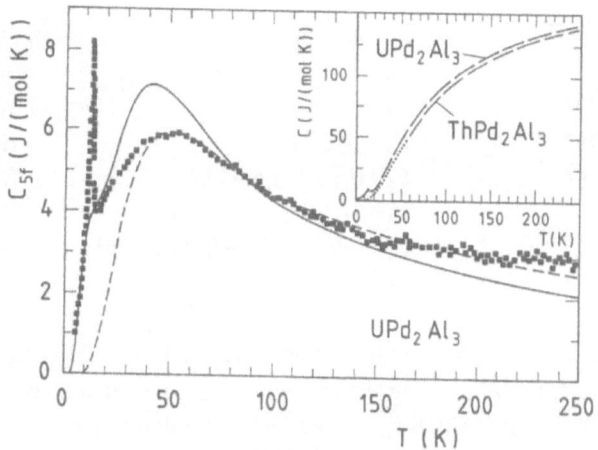

Figure 1. Temperature dependence of the 5f contribution to the specific heat of UPd₂Al₃ as C(UPd₂Al₃) - C(ThPd₂Al₃). Solid and dashed curves represent CF driven Schottky anomalies (see text). The inset shows the specific heat of UPd₂Al₃ and ThPd₂Al₃ vs T /11/.

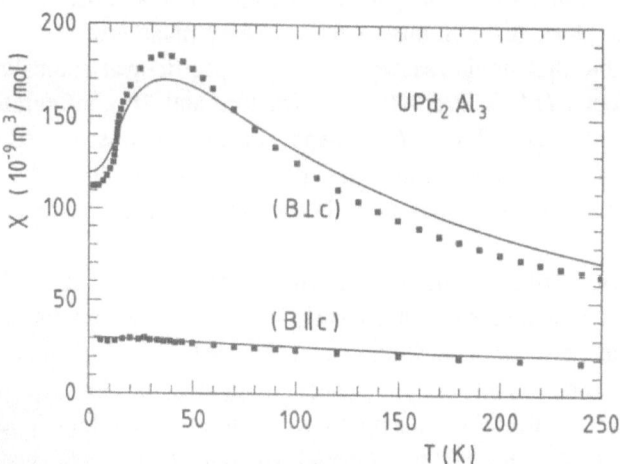

Figure 2. Temperature dependence of $\chi_\perp$ (B⊥c) and $\chi_\parallel$ (B∥c). The solid lines show CF calculations for both directions (see text) /11/.

interactions were taken into account in a mean-field approximation by introducing two mean-field fit parameters $\lambda_{\perp/\parallel}$ as follows: $(\chi_m)^{-1}=(\chi_{CEF})^{-1}-\lambda$. Starting from the $5f^3$-configuration of trivalent Uranium we found it impossible to reproduce the two salient features of the susceptibility data: the maximum in $\chi_\perp(T)$ and the small and non-diverging $\chi_\parallel(T)$. These features could, however, be reproduced by assuming the $5f^2$-configuration of tetravalent Uranium and we, therefore, exclude the former from consideration and concentrate on the latter. For the $5f^2$ configuration our best fit to the susceptibility and specific-heat data was obtained with the parameters of Table 1.

Table 1. Crystal field parameters used for the fit of the specific-heat and susceptibility data of UPd_2Al_3.

CF operators	eigenfunctions	eigenvalues
$B_2'=8.74K$	$\Gamma_4=1/\sqrt{2}[\,\mid -3\rangle - \mid 3\rangle\,]$	0K
$B_4=57.5mK$	$\Gamma_1=\mid 0\rangle$	32K
$B_6=3.10mK$	$\Gamma_6=\mid \pm1\rangle$	109K
$B_6'=38.3mK$	$\Gamma_5=0.29\mid \pm4\rangle -0.96\mid \mp2\rangle$	168K
	$\Gamma_3=1/\sqrt{2}[\,\mid -3\rangle + \mid 3\rangle\,]$	290K
	$\Gamma_5=0.96\mid \pm4\rangle +0.29\mid \mp2\rangle$	564K

The fitted values of the mean-field parameters, $\lambda_\parallel=2.8\cdot10^7$mole/m^3 and $\lambda_\perp=2.8\cdot10^6$mole/m^3, are in good agreement with a rough estimate according to the simple relation $\lambda_{\perp/\parallel}=1/\chi_{\perp/\parallel}(T_N)$ which leads to $\lambda_\parallel=3\cdot10^7$mole/m^3 and $\lambda_\perp=7\cdot10^6$mole/m^3. The same set of CF parameters used for the specific-heat data gives the solid line in Fig. 1. The entropies of both the CF fit and the measured specific heat reach the same value at about 90K. We think that magnetic interactions, essential for establishing magnetic order in a singlet-ground-state system, account for the differences observed at low temperatures. With two lowest lying singlets the susceptibility for $T\rightarrow0$ is driven by van-Vleck contributions. The maximum in $\chi_\perp(T)$ is caused by successive thermal population of excited levels. Expanding our analysis by including finite magnetic fields /13/ we find that for $B\perp c$ the eigenvalues strongly depend on magnetic field. We suggest that this produces some kind of instability of the magnetic ordering and that this, in turn, leads to the complex magnetic phase diagram observed in the easy plane /10/.

It is remarkable that similar susceptibility and specific-heat anomalies as in UPd_2Al_3 have also been found /14/ in $PrNi_5$ (CaCu$_5$ structure and $4f^2$ configuration of Pr^{3+}) and have likewise been ascribed to the influence of the same sequence of CEF levels. Furthermore, a singlet ground state was proposed for another HF superconductor, URu_2Si_2, which presents a similar peak in the susceptibility along the easy direction /15/16/17/18/.

We now turn to the Kondo effect. The negative temperature derivative of the resistivity at high temperatures which is clearly seen in the raw data for UPd_2Al_3 and, after substraction of the phonon contribution, for UNi_2Al_3, suggest some kind of Kondo interaction. Thus, one would at first glance interpretate the decrease of $\rho(T)$ to lower temperature to the transition from a single-ion Kondo regime at high T to a Kondo-lattice regime at low T in analogy to the Ce-based HF compounds. The investigation of the pseudo-binary system $U(Pd_{1-x}Ni_x)_2Al_3$ sheds more light on this question /19/. We have observed that in the whole concentration range, both maxima in ρ and χ remain well resolved and shift toward higher temperature with increasing Ni-content (Fig. 3a). This is in contrast to the typical behavior in Ce-based systems, where alloying usually results in a large increase of the residual resistivity and a strong reduction or even disappearance of the coherence maximum in ρ, see e.g. the $Ce(Cu_{1-x}Ni_x)_2Ge_2$ system /20/. Thus, either the coherence in U-based HF compounds is less sensitive to disorder, or the maximum in $\rho(T)$ has another origin.

Remarkably, although the temperatures of the maxima increase by a factor 3 on going from the Pd- to the Ni-compound, the ratio $T_{\rho m}/T_{\chi m}$ of the maximum positions of both resistivity and susceptibility remains unchanged (Fig. 3a), indicating a common origin for both phenomena, i.e. CF effects. This is supported by an analysis of the 5f-derived entropy at the temperature of the maximum in $\chi(T)$, $S_{5f}(T_{\chi m})$, which also remains constant over the whole concentration range. One may ask whether the maximum in $\chi(T)$ could be

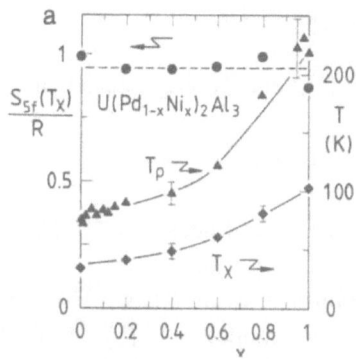 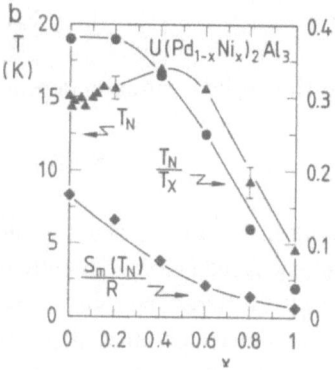

Figure 3. Left frame (a): Concentration dependence of the temperatures of the maxima in $\rho(T)$ and $\chi(T)$, $T_{\rho m}$ (▲) and $T_{\chi m}$ (◆), as well as of the entropy at T_χ, $S_{5f}(T_{\chi m})$ (●). Solid and broken lines are guides to the eye. Right frame (b): Concentration dependence of T_N (▲) as well as of the ratio $T_N/T_{\chi m}$ (●) and of the entropy at T_N, $S_m(T_N)$ (◆). Solid lines are guides to the eye /19/.

related to another phenomenon, e.g. to the metamagnetic transition observed at $B=18$ Tesla in UPd_2Al_3 /15/ (and at 38 Tesla in URu_2Si_2 /16/). However, it is rather difficult to understand how a purely magnetic inter site interaction could lead to a maximum in $\rho(T)$ at a temperature twice as large as the temperature of the maximum in $\chi(T)$. Another explanation would involve the Kondo effect of a magnetic ion with a large ground state degeneracy since $S_{5f}(T_{\chi m})$ is already rather large $\sim Rln3$. But for this case the theory /21/ predicts $T_K \approx 2 \cdot T_{\chi m}$ and one would obtain $T_K \approx 70K$ for UPd_2Al_3. In this case it is difficult to understand how a large moment of $0.85\mu_B$ can still be formed at $T_N=14.5K$. On the other hand, no difficulties have to be encountered, if one assumes that the maxima in $\chi(T)$ and $\rho(T)$ are mainly related to CF effects in the presence of strong 5f-conduction electron hybridization. Calculations of the resistivity for a combined action of CF and Kondo effect show that, if $T_K/\Delta \geq 0.1$ (Δ being the CF splitting), the Kondo coherence peak merges with the CF derived peak (see e.g. Ref. 22). This seems to be the case for UPd_2Al_3, since the model for the CF level scheme gives for the first excited level $\Delta \approx 33K$ and, from the reduction of the ordered moment, one expects $T_K \approx T_N \approx 15K$.

Another interesting scaling behaviour can be observed for the AF ordering. In Fig. 3b we have plotted the concentration dependence of T_N, of the ratio T_N/T_{xm} and of the "magnetic inter site" contribution to the 5f-derived entropy at T_N, defined by $S_m(T_N) = S_{5f}(T_N) - \gamma T_N$. Whereas T_N vs x shows a maximum around x=0.4, both T_N/T_{xm} vs x and $S_m(T_N)$ vs x decrease monotonically with increasing Ni content. In the pure compounds, $S_m(T_N)$ scales with the values of the ordered moments. In a simple phenomenological approach, the ordered moment μ_s should be roughly proportional to the ratio between the energy scale $k_B T_{int}$ of the inter site interaction responsible for the AF ordering and the energy scale $k_B T_{red}$ of the intrasite interaction responsible for the moment reduction. The reduction of T_N in UNi_2Al_3 is certainly not due to a frustration problem since in that case one would expect a shift of the magnetic entropy to lower temperature but not a reduction as observed. Then in a first approximation T_{int} should be proportional to T_N and we obtain

$$\mu_s \sim S_m(T_N) \sim T_{int}/T_{red} \sim T_N/T_{red} \tag{3}$$

Now, the similar concentration dependence of $S_m(T_N)$ and T_N/T_{xm} means that

$$S_m(T_N) \sim T_N/T_{red} \sim T_N/T_{xm} \tag{4}$$

The proportionality of T_{red} and T_{xm} suggests that either T_K scales with the CF splitting or that the moment reduction is directly connected to CF effects.

Somewhat different behavior between Uranium- and Cerium-based HF compounds is seen also in the temperature dependence of the thermopower, S(T). In Ce-based HF compounds, one often observes a negative minimum at the characteristic Kondo lattice temperature but a positive maximum at higher temperatures related to CF effects. In U-based HF compounds, there is no common behavior at low temperatures, but at high temperatures (above 50K) the absolute value of S(T) generally increases steadily with increasing temperatures. This was also observed in UNi_2Al_3 and UPd_2Al_3 (Fig. 4a) (for more details see /23/). In both compounds, S(T) is positive and increases monotonically with the temperature assuming large values above 50K. Small anomalies are visible at the ordering temperature. In Fig. 4b, we have plotted T/S vs T. For T > 50K we get a straight line, indicating that the thermopower can be phenomenologically described by the general expression S(T)=AT(1+T/T'). In Ref. 23, it was shown that this is also the case for other U-based heavy fermion compounds and that the characteristic temperature T' scales approximately with the temperature of the maximum in the resistivity. This could be an indication that CF effects are also manifested in the temperature dependence of the thermopower. Since the U-based HF compounds seem to exhibit either a $5f^2$ or a $5f^3$ configuration the number of CF levels and thus of CF transitions is larger and this may be responsible for the larger thermopower at high temperature in these compounds compared to the Ce-based compounds.

SUPERCONDUCTING STATE

Some of the most fascinating aspects of the HF superconductors were revealed in specific- heat C(T) measurements: e.g. the giant jump at T_c in the specific heat of $CeCu_2Si_2$ proved that the Cooper pairs are formed by the heavy electrons, the double transition in the specific heat of UPt$_3$ was the definitive evidence for a complex superconducting phase diagram and the power laws observed in C(T) below T$_c$ stimulated the speculation about highly anisotropic order parameters with nodes or lines of zero gap on the Fermi surface /3/. Preliminary measurements on UPd$_2$Al$_3$ revealed no spectacular behavior: Only one sharp (ΔT$_c$ < 0.1K) transition at T$_c$, a linear decrease of C/T with T below T$_c$ with some evidence for a higher power law at the lowest temperatures and a residual γT term which was attributed to defects in the sample /5/. The reduced jump height at T$_c$, ΔC/C$_n$(T$_c$) $\approx$ 1.2, was found to be close to the BCS value. In UNi$_2$Al$_3$, the transition was found to be much broader and the jump height smaller due to the lower quality of the sample. Since these results do not reflect the intrinsic properties of this compound, we shall not discuss UNi$_2$Al$_3$ further here.

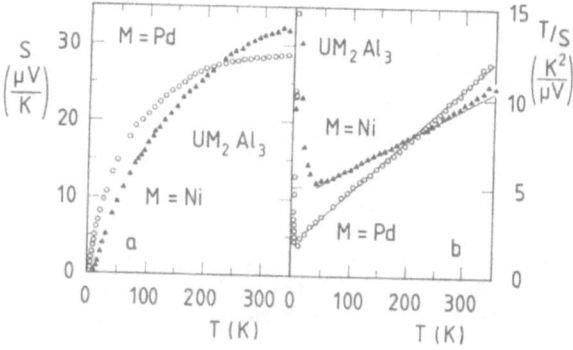

Figure 4. a: Thermopower S(T) vs T of polycrystalline UPd$_2$Al$_3$ (o) and UNi$_2$Al$_3$ (▲). **b**: T/S vs T in polycrystalline UPd$_2$Al$_3$ and UNi$_2$Al$_3$ /23/.

We have now performed a more detailed investigation of the specific heat of UPd$_2$Al$_3$ below 3K /24/ (Fig. 5a). In general, the previous results were confirmed, but since the experiments were extended to lower temperature, down to 0.18K, the low-T behavior can now be studied more precisely. For all values of the applied magnetic field, the specific heat could be well fitted, in the temperature range 0.18K < T < 0.6K, to the expression C = a/T^2 + γT + bT^n where a, γ and b are field-dependent coefficients. The first term represents the contribution of the nuclear moments of ^{27}Al due to the Zeeman splitting in a magnetic field. For high field (4T and 8T) where this term is dominating, the calculated value of the field at the Al-site coincides within 6% with the external field, indicating that the net transfered hyperfine field is nearly zero, i.e. that the ^{27}Al atoms occupy a high-symmetry site in the AF structure. This is in accordance with the results of neutron diffraction measurements /8/. The contribution of the other isotopes as well as a quadrupolar effect can be neglected within experimental resolution.

The second term γT represents the contribution of unpaired electrons. For other samples this term (taken at B=0) was found to increase with increasing width of the superconducting transition. This suggests that γT(B=0) is due to pairbreaking caused by some defects. Below B$_{c2}$ this term increases linearly with applied field.

The third term represents the quasiparticle contribution of the superconductor. The best fit to the data was obtained with n=2.9±0.1 independent of the applied field. Wealso tried an exponential function, but the quality of the fit was much worse. A very similar term was also observed in $CeCu_2Si_2$ /25/ and UBe_{13} /26/. For $0.3 < T/T_c < 1$ $C(T)$ increases linearly with T like in $CeCu_2Si_2$, UPt_3 /27/ and URu_2Si_2/28/. Thus, in the superconducting state the specific heat of UPd_2Al_3 reveals a profound similarity to that of other HF superconductors.

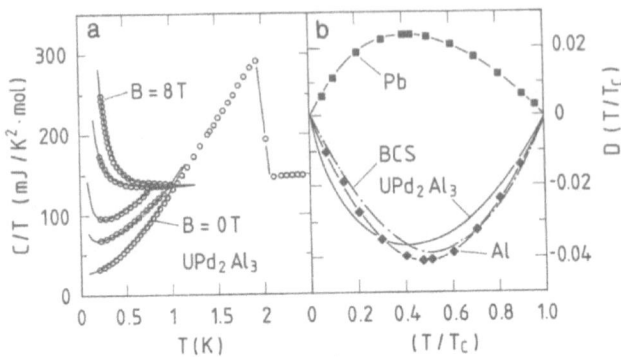

Figure 5. a: Temperature dependence of the specific heat of UPd_2Al_3 below 3.5K for B=0, 1, 2, 4, and 8 Tesla (from bottom to top). The lines correspond to the fits described in the text. **b:** Deviation function $D(T/T_c)$ vs $(T/T_c)^2$ for UPd_2Al_3, Al, Pb and the BCS-theory.

From the specific heat in the normal and in the superconducting state, we determined the thermodynamic field B_c and the deviation function $D(t)=b-(1-t^2)$ where $t=T/T_c$ and $b=B_c(T)/B_c(0)$. UPd_2Al_3 is the first HF-superconductor where this can safely be done because its specific heat in the normal state is nearly field-independent and T_c is rather large. In Fig. 5b, we compare the deviation function of UPd_2Al_3 with the deviation function of a weak coupling superconductor, Al, (ii) of a strong coupling superconductor, Pb and (iii) the BCS result. The curve for UPd_2Al_3 is very close to the results for Al and the universal BCS curve. Only at low temperature, a significant deviation can be seen. This implies that the excitation spectrum of the superconductor UPd_2Al_3 is, at least for $T > 0.3T_c$, quite similar to that of a classical weak-coupling superconductor.

Finally, we present new measurements of the thermal conductivity, $\kappa(T)$, performed on a polycrystalline sample of UPd_2Al_3 (Fig. 6). In this sample the low temperature thermal conductivity is dominated by the electronic contribution, since the $\kappa\rho/T$ ratio comes close to the Lorenz number. This is a further indication for a large electronic mean free path in this system. Correspondingly $\kappa(T)$ increases strongly in a field larger than B_{c2}. By extrapolating κ/T vs T^2 for $T\rightarrow 0$, we extracted the electronic contribution $\kappa_e(T)$. We notice that the ratio of κ_e in the superconducting and in the normal state is the same as the ratio in the electronic specific heat coefficient γ. Since the samples used in both experiments are comparable this would suggest that the electrons responsible for the residual γ term have the same heavy mass as the electrons which form the Cooper pairs, supporting a pairbreaking origin for the unpaired electrons.

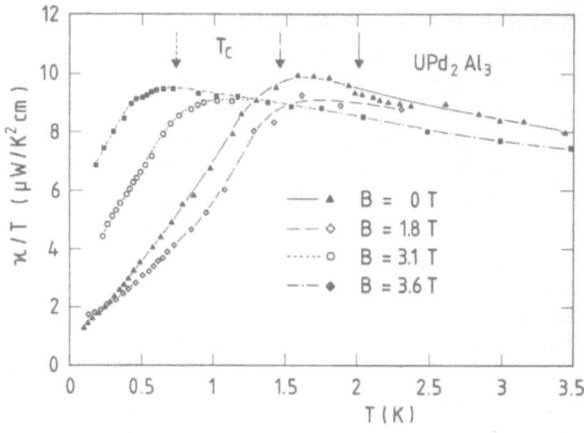

Figure 6. Temperature dependence of κ/T for B=0 (▲), B=1.8T (◇), B=3.1T (○) and B=3.6T (♦).

CONCLUSION

In summary, we observe in UPd_2Al_3 between 30K and 100K pronounced maxima in the temperature dependence of the easy plane susceptibility, of the resistivity and of the 5f-derived specific heat. These anomalies can be described surprisingly well by a CF model for $5f^2$ configuration (tetravalent Uranium). A comparison of UPd_2Al_3 and UNi_2Al_3 suggests in the latter one an increase by a factor 3 in the crystal field splitting and a similar increase in the energy scale of the interaction responsible for the reduction of the ordered moment. The power laws observed below T_c in the specific heat of UPd_2Al_3 are very similar to those observed in the other HF superconductors. The temperature dependence of the thermodynamical field determined from the specific heat is close to the results of the BCS theory, implying that the excitation spectrum of the superconductor UPd_2Al_3 is, at least for $T > 0.3T_c$, quite similar to that of a classical weak coupling superconductor.

REFERENCES

/1/   K. Andres et al., Phys. Rev. Lett. **35**, 1779 (1975).
/2/   F. Steglich et al., Phys. Rev. Lett. **43**, 1892 (1979).
/3/   For a recent review, see: N. Grewe, F. Steglich, in: Handbook on the physics and chemistry of rare earths, K.A. Gschneidner Jr., L. Eyring (eds.), Vol. 14, Chapt. 97.
/4/   C. Geibel et al., Z. Phys. B **83**, 305 (1991).
/5/   C. Geibel et al., Z. Phys. B **84**, 1 (1991).
/6/   F. Steglich et al., in: Frontiers in solid state sciences, L.C. Gupta, M.S. Multani (eds.), World Scientific Publishing Company, in press.
/7/   A. Amato et al., Z. Phys. B **86**, 159 (1992).
/8/   A. Krimmel et al., Z. Phys. B **86**, 161 (1992).
/9/   F. Steglich et al., Physica C **185-189**, 379 (1991).

/10/ A. Grauel et al., Phys. Rev. B **46**, 5818 (1992).

/11/ A. Böhm et al., Int. J. of Modern Physics, in press.

/12/ C. Geibel et al., Physica C **185-189**, 2651 (1991).

/13/ A. Böhm et al., to be published.

/14/ K. Andres et al., Phys. Rev. Lett. **52**, 1551 (1984).

/15/ R. Felten, PhD Thesis, TH Darmstadt (1987).

/16/ G.J. Nieuwenhuys, Phys. Rev. B **35**, 5260 (1987).

/17/ R.J. Radwanski, J. Magn. Magn. Mat. **103**, L1 (1992).

/18/ C. Broholm et al., Phys. Rev. B **43**, 12809 (1991).

/19/ C. Schank et al., Int. J. of Modern Physics.

/20/ A. Loidl et al., Ann. Physik **1**, 78 (1992).

/21/ V.T. Rajan, Phys. Rev. Lett. **51**, 308 (1983).

/22/ see, e.g.: Y. Lassailly et al., J. Magn. Magn. Mat. **47&48**, 501 (1985).

/23/ A. Grauel et al., Int. J. of Modern Physics, in press.

/24/ R. Caspary et al., to be published.

/25/ F. Steglich, Vol. 62 of: Springer Series in Solid-State Sciences, p. 23 (1985).

/26/ A. Ravex et al., J. Magn. Magn. Mat. **63&64**, 400 (1987).

/27/ A. Sulpice et al., J. Low Temp. Phys. **62**, 39 (1986).

/28/ W. Schlabitz et al., Z. Phys. B **62**, 171 (1986).

PRESSURE-INDUCED SUPERCONDUCTIVITY

OF $CeCu_2Ge_2$

D. Jaccard

Université de Genève, DPMC, 24, quai E.-Amsermet
CH 1211 Genève 4, Switzerland

Systematic trends observed in the thermopower of heavy fermion compounds have led us to study $CeCu_2Ge_2$ under high pressure. As expected, the transport properties of this compound above 70 kbar were found to be quite similar to the normal pressure ones of $CeCu_2Si_2$. Around this pressure, magnetic ordering vanishes and superconductivity appears. At 101 kbar, the transition temperature $T_c \approx 0.64K$ and the critical field $B_{c2}(0) \approx 2T(1)$.

REFERENCE

(1) D. Jaccard, K. Behnia and J. Sierro, Phys. Lett A163, 475 (1992)

WEAK MAGNETISM AND MAGNETIC PHASE TRANSITIONS

IN MIXED COMPOUND-U(Ru$_{1-x}$Rh$_x$)$_2$Si$_2$

Y. Miyako

Department of Earth and Space Science
Faculty of Science, Osaka University
Toyonaka 560, Osaka, Japan

Abstract

The magnetic properties of URu$_2$Si$_2$ and mixed compound U(Ru$_{1-x}$Rh$_x$)$_2$Si$_2$ are reviewed.   Weak antiferromagnetic order was found by neutron diffraction in URu$_2$Si$_2$ and random order was suggested by μSR while no magnetic order was detected in NMR.

U(Ru$_{1-x}$Rh$_x$)$_2$Si$_2$ shows various magnetic properties depending on the local environment of uranium from heavy fermion properties to devil's staircase type magnetic phase transitions.

Introduction

RM$_2$Si$_2$(R = U and Ce, M =   3d, 4d and 5d transition metals) ternary uranium and cerium compounds have a variety of magnetic properties.URu$_2$Si$_2$[1] and CeRu$_2$Si$_2$[2] are known as heavy fermion compound and UNi$_2$Si$_2$[3] and CeRh$_2$Si$_2$[4] show magnetic phase transition with ordered moment of ～ 3 μ$_B$. Uranium compound[5][6] as well as cerium[7][8] changes the magnetic properties with kinds of atoms of environment.

URu$_2$Si$_2$ is a canonical heavy fermion compound which exhibits superconductivity[1] and URh$_2$Si$_2$ is a simple collinear antiferromagnet polarized along the c-axis[5].   By mixing two compounds , U(Ru$_{1-x}$Rh$_x$)$_2$Si$_2$ displays different kinds of magnetic properties.   U(Ru$_{1-x}$Rh$_x$)$_2$Si$_2$ is a heavy fermion in

$0 < x < 0.08$ and a crossover occurs around $x = 0.1$ from delocalized to localized f-electron state.

For $0.2 < x < 0.35$, the mixed compound shows successive magnetic phase transitions.

Valence state of Uranium

Heavy fermion compound, URu_2Si_2, has a large anisotropy which is considered to be originated in the crystalline field, although the valence state of uranium is not so well defined as cerium. In the case of $CeRu_2Si_2$, the crystalline field splitting between the ground and excited doublets was estimated to be 220 K from the analysis of a Schottky anomaly observed in the specific heat measurement[9]. Uranium compounds seem to be more complex than cerium compounds due to the extent of f-orbits of uranium atom. UPt_2Si_2[10] and UPd_3[11] are the compounds where the crystalline field splittings are observed by neutron inelastic scattering. The magnetic properties of UPt_2Si_2 are explained by a simple crystalline field model assuming U^{4+}, which are the magnetic anisotropy of the susceptibility[12], the high field magnetization[13] and the specific heat[13].

The crystalline field model was applied to URu_2Si_2[12], but the low temperature properties below 100K could not be expressed even qualitatively. The low temperature properties of URu_2Si_2 are considered to be a coherent Kondo state developed as a result of the competitions among Kondo couplings, RKKY interactions and the crystalline field effect.

In the specific heat measurements, the observation of a Schottky type anomaly in 10 - 50 K is not a few in cerium and uranium heavy fermion compounds. UPt_2Si_2 exhibits the antiferromagnetic phase transition at 35 K with the ordered moment of $1.67~\mu_B$. The magnetic specific heat anomaly is analyzed by the crystalline field model[14].

The Schottky anomaly in URu_2Si_2 was observed at 32 K and was fitted by the doublet - doublet crystalline field splitting[15]. The doublet ground state contradicts the prediction of neutron scattering experiment[16].

The Schottky anomaly in URu_2Si_2 is also interpreted by a singlet-doublet(85K) crystalline field splitting model as shown in Fig.1[17] and the entropy change is given in Fig.2 as a function of temperature.

The best fitting (solid line in Fig.1) to the experimental data is obtained assuming a singlet ground and doublet excited level scheme with a splitting of 85 K[17]. Dotted and dush-dot curves are calculations that are a singlet-singlet(42K)-singlet-(170K)-doublet(550K) model by Nieuwenhuys[12] and a doublet-doublet(75K) model by Renker et al.[15], respectively.

The theoretical prediction of Kondo model[18] is smaller than the observed Schottly anomaly in URu_2Si_2. Similar large Schottky anomaly is observed even in UAl_2 with cubic symmetry[19].

These results suggest that the interpretation of Schottky anomaly in URu_2Si_2 has ambiguity because of complex f-electronic state. In the case of $CeRu_2Si_2$, a Schottky anomaly around 10K was explained as Kondo anomaly of $S = 1/2$[18] with the Kondo temperature of 24.4 K.

In a dilute compound $La_{0.95}U_{0.05}Ru_2Si_2$, the susceptibilty in Fig.3 increases down to 4.2 K, which suggests the contribution of a Kramers doublet ground state. The electronic state of uranium varies from compound to compound and seems to be sensitive to local environment of uranium.

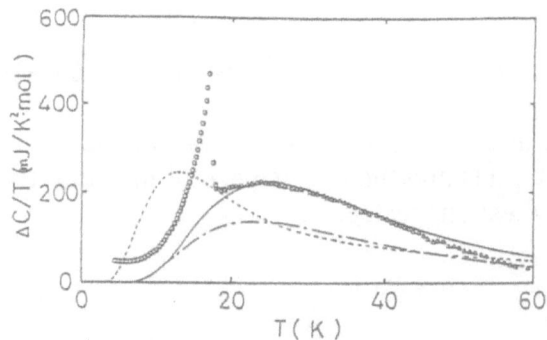

Figure 1. The specific heat of f-electron part in URu_2Si_2.
Lattice part is subtracted using the specific heat of $LaRu_2Si_2$ making a correction of mass difference. after [18].

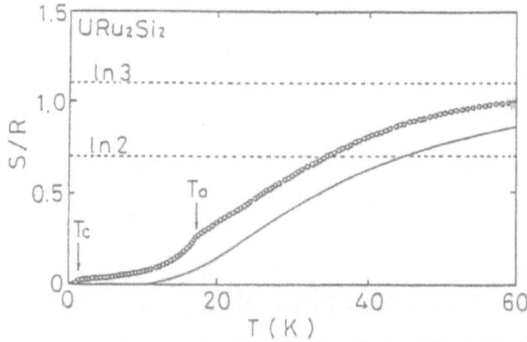

Figure 2. The entropy of f-electrons in URu_2Si_2 as a function of temperature. after [18].

Weak Magnetism

Heavy fermions have a variety of low temperature properties as the result of the delicate balance of the competition among the Kondo coupling, the RKKY interaction and the crystalline field effect. When the Kondo coupling is dominant, a coherent Kondo state, which is an antiferromagnetically correlated Fermi liquid, evolves at low temperatures.

By doping a few percent of impurities, the coherent Kondo state is easily collapsed and magnetic order appears.

URu_2Si_2 and UPt_3 [20] are canonical compounds which show a superconducting phase transition and the antiferromagnetic order persists in their superconducting state.

URu_2Si_2

URu_2Si_2 is shown to be a Kondo lattice system from the resistivity at high temperatures[21].

Antiferromagnetic and superconducting phase transitions in URu_2Si_2 were found by Schlabitz et al.[1] and the coexistence of the antiferromagnetic order and superconductivity was suggested.

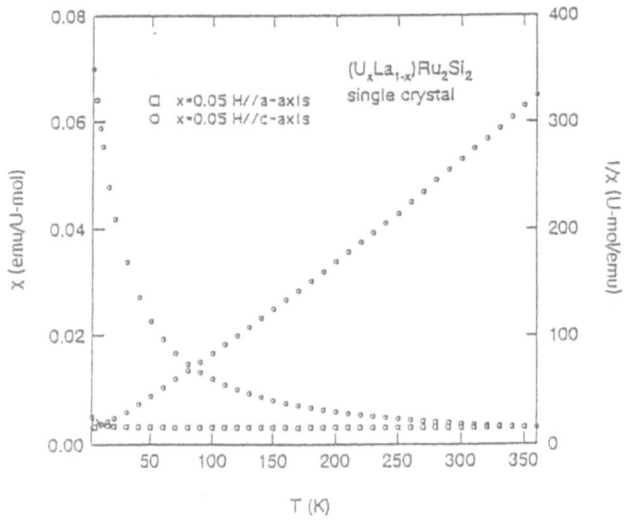

Figure 3. The susceptibility of $La_{0.95}U_{0.05}Ru_2Si_2$.

Neutron scattering experiment[16] confirmed the coexistence and showed the small ordered moment of 0.03 μ_B polarized along the tetragonal c-axis. However, the magnetic Bragg peak seems to appear from above about 30 K. µSR experiment also showed the evidence of magnetic ordering below about 30 K[22].

Contrary to the above early experiments, recent X-ray[23] and neutron scattering[24] experiments have demonstrated the abrupt increase of the magnetic Bragg reflection from about 17 K, where the specific heat anomaly is observed. In these experiments, it was stressed that good sample quality provides the rapid appearance of the Bragg reflections around 17 K.

However, there is still mystery that the sharp specific heat anomaly at 17 K is insensitive to sample preparation. The entropy change due to the sharp specific heat anomaly is $\sim 0.4\,R\ln 2$. This large entropy change does not seem to be explained by the antiferromagnetic order with small ordered moment($0.03\,\mu_B$). Moreover, no magnetic order is found in NMR although the internal field on ^{29}Si site(4kOe) may be too small to detect the magnetic order. In NMR experiment on ^{29}Si in URu_2Si_2 by Kohori et al.[25], the line width was about 150 Oe, which was attributed mainly to the distribution of the Knight shift. In the powdered sample, the c-axis of crushed microcrystals is roughly aligned under the external magnetic field due to the strong anisotropy of the susceptibility. No broadening of the line width was observed at T_0, although the transferred hyperfine field from uranium moments does not cancel at the Si sites owing to the low symmetry.

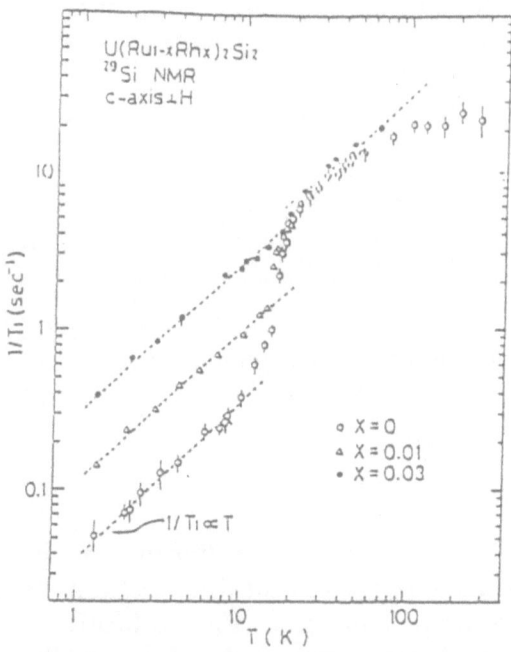

Figure 4. Temperature dependence of $1/T_1$ of ^{29}Si in $U(Ru_{1-x}Rh_x)_2Si_2$ for $0<x<0.03$ ($H\perp$ c-axis). after [25].

The nuclear spin lattice relaxation time, T_1, becomes longer below T_0 as shown in Fig.4. $1/T_1$ is proportional to temperature T below T_0, predicting the Fermi liquid state. The temperature dependence of $1/T_1$ predicts the creation of the Fermi surface, which stems from the Kondo coherent state.

Maple and his collaborators[26] proposed a partial gap opening of the Fermi surface at T_0 by the formation of spin density wave(SDW) or charge density wave(CDW). They suggested that SDW or CDW removes about 40 % of the Fermi surface by estimating the elecronic specific heat coefficient from extrapolating the slope of C/T vs T^2 curve from above and below T_0.

They also estimated an energy gap , Δ , to be $\sim$ 11 meV by fitting the specific heat data to the function of A exp($-\Delta/T$) where A and Δ are fitting parameters.

The temperature dependence of the resistivity of URu_2Si_2 seems to support the gap opening on the Fermi surface[22]. However, the shoulder of the specific heat below T_o cannot be expressed by an exponential function for a partial gap opening of the Fermi surface.

It is a very delicate problem whether a partial gap opening occurs by SDW transition at T_o in URu_2Si_2 or not.

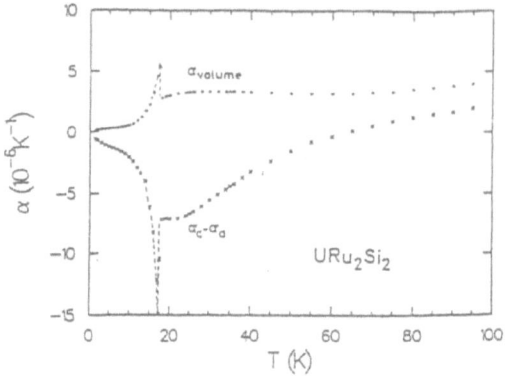

Figure 5. The volume thremal -expansion coefficient α_{volume} and the temperature dependence of the c/a ratio $\alpha_T = \alpha_c - \alpha_a$. after [30]

According to inelastic neutron scattering by Broholm et al.[16], the splitting between the ground and the first excited levels increases rapidly below T_o like an order parameter. Holland-Moritz and his colaborators[28] carried out neutron scattering experiment and showed the appearance of inelastic scattering below about 35 K and suggested the possibility of quadrupolar ordering.

de Visser et al. measured the thermal expansion coefficients along the three crystallographic axes of the tetragonal URu_2Si_2[29] and found an anomalous peak at T_o as shown in Fig.5.

The temperature dependence of the volume thermal expansion coefficient of URu_2Si_2 below T_o is different from the case of Cr metal[30].

Magnetization, M, is given as a function of an external magnetic field, H, in a paramagnetic state that $M = \chi_0 H + \chi_2 H^3$. Here, χ_0 and χ_2 are the (linear)susceptibility and the nonlinear susceptibility, respectively. Nonlinear susceptibilty is a sensitive probe for the study of the phase transition related to four spin correlations. Morin and Schmit [31] proposed the study by nonlinear susceptibility of quadrupolar ordering of localized f-electrons in rare earth compounds.

Fig.6 (a) and (b) are the temperature dependence of the nonlinear susceptibility of URu_2Si_2 and $U(Ru_{0.99}Rh_{0.01})_2Si_2$. χ_2 has a positive sign and a sharp peak at T_0 which corresponds to the peak of $-d(c/a)/dT$ where c/a is the ratio of the lattice constant of URu_2Si_2. We calculated the derivative of c/a with respect to temperature using the temperature dependence of c/a measured by de Visser[29].

The positive divergent behavior can not be explained by magnetic phase transitions and suggests the possibility of quadrupolar ordering[32][33][34], although the ordering temperature is too high.

At present, the origin of the phase transition accompanied by lattice instability is still a puzzle.

By replacing Ru atoms with a few percent Rh, the specific heat anomaly in Fig.7 becomes smaller with decreasing T_0 and vanishes above around 6% Rh concentration. The slope of C/T vs T^2 is independent of x above T_0 for $0 < x < 0.06$ of $U(Ru_{1-x}Rh_x)_2Si_2$.

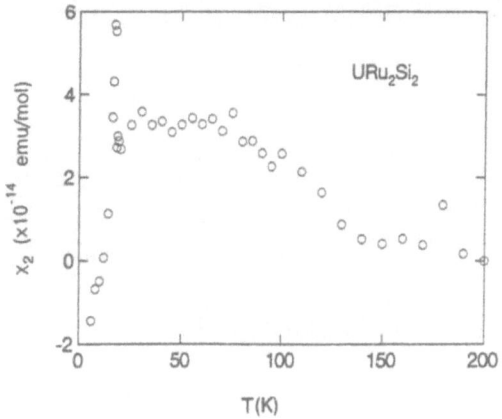

a) The nonlinear susceptibility
of URu_2Si_2.

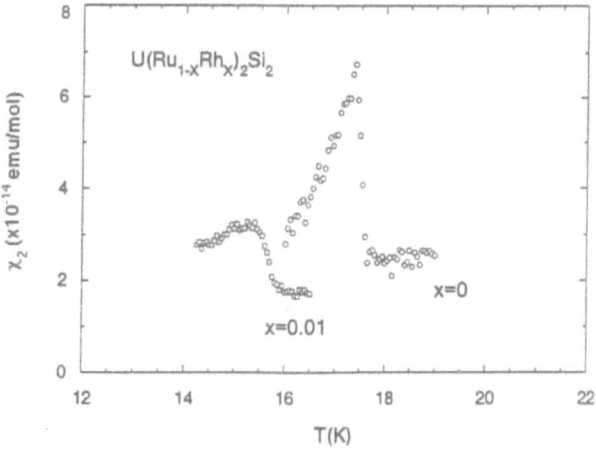

b) The nonlinear susceptibility
of URu_2Si_2 and of $U(Ru_{1-x}Rh_x)_2Si_2$.

Figure 6

An external magnetic field reduces the Kondo screening and a metamagnetic like transition occurs at $H_M=358$ kO$_e$ in $URu_2 Si_2$ [35]. Fig.8 shows the H_M versus T_0 curves as a function of x in $U(Ru_{1-x} Rh_x)_2Si_2$. H_M decreases with x having the minimum value at around x=0.08 and start to increase with x up to x=0.2. This is clearly shown in Fig.9.

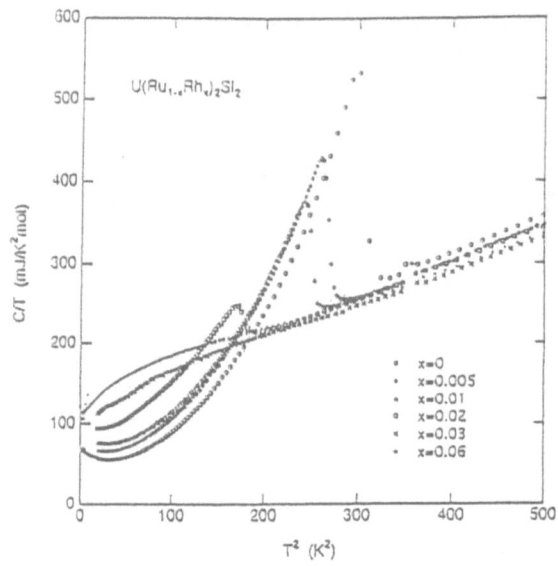

Figure 7. The specific heat of f-electrons in $U(Ru_{1-x}Rh_x)_2Si_2$.

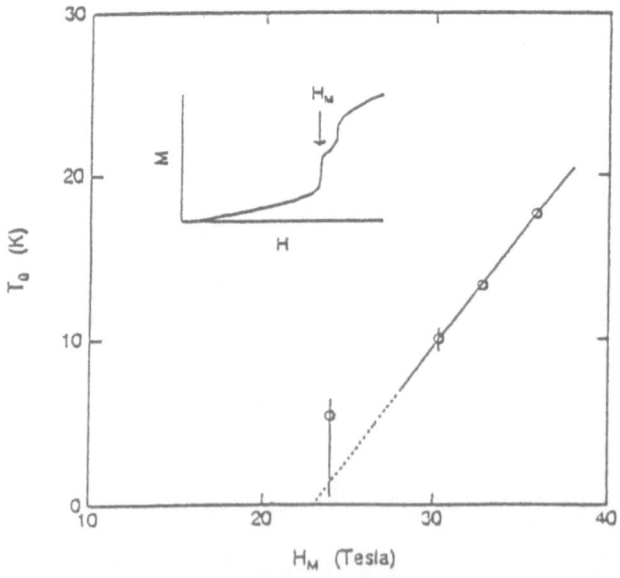

Figure 8. T_0 versus H_M curve in $U(Ru_{1-x}Rh_x)_2Si_2$.
T_0 and H_M are defined in the text.

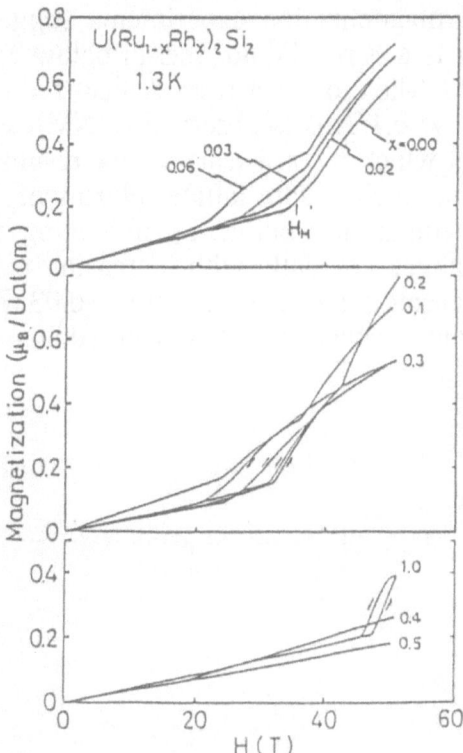

Figure 9. High field magnetization curve of $U(Ru_{1-x}Rh_x)_2Si_2$.

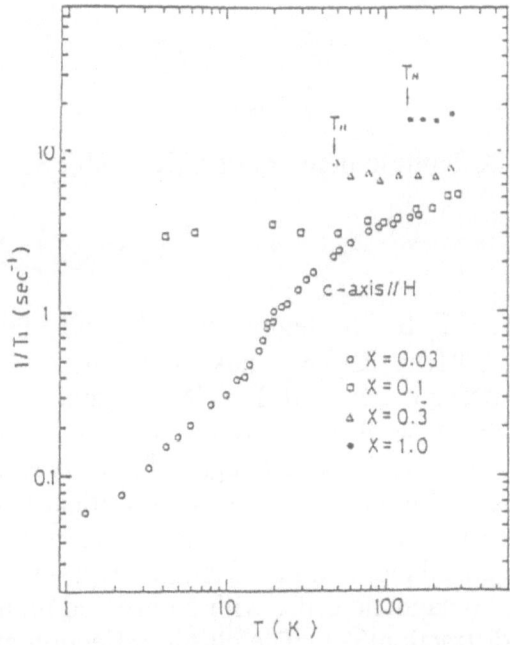

Figure 10. Temperature dependence of $1/T_1$ of ^{29}Si in $U(Ru_{1-x}Rh_x)_2Si_2$ for 0.03<x<1.0 (H⊥c-axis). after [26].

The specific heat magnetization experiments suggest that the coherent Kondo state, which is a Fermi liquid, sets in below T_o for $U(Ru_{1-x}Rhx)_2Si_2$ with x less than 0.08. The crossover from delocalized to localized f-electron state occurs around x=0.1.This is observed in NMR experiment on ^{29}Si in $U(Ru_{1-x}Rh_x)_2Si_2$[25], which is consistent with the results of neutron scattering experiment[36]. The nuclear spin lattice relaxation, $1/T$, of ^{29}Si varies in proportion to the temperature below T_o, indicating a Fermi liquid state. Above T_o, $1/T_1$ changes gradually with increasing temperature as seen in Fig.4 and has almost same value for x=0, 0.01 and 0.03 within the experimental accuracy, which attains a constant value around 100K.

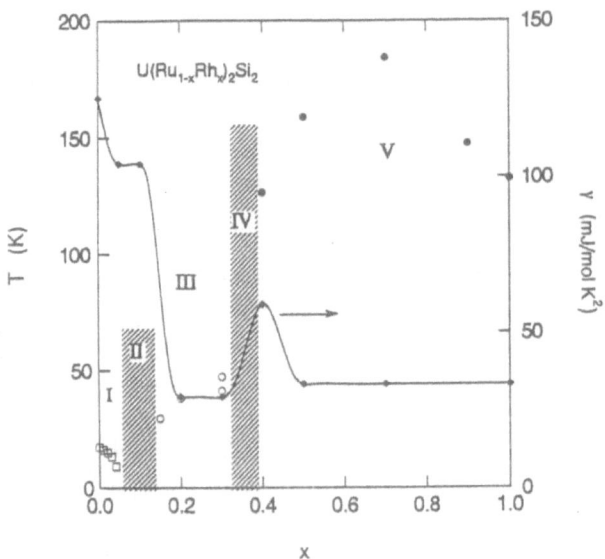

Figure 11. X-T phase diagram of $U(Ru_{1-x}Rh_x)_2Si_2$. after [33].

Fig.10 shows that $1/T_1$ has no temperature dependence for x=0.1,0.3 and 1.0. For x>0.1, $U(Ru_{1-x}Rh_x)_2Si_2$ has a localized moment and an antiferromagnetic order occurs. At T_N, NMR signal becomes too broad to observe due to the increase of the internal field. x=0.1 compound shows no magnetic transition down to 4.2K, although $1/T_1$ behaves like the temperature dependence of localized moment. x=0.1 is the critical concentration for Rh to separate the low temperature phase into the coherent Kondo and magnetic ordered state of localized moment in x-T phase diagram as shown in Fig.11.

In UPt_3, as antiferromagnetic order with an ordered moment of $0.02\mu_B$ was found by neutron diffraction[37]. The elastic reflection at q=($\pm1/201$) has a width broader than the spectrometer resolution and no specific heat anomaly is observed for the antiferromagnetic order.

However, UPt_3 with 5% Pd impurities orders at 5K exhibiting a sharp specific heat anomaly [38]. NMR signal was observed in the ordered state in $U(Pt_{0.95}Pd_{0.05})_3$ and the isotropic hyperfine field was estimated to be -85 kOe/μ_B.

If we assume the same hyperfine coupling constant in UPt_3 as in $U(Pt_{0.95}Pd_{0.05})_3$, the Knight shift of 900 Oe is expected in the antiferromagnetic ordered state $(0.02\mu_B)$ in UPt_3. No sign of the magnetic ordering was observed in NMR experiment in pure UPt_3[39], although μSR experiment suggested the static random order below about 5K[40].

The magnetic moments in UPt_3 may be fluctuating with the time scale of $10^{-12} < t < 10^{-6}$.

Successive Magnetic Phase Transitions

The mixed compound $U(Ru_{1-x}Rh_x)_2Si_2$ displays a variety of magnetic properties as a function of x. A magnetic phase diagram of Fig.11 has been obtained from susceptibility, magnetization, specific heat and neutron scattering experiments[32]. The effective magnetic moment estimated from high temperature susceptibility and the Weiss constant Ⓗ are given in Fig. 12.

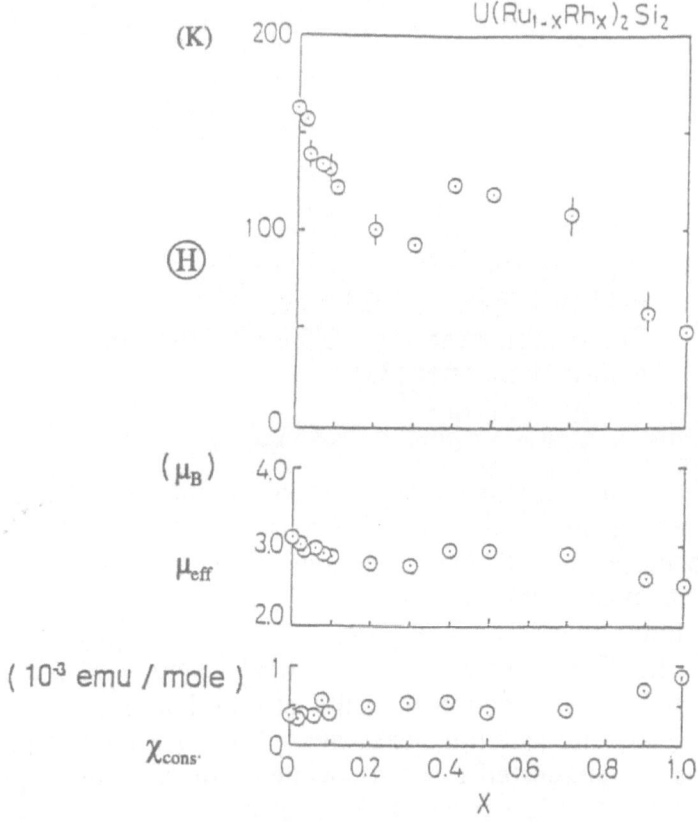

Figure 12. Weiss constant Ⓗ, effective magnetic moment μ_{eff} and the susceptibility independent of temperature χ_{cons}

x=0.1 area hatched in Fig.11 is a crossover region from delocalized heavy fermion to localized f-electron state. In region III, the mixed compound of x=0.25,0.3 and 0.35 exhibits successive magnetic phase transitions[39].

The specific heat and the susceptibility of $U(Ru_{0.7}Rh_{0.3})_2Si_2$ are shown in Figs.13 and 14 [32]. The three successive magnetic phase transitions are indicated by T_o^1, T_o^2 and T_o^3. The magnetic structure of these three phases are determined by neutron scattering experiment[32].

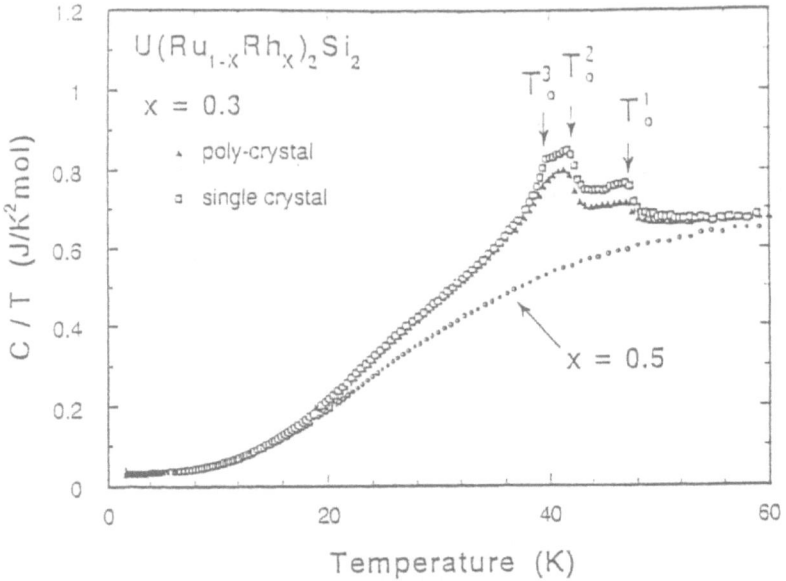

Figure 13. The specific heat of $U(Ru_{1-x}Rh_x)_2Si_2$ for x = 0.3 single and polysrystals and x = 0.5 polycrystal.
The specific heat of x = 0.5 is used to estimate lattice contribution. after [33].

Fig.15 is the temperature dependence of the magnetic Bragg reflections whose indices are indicated. The reflections consist of clear three peaks. In addition to these three Bragg reflections, three are some other magnetic scattering which are weak but significant.

Fig.16 shows the scattering spectrum in the reciprocal space region between (1/21/2 1/2) and (1/2 1/2 1/2 ± 1/6) at several temperatures between 39.8 and 43K. The reflection intensity of (1/2 1/2 1/2) peak decreases at around 43K and the satellite peaks at (1/2 1/2 1/2±1/6) appear. Their modulations are suggested to be associated with a polarization along the c-axis.

The peak position moves from near (1/2 1/2 1/2) to (1/2 1/2 1/2±1/6) with decreasing temperature, which suggests devil's staircase type phase transitions.

Below the temperature (43K) where the satellites (1/2 1/2 1/2±1/6) appear, (1/2 1/2 1/2) peak still remains with strong scattering intensity and there is a thermal hysteresis below 43K as seen in Fig.15. Similar behavior has been observed in the temperature dependence of the modulation wave vector along the c-axis of holmium[42].

This may be due to pinning by defects in bond random system, which is similar to lock in transition in holmium.

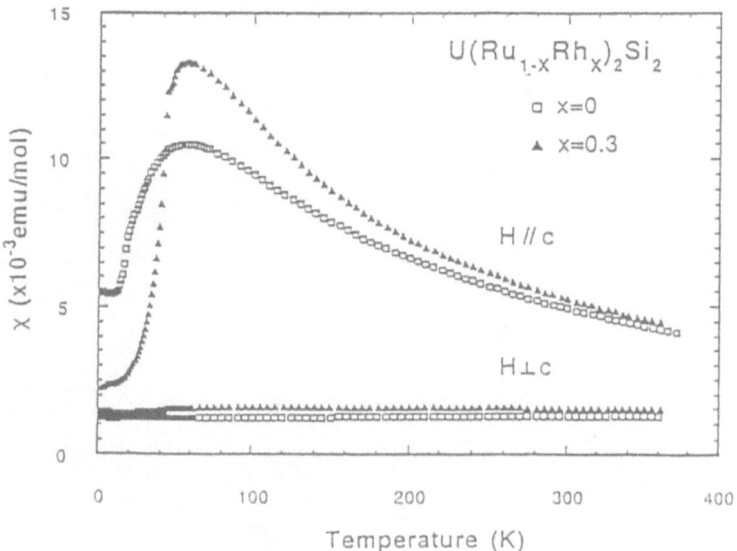

Figure 14. The susceptibility of URu_2Si_2 and $U(Ru_{0.7}Rh_{0.3})_2Si_2$

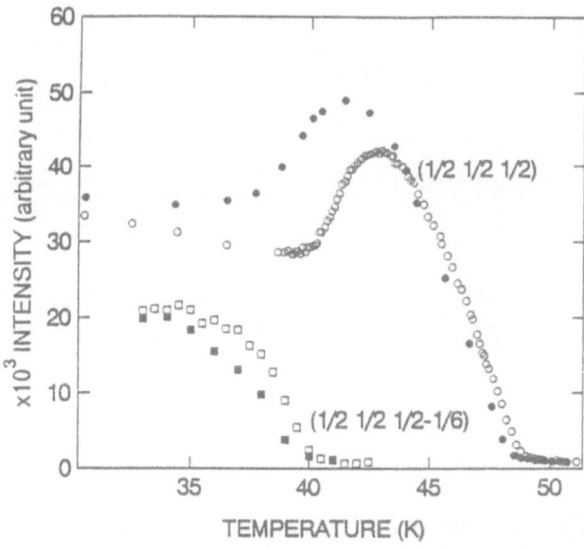

Figure 15. Thermal hysteresis of the intensities of the (1/2 1/2 1/2) and the (1/2 1/2 1/2 - 1/6) reflections. Directions of the thermal scanning are indicated by the arrows. after [42].

Devil's staircase type magnetic phase transitions were observed in erbium [43] and holmium [44] in addition to a classical CeSb [45]. To explain devil's staircase magnetic phase transitions, Bak proposed two theoretical models, which are Ising model with long range repulsive force [46] and an axial next nearest neighbor Ising model (ANNNI model)[47]. These simple models qualitatively explain the devil's staircase phenomenon, which is the transitions that the magnetic wave vector passes a discrete series of commensurate values as observed in holmium under an applied magnetic field [41].

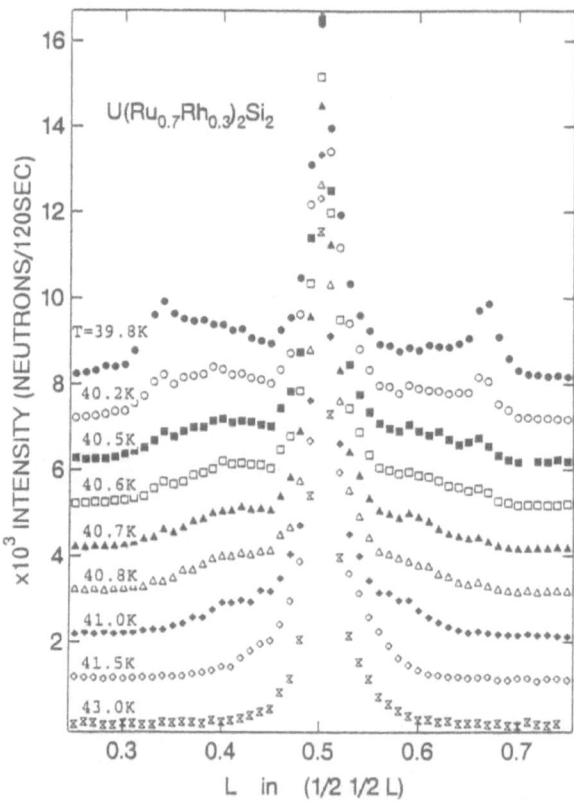

Figure 16. Diffraction spectrum between (1/2 1/2 1/2) and (1/2 1/2 1/2 + 1/6) reciprocal space region at several temperatures. after [42].

Very recently, the magnetic structure of UNi_2Si_2 was determined by neutron diffraction [3] and the wave vector of incommensurate phase changes smoothly with decreasing temperature, leading to a commensurate phase. $U(Ru_{1-x}Rh_x)_2Si_2$ and UNi_2Si_2 doped with impurities are new candidates to show devil's staircase transitions.

References

[1] W. Schlabitz, J. Baumann, B. Pollit, U. Rauschwalbe, H. M. Mayer, U. Ahlheim and C. D. Bredl, Z. Phys. B 62 (1986) 171

[2] A. Amato, D. Jaccard, J. Sierro, P. Haen, P. Lejay and J. Flouquet, J. Low Temp. Phys., 77 (1989) 195

[3] H. Lin, L. Rebelsky, M. F. Collins, J. D. Garrett and W. J. L. Buyers, Phys. Rev. B 43 (1991) 13232

[4] B. Lloret, B. Chevallier, B. Buffat, J. Etourneau, S. Quewzel, A. Lamharrar, J. Rossat-Mignot, R. Calemczuk and E. Bonjour, J. Mag. Mag. Mater. 63&64 (1987) 85

[5] H. Ptasiewicz-Bak, J. Leciejewicz and A. Zygmunt, J. Phys., F 11 (1981) 1225

[6] L. Chetmicki, J.Leciejewicz and A. Zygmunt, J. Phys. Chem. Solids 46 (1985) 529

[7] B. H. Grier, J. M. Lawrence, V. Murgai and R. D. Parks, Phys. Rev.,B 29(1984)2664

[8] A. Severing, E. Holland-Moritz and B. Frick, Phys. Rev.,B 39(1989)4164

[9] M. J. Besnus, J. P. Kappler, P. Lehmann and A. Meyer, Solid State Commun. 55 (1985) 779

[10] R. A. Steeman, E. Frikkee, C. van Dijk, G. J. Nieuwenhuys, and A. A. Menovsky, J. Mag. Mag. Mater., 76 & 77 (1988) 435

[11] W. J. L. Buyers, A. F. Murray, T. M. Holden, E. C. Svensson, P. de V. Duplessis, G. H. Lander and O. Vogt, Physica B 102(1980)291

[12] G. J. Nieuwenhuys, Phys. Rev. B 35 (1987) 5260

[13] H. Amitsuka, T. Sakakibara, K.Sugiyama, T. Ikeda, Y. Miyako , M. Date and A. Yamagishi, Physica B 177 (1992) 173

[14] R. A. Steeman, E. Frikkee, S. A. M. Mentink, A. A. Menovsky, G. J. Nieuwenhuys and J. A. Mydosh, J. Phys., C 2(1990) 4059

[15] B. Renker, F. Gompf, E. Gering, P. Frings, H. Rietschel, R. Felten, F. Steglich and G. Weber, Physica B 148 (1987) 41

[16] C. Brohlm, J. K. Kjems, W. J. L. Buyers, P. Mathews, T. M. Palstra, A. A. Menovsky and J. A. Mydosh, Phys. Rev. Lett., 58 (1987)1467

[17] H. Amitsuka, a private communication.

[18] V. Rajan and J. H. Lowenstein, Phys. Rev. Lett., 49(1982)497

[19] T. Kuwai and Y. Miyako, to be published.

[20] G. R. Stewart, Z. Fisk, J. O. Willis and J. L. Smith, Phys. Rev. Lett., 52 (1984) 679

[21] J. Schoenes, C. Schonenberger, J. J. M. Franse and A. A. Menovsky, Phys. Rev., B 35(1987)5375

[22] D. E. MacLaughlin, D. W. Cooke, R. H. Heffner, R. H. Hutson, M. W. McElfresh, M. E. Schllaci, H. D. Rempp, J. L. Smith, J. O. Willis, E.Zirngieble, C. Boekema, R. L. Licht, J. Oostens, Phys. Rev. B 37(1988)3153

[23] E. D. Isaacs, D. B. Mcwhan, R. N. Keinman, D. J. Bishop, G. E. Ice, P. Zschack, B. D. Gaulin, T. E. Mason, J. D. Garrett and W. J. L. Buyers, Phys. Rev. Lett., 65(1990)3185

[24] T. E. Mason, B. D. Gaulin, J. D. Garrett, Z. Tun, W. J. L. Buyers and
    E. D. Isaacs,   Phys. Rev. Lett.,   65(1990)3189

[25] Y. Kohori, Y. Noguchi, T. Kohara, K. Asayama, H. Amitsuka and
    Y. Miyako,   Solid State Commun. 82 (1992) 479

[26] M. B. Maple, J. W. Chen, Y. Dalichaouch, T. Kohara, C. Rossel and
    M. S. Torikachvili,   Phys. Rev. Lett.,   56 (1986) 185

[27] T. T. M. Palstra, A. A. Menovsky and J. A. Mydosh,
    Phys. Rev., B 33 (1986) 6527

[28] E. Holland-Moritz, W. Schlabitz, M. Loewenhaupt and U. Walter,
    Phys. Rev., B   (1989)551

[29] A. de Visser, F. E. Kayzel, A. A. Menovsky, J. J. M. Franse, J. van den
    Berg and G. J. Nieuwenhuys,   Phys. Rev. , B 34 (1986) 8168

[30] E. W. Lee and M. A. Asgar,   Phys. Rev. Lett.,   22 (1969) 1436

[31] P. Morin and D. Schmitt,   Phys. Rev., B 23 (1981) 5936

[32] Y. Miyako, S. Kawarazaki, H. Amitsuka, C. C. Paulsen and
    K. Hasselbach,   J. Appl. Phys., 70 (1991) 5791

[33] A. P. Ramirez, P. Coleman, P. Chandra, E. Bruck, A. A. Menovsky,
    Z. Fisk and E. Bucher,   Phys. Rev. Lett., 68(1992)2680

[34] Y. Miyako, H. Amitsuka, S. Kunii and T. Kasuya, Physica (1993)
    to be published.

[35] F. R. deBoer, J. J. M. Pastra, U. Rauchschwalbe, W. Schlabitz,
    F.Steglich and A. de Visser,   Physica B 138(1986)1

[36] P. Burlet, F. Bourdarot, S. Quezel, J. Rossat-Mignot, P. Lejey,
    B. Chevalier and H. Hickey,   J. Mag. Mag. Mater., 108 (1992) 202

[37] G. Aeppli, E. Bucher, A. I. Goldman, S. Shirane and C. Broholm,
    J. Mag. Mag. Mater.,   76&77 (1988) 385

[38] J. J. M. Franse, A. de Visser, A. Menovsky and P. H. Frings,
    J. Mag. Mag. Mater.,   52(1985)61

[39] Y. Kohori, M. Kyogaku, T. Kohara, K. Asayama, H. Amitsuka and
    Y. Miyako,   J. Mag. Mag. Mater.,   90&91 (1990) 510

[40] D. W. Cooke, R. H. Heffner, R. L. Hutson, M. E. Schillaci and J. L.
    Smith,
    Hyperfine Interaction   31 (1986) 425

[41] S. Kawarazaki, T. Taniguchi, H. Iwabuchi, Y. Miyako, H. Amitsuka and
    T. Sakakibara,   Phys. Lett.   A   160 (1991) 103

[42] D. Gibbs, D. E. Moncton, K. L. D'Amico, J. Bohr and B. H. Grier,
    Phys. Rev. Lett.,   55(1985)234

[43] M. Habenschuss, C. Stassis, S. K. Sinha, H. W. Deckman and
    F. H. Spedding,   Phys. Rev., B   10 (1974) 1020

[44] R. A. Cowley, D. A. Jehan, D. F. McMarrow and G. J. McIntyre,
    Phys. Rev. Lett.,   66 (1991) 1521

[45] J. Rossat-Minod, P. Burlet, J. Villain, H. Bartholin, Wang Tcheng-Si,
    D. Florence and O. Vogt,   Phys. Rev., B   16 (1977) 440

[46] P. Bak and R. Bruinsma,   Phys. Rev. Lett.,   49 (1982) 249

[47] P. Bak and J. von Boehm,   Phys. Rev., B 21 (1980) 5297

TRANSPORT AND THERMAL PROPERTIES OF SOME SELECTED
HEAVY-FERMION MATERIALS:
PROBING THE ELECTRONIC INSTABILITY

Anne de Visser

Van der Waals - Zeeman Laboratorium
Universiteit van Amsterdam
Valckenierstraat 65
1018 XE Amsterdam
The Netherlands

1. INTRODUCTION

In the past decade it has been recognized that the delocalization of 4f or 5f electrons in cerium or uranium intermetallics occasionally gives rise to the low-temperature formation of a strongly correlated electron band close to the Fermi level. The unprecedented strong renormalization is reflected in a Fermi-liquid quasiparticle mass of the order of 100 times the free electron mass. Concurrently, the Fermi-liquid temperature is strongly renormalized: $T_F \sim$ 100 K. The enormous effective mass is built up by a variety of competing electron-electron interactions, of which the on-site Kondo effect and the inter-site Ruderman-Kittel-Kasuya-Yosida (RKKY) interactions are thought to play the leading parts. The stabilization of the heavy-fermion state is attended by striking anomalies in the thermal and transport properties. As to the thermal properties, spin degrees of freedom turn up predominantly in the low-temperature entropy and enhance the electronic specific heat accordingly, while the strains are tightly coupled to the heavy-electron bands via anomalously large Grüneisen parameters giving rise to large coefficients of thermal expansion. As far as the transport properties are concerned, a rapid variation of the scattering processes takes place at low temperatures, as a consequence of the competing on-site and inter-site interactions. This is manifested in the electrical resistivity as a crossover from the Kondo-esque increase to the Fermi-liquid-like decrease in the coherent state, at lowering the temperature. At studying the physics of correlated electrons in heavy-fermion materials, the influence of clean external parameters, e.g. pressure and magnetic field, has received a wide attention. The pressure effects on the heavy-fermion bands are unusually large because of the strong hybridization. However, the influence of a magnetic field is relatively small, so that very strong magnetic fields are needed in order to investigate the heavy-fermion state.

Transport and Thermal Properties of f-Electron Systems
Edited by G. Oomi *et al.*, Plenum Press, New York, 1993

Electronic instabilities in the heavy-electron liquid at low temperatures are mostly of antiferromagnetic nature. However, for a few intriguing materials a superconducting ground state coexists with an antiferromagnetic one. The non-standard-BCS thermal and transport properties of some of these heavy-fermion superconductors provide strong evidence for an unconventional Cooper state and have led to speculations upon electron-electron mediated superconductivity.

The purpose of the present paper is to high-light a number of salient properties of heavy-fermion systems, as detected by low-temperature transport and thermal measurements. Special emphasis is given to the phase transitions that take place in the heavy-electron liquid, e.g. long-range antiferromagnetism, unconventional superconductivity, pseudo-metamagnetism and gap-formation. These phenomena are elucidated by experimental studies on exemplary systems, such as $U(Pt,Pd)_3$, $(Ce,La)Ru_2Si_2$, UPd_2Al_3 and CeNiSn. Furthermore, in a few special cases the behaviour under extreme conditions (high pressures and high-magnetic fields) is discussed.

2. THE SUPERCONDUCTING AND MAGNETIC INSTABILITY IN $U(Pt,Pd)_3$

Many intriguing aspects of heavy-fermion physics can be illustrated by experimental studies on the pseudobinary series $U(Pt,Pd)_3$ [1]. Thermal, magnetic and transport studies on the renowned 5f-electron compound UPt_3 [2] have revealed various anomalous properties at low temperatures, i.e. a large coefficient of the linear term in the electronic specific heat, $c(T)$ ($\gamma = 420$ mJ/molK^2), an enhanced low-field magnetic susceptibility, $\chi(T)$, with a pronounced maximum at $T_{max} = 18$ K, and a strongly temperature dependent electrical resistivity, $\rho(T)$, which exhibits a sharp decrease near 10 K, and passes to a Fermi-liquid AT^2 regime ($T < 1.5$ K), with a coefficient A enhanced two orders of magnitude over that of a normal metal. These anomalous low-temperature properties are reputed as hints for pronounced spin-fluctuation phenomena. Studies of the thermal and transport properties in an external magnetic field provide evidence that the electronic correlations are primarily of antiferromagnetic character.

Solid evidence for antiferromagnetic spin-fluctuation phenomena comes from inelastic neutron-scattering experiments [3]. The fluctuation spectrum is quite complex, as different energy scales are present. Experiments on polycrystalline samples yield a quasi-elastic contribution centered at 10 meV, which is related to the fluctuating local f-moment. The size of the fluctuating moment is of the same order as the effective moment deduced from the high-temperature Curie constant. Experiments on single-crystalline samples reveal a response centered at 6-8 meV, which reveals antiferromagnetic short-range order between nearest neighbour uranium atoms located in adjacent planes (UPt_3 has a hexagonal closed-packed type structure). The antiferromagnetic correlations disappear above T_{max}, whereas in-plane ferromagnetic correlations are present up to about 150 K. At yet a lower energy (0.5 meV), a second type of antiferromagnetic (basal-plane) correlations is found, and, surprisingly, also a very weak long-range magnetic moment appears with a Néel temperature $T_N = 5$ K. The size of the ordered moment amounts to 0.02 ± 0.01 μ_B/U-atom, and is directed along the b-axis. To what extent this extremely small ordered moment is intrinsic to UPt_3 is still an open question, as one cannot exclude that it emerges from crystalline imperfections.

The thermodynamic and spectroscopic studies on UPt_3 clearly demonstrate the proximity of an antiferromagnetic instability. Having noticed the corresponding spin-fluctuation phenomena, which constitute the anomalous low-temperature behaviour of

UPt$_3$, next two extraordinary findings will be discussed: the appearance of antiferromagnetic order with a substantial moment by alloying with Pd [1] and the superconducting instability at $T_c = 0.5$ K [4].

2.1 Long-range Antiferromagnetism in U(Pt,Pd)$_3$

By alloying UPt$_3$ with iso-electronic Pd, additional large anomalies appear in the specific heat (see fig.1) [1] and the electrical resistivity [5]. The λ-like anomaly in $c(T)$ and the Cr-type anomaly in $\rho(T)$ give evidence for an antiferromagnetic transition of the spin-density-wave type. Neutron-diffraction experiments [6] on a single-crystalline sample with composition U(Pt$_{0.95}$Pd$_{0.05}$)$_3$ ($T_N = 5.8$ K) confirmed the antiferromagnetic order. The ordered moment equals 0.6 ± 0.2 μ_B/ U-atom and points along the b-axis. The magnetic structure consists of a doubling of the nuclear cell along the b-axis, just as is the case for the weak antiferromagnetic order observed in pure UPt$_3$.

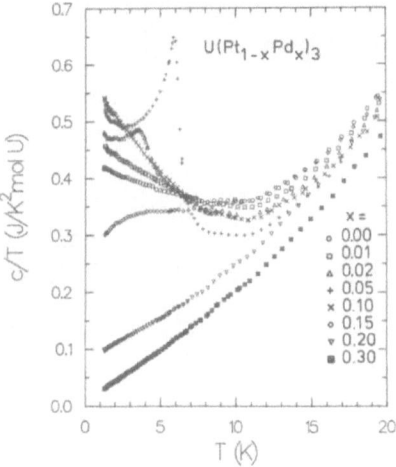

Figure 1. Specific heat of U(Pt$_x$Pd$_{1-x}$)$_3$ in a plot of c/T versus T, for $x \leq 0.30$ (after Ref.1). The peaks at $T_N = 3.6$ K for $x = 0.02$ and $T_N = 5.8$ K for $x = 0.05$ indicate long-range antiferromagnetic order.

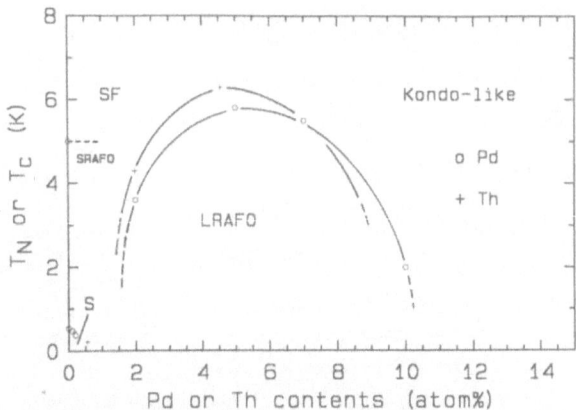

Figure 2. Superconducting and magnetic phase diagram for U(Pt,Pd)$_3$ (o) and (U,Th)Pt$_3$ (+) (after Ref.7). S = superconductivity; LRAFO = long-range antiferromagnetic order; SRAFO = (short-range) small-moment antiferromagnetic order; SF = spin fluctuations.

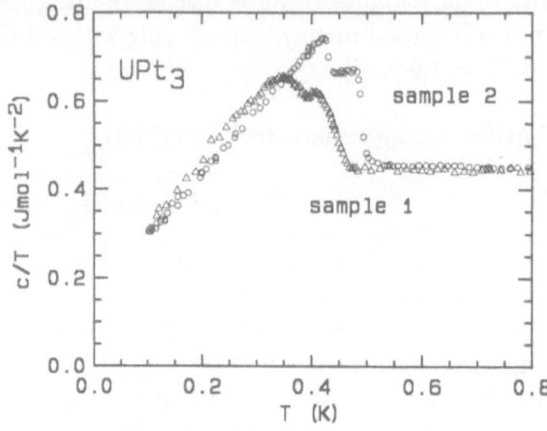

Figure 3. The double anomaly in the specific heat of single-crystalline UPt3 at the superconducting transition, in a plot of c/T versus T. Data are from Ref.16 (sample 1) and Ref.15 (sample 2).

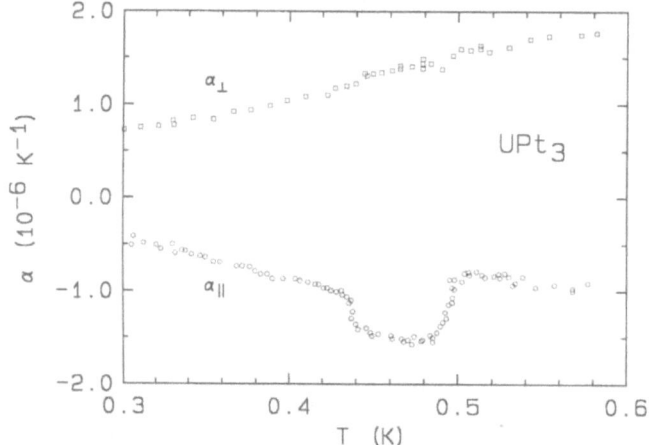

Figure 4. The coefficient of linear thermal expansion of single-crystalline UPt3 versus temperature, along (○) and perpendicular to (□) the hexagonal axis. After Ref.17.

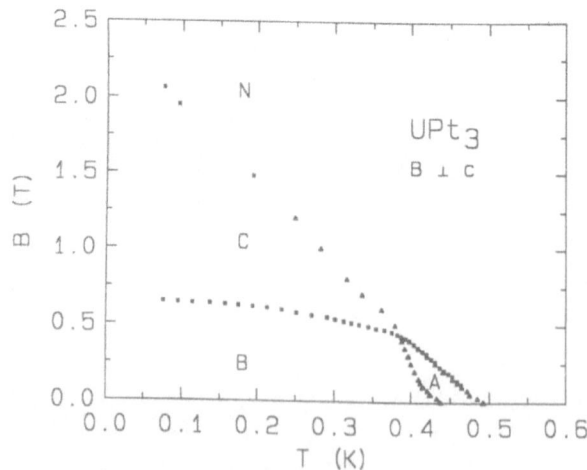

Figure 5. Superconducting phase diagram of UPt3 for $B \perp c$, determined from magnetostriction (■) and thermal expansion (△) data (after Ref.18). The superconducting phases are labeled A, B and C. N denotes the normal phase.

By tracing the Néel temperatures, deduced from the $c(T)$ and $\rho(T)$ data, as function of Pd concentration, the antiferromagnetic phase boundary for the $U(Pt_{1-x}Pd_x)_3$ series is determined (fig.2) [7]. Apparently, the long-range antiferromagnetic order, with fairly large moments, appears in a very limited concentration range, $0.02 \leq x \leq 0.07$, and seems to be unrelated to the reduced moment antiferromagnetism observed for pure UPt_3. It is remarkable that an almost similar magnetic phase diagram is obtained by substituting U by Th (see fig.2) [7]. This proofs that the tendency of the f-moments to localize is not dominated by the volume of these pseudo-binary compounds (the molar volume decreases by alloying with Pd, while it increases by alloying with Th). Other investigations have shown that long-range antiferromagnetic order also appears when UPt_3 is doped with 5% Au, while substitutions of 5 at% Ir, Rh, Y, Ce or Os do not lead to magnetic order [8,9]. Possibly a shape effect (a subtle change in the c/a ratio) is the relevant parameter here. The reduced entropy involved in the ordering (0.1Rln2 for 5 at% Pd) and the enhanced (with respect to pure UPt_3) low-temperature c/T-values (fig.1), demonstrate that only a part of the fluctuating moment orders. The heavy-electron behaviour is not removed by the ordering; it persists in the ordered state. The antiferromagnetic coupling is relatively strong, as a substantial magnetic field is needed to suppress the antiferromagnetic order. In the case of the $U(Pt_{0.95}Pd_{0.05})_3$ compound, specific-heat measurements in field revealed that a field of 13 T (a-axis) or 12 T (b-axis) directed in the easy plane for magnetization is needed to suppress the antiferromagnetic state completely [10].

2.2 The Superconducting Phases of UPt_3

The occurrence of superconductivity in the strongly correlated electron system UPt_3 is highly remarkable, the more because the alloying studies with Pd indisputably reveal the proximity of a magnetic instability. Therefore, it has been suggested that electron-electron interactions mediate superconductivity, instead of the conventional electron-phonon interaction. This issue is, however, not easily accessible by experiments, and constructive evidence for it is lacking, albeit that hydrostatic pressure experiments [11] have revealed a correlation between T_c and the spin fluctuation temperature (T_{sf}). As far as the superconducting order parameter is concerned, measurements of the thermal and transport properties yield strong evidence for an unconventional (L $\neq 0$) Cooper state and a gap-function with reduced symmetry. The electronic activation energy, determined by various techniques, e.g. sound attenuation [12] and London penetration depth [13], exhibits a power-law temperature dependence and a strongly anisotropic response, indicating a hybrid gap with line nodes at the equator and point nodes at the poles. However, as impurity scattering may contribute significantly, the gap function has not been determined unambiguously so far.

Recently, another type of evidence for unconventional superconductivity in UPt_3 has been put forward, namely the occurrence of additional anomalies in the thermo-dynamic quantities of the superconducting phase. High-precision measurements of the specific heat [14,15,16] (see fig.3), the thermal expansion (see fig.4) [17,18] and the sound velocity [19], on high-quality single-crystalline samples, clearly identified a second phase transition $\sim$ 50 mK below the superconducting one. The small temperature difference $(T_c/10)$ between the two transitions, strongly suggests that the superconducting phase transition is split. From an analysis of the data, using phenomenological Ginzburg-Landau theory [20], it has been inferred that the splitting might arise from the lifting of the degeneracy of a superconducting *vector* order parameter by a symmetry breaking field. Within this analysis, the unconventional nature of the order parameter is irrefutable. However, the model allows not for an unambiguous determination of the angular momentum of the Cooper pair. A study of the double superconducting transition in a magnetic field revealed a remarkable phase diagram with three distinctly different superconducting phases that meet at a tetracritical point (fig.5) [15,17-19]. From the theoretical side much effort is put in a phenomenological description of the superconducting phase diagram using Ginzburg-Landau theory [20]. In the most appealing scenario the symmetry breaking field is supposed to originate form the extremely small antiferromagnetic moment that is found below $T_N = 5$ K. However, the available Ginzburg-Landau models are inadequate at several points [21], in particular as to the topology of the phase diagram, which appeared to be rather isotropic with respect to the magnetic field orientation. In order to gain more insight in the puzzling aspects of the superconducting phase diagram, the current experiments aim at probing directly the coupling of the superconducting order parameter to the symmetry breaking field, via microscopic techniques, e.g. neutron scattering. Besides, much experimental work is directed towards a study of the superconducting phases under pressure. In particular, uniaxial pressure might be a very helpful external parameter, as a further reduction of the crystalline symmetry can be imposed.

3. COMPETITION BETWEEN RKKY AND KONDO EFFECT IN U(Pt,Pd)$_3$

Evidence for competing electronic interactions in heavy-fermion systems has for the greater part been gathered by alloying studies, i.e. by progressive replacements of one of the constituents. Detailed measurements of the transport, magnetic and thermal properties along such a series yield often distinctly different regimes. In this respect pseudobinary U(Pt$_{1-x}$Pd$_x$)$_3$ is an exemplary system [1]. As discussed in section 2, the low-temperature properties of UPt$_3$ are dominated by antiferromagnetic interactions. However, by substituting small amounts of Pt by Pd, a crossover to a regime dominated by Kondo fluctuations is observed. This change in regime manifests itself most clearly in the electrical resistivity (see fig.6) [5]. For pure UPt$_3$, the gradual drop of $\rho(T)$ with decreasing temperature is ascribed to the stabilization of antiferromagnetic correlations, while for a Pd contents of only 10 at% a Kondo-like upturn appears. The maximum in the magnetic susceptibility at $T_{max} = 18$ K, and the metamagnetic-like transition at a field of 21 T ($T \leq T_{max}$), characteristic of pure UPt$_3$, both shift towards lower energies on alloying, and are no longer observed for $x = 0.10$ [1], lending further support for a suppression of the antiferromagnetic correlations. Large variations are also observed in the thermal properties. The γ-value passes through a pronounced maximum near $x = 0.10$ (see fig.1), while $\partial\gamma/\partial B$ for $B \rightarrow 0$ changes sign between $x = 0.07$ and $x = 0.10$ (as $\partial\rho/\partial B$ does at low temperatures). In the same Pd concentration range the coefficient of volume expansion ($\alpha_v(T)$) changes sign [22] and the Grüneisen parameter, $\Gamma(T \rightarrow 0)$, shows a huge drop from ~ 60 for pure UPt$_3$ to -300 for $x = 0.15$ [22]. This salient change in regime clearly indicates the presence of competing electronic interactions and is more generally attributed to a competition between the RKKY (T_{RKKY}) and Kondo-effect (T_K). However, the variation of T_{RKKY} and T_K with Pd content is not easily determined in the crossover regime [23], as no clear-cut definitions are at hand. Furthermore, as it are the very same electrons that take part in both phenomena, the separation of both contributions is thwarted. A next complication arises from the part of the fluctuating moment that orders antiferromagnetically (see section 2.1), thus giving rise to another superimposed contribution. In order to unravel the composite low-temperature properties of the U(Pt,Pd)$_3$ system, inelastic-neutron scattering experiments will be extremely helpful.

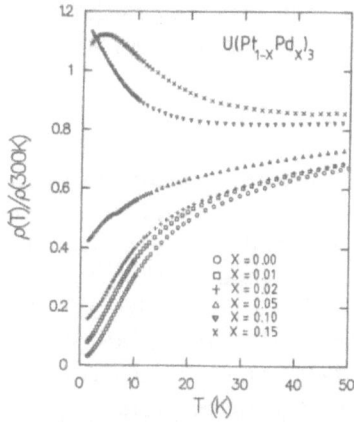

Figure 6. Electrical resistivity versus temperature for polycrystalline $U(Pt_{1-x}Pd_x)_3$ for $x \leq 0.15$. The resistivity values are normalized to 1 at $T = 300$ K (after Ref.5).

4. PSEUDO-METAMAGNETISM IN U(Pt,Pd)₃

Besides the superconducting and magnetic instabilities, the electron liquid of $U(Pt,Pd)_3$ exhibits a third electronic transition, albeit in very strong magnetic fields. This transition is often referred to as the pseudo-metamagnetic or metamagnetic-like transition and occurs at liquid helium temperatures ($T < T_{max}$) at a threshold field $B^* \approx 21$ T directed in the hexagonal plane [24]. The transition appears as a gradual increase in the magnetization [24] and as a sharp maximum in the magnetoresistance [25]. Note that the magnetic energy $\mu_B B^*$ is almost equal to the thermal energy $k_B T_{max}$. Taking into account the variety of thermal, magnetic, transport and alloying studies performed on UPt₃ [2], it is plausible that the 21 T anomaly is connected with a strong reduction of the antiferromagnetic intersite correlations. It is remarkable that the magnetization increase at B^* amounts to 0.6 μ_B/U-atom, which is equal to the size of the ordered moment in the 5 at% Pd compound, suggesting a common origin of the field-induced and alloying-induced moment. Specific-heat measurements in very strong magnetic fields ($B \leq 24.5$ T) [26] have shown that fields much larger than B^* are required in order to suppress the heavy-fermion state. At increasing the magnetic field, the γ-value initially increases and passes through a pronounced maximum at B^*, where the field-induced quasiparticle-mass enhancement amounts to 1.4 times the zero-field value. For fields $B > B^*$, the γ-value drops only slowly and is still larger than the zero-field value at the maximum applied field. This indicates that correlated electron phenomena, probably of the Kondo-type, persist in very strong magnetic fields. Note that the transition at B^* is not a true phase transition, but rather indicates a rapid crossover.

By substituting Pt by Pd in UPt₃, B^* decreases, which supports the claim that the intersite fluctuations become weaker (section 3). For a 10 at% Pd compound the metamagnetic-like transition is no longer observed. Hence the high-field experiments yield important information about the heavy-fermion state. Recently, high-magnetic fields were combined with high-pressures, in order to study several aspects of the pseudo-metamagnetism in the U(Pt,Pd)₃ system. The issue of one-parameter scaling was addressed by high-field high-pressure experiments on UPt₃ [27], while for

U($Pt_{0.95}Pd_{0.05}$)$_3$ the magnetic phase diagram in the $B-T$ plane was investigated [28]. In the following sections a brief account of these studies is presented.

4.1 One-Parameter Scaling in UPt$_3$

The heavy-electron bands are very sensitive to volume (and shape) effects, as follows from the extremely large thermal Grüneisen parameters, $\Gamma_T = -\partial \ln T^*/\partial \ln V$, that are roughly two orders of magnitude larger than for ordinary metals [29,30]. Here T^* ($\propto T_{max}$) is the characteristic temperature of the heavy-fermion resonance which is of the order of several Kelvin. In the case of UPt$_3$, Γ_T amounts to 60 [29]. The close connection between the thermal and magnetic energy scales ($k_B T_{max} \approx \mu_B B^*$) suggests that the relevant free energy term can be written as $F = F(T/T^*(V), B/B^*(V))$. This implies that the thermal and magnetic properties can be scaled by one single volume-dependent energy parameter and that the thermal and magnetic Grüneisen parameters $\Gamma_B = -\partial \ln B^*/\partial \ln V$ are equal, $\Gamma_T = \Gamma_B$. In order to verify the scaling anzats for B^*, a comparison of high-field magnetization and magnetostriction data was made [29]. However, Γ_B can also be measured directly in a high-field high-pressure experiment. This was achieved by measuring the longitudinal magnetoresistance of a single-crystalline UPt$_3$ sample ($B\|a\|I$) up to a field of 28 T under hydrostatic pressures up to 5 kbar [27]. The results are shown in fig.7. A large shift of the maximum in the magnetoresistance towards higher fields with increasing pressure is observed. The pressure variation of B^* is also plotted in fig.7, from which it follows that dB^*/dp is constant over the pressure range 0-5 kbar and amounts to 0.60 T/kbar. Consequently, the magnetic Grüneisen parameter Γ_B equals 59. The thermal Grüneisen parameter of UPt$_3$ has been determined in several ways [27]. From a combination of thermal-expansion and specific-heat measurements a value for Γ_T of 71 results. Pressure experiments yield values of 52 (from the suppression of the coefficient of the T^2 term in the resistivity, where $A \propto 1/(T^*)^2$), 58 (from the pressure induced shift of $T_{max} \propto T^*$ in the susceptibility) and 55 (from specific heat measurements under pressure, where $\gamma \propto 1/(T^*)$). Hence it is readily concluded that $\Gamma_T \simeq \Gamma_B$.

The presence of one single energy scale that governs the low-temperature thermal, magnetic and transport properties, is well established experimentally for UPt$_3$. However, the one-parameter scaling is not easily reconciled with the notion that competing magnetic interactions, namely the Kondo screening and the RKKY exchange, are at the basis of the quasiparticle formation, as this brings two energy scales into the problem. The scaling via B^* suggests that the characteristic energy scale is mainly determined by the intersite interactions. However, the complex nature of f-electron screening and f-electron exchange in heavy-fermion compounds likely implies that the on-site and inter-site interactions are intimately connected.

4.2 The Magnetic Phase Diagram of U($Pt_{0.95}Pd_{0.05}$)$_3$

In the following a discussion is presented of the magnetic phase diagram of the 5% Pd compound (fig.8). The antiferromagnetic phase boundary for a field directed in the easy plane for magnetization has been detected by specific-heat experiments in applied magnetic fields [10]. For $B\|a$ and $B\|b$ the suppression of the long-range magnetic order occurs at $B_c = 13$ T and $B_c = 12$ T for $T \to 0$, respectively. Magnetization measurements performed in the temperature range 1.3 K $< T < 20$ K indicated, however, the presence of a weakly temperature-dependent second phase transition at $B^* =$ 12-13 T [31]. B^* is interpreted as the threshold field for the suppression of the antiferromagnetic intersite correlations, i.e. the same phenomenon that takes place in pure UPt$_3$ at $B^* = 21$ T. The pseudo-metamagnetism in the 5 at% Pd compound, is related to the fluctuating f-moment (the part that does not order antiferromagnetically). The

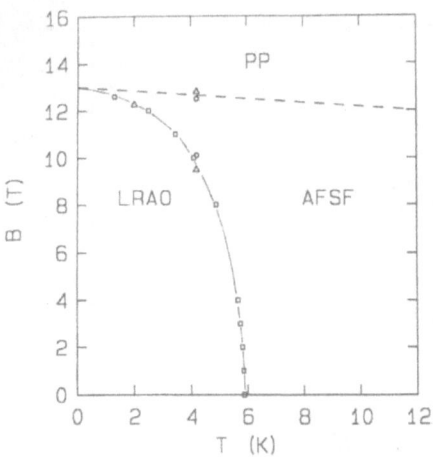

Figure 7. Left frame: High-field magnetoresistance of UPt_3 ($B\|I\|a$) at $T = 2.0$ K under hydrostatic pressures up to 5.1 kbar as indicated (after Ref.27). Right frame: Pressure variation of the metamagnetic threshold field B^* of UPt_3 at $T = 2.0$ K. B^* increases at a constant rate of 0.60 T/kbar (solid line) (after Ref.27).

fact that electron correlations still play an important role below T_N, is demonstrated by the large γ-value, which is enhanced considerably ($\sim 50\%$) with respect to pure UPt_3 (see fig.1).

Surprisingly, the extrapolation of both phase lines (fig.8) suggests that they meet at $T = 0$ K. Apparently, the magnetic energies for suppression of the antiferromagnetic order ($\mu_B B_c$) at 0 K and the antiferromagnetic intersite correlations ($\mu_B B^*$) are approximately equal. On the other hand, an intriguing question is whether the metamagnetic-like transition still exists at $T = 0$ K. While the anomaly at B^* becomes more pronounced in pure UPt_3 when the temperature is decreased, the sharp anomaly at B_c hampers the observation of B^* in the case of the 5% Pd compound. Interestingly, for the heavy-fermion antiferromagnets $Ce_{0.90}La_{0.10}Ru_2Si_2$ ($T_N = 2.7$ K) and $Ce_{0.87}La_{0.13}Ru_2Si_2$ ($T_N = 3.8$ K) a similar magnetic phase diagram has been reported [32], i.e. the boundary for the metamagnetic-like transition (for both compounds at $B^* \simeq 3.5$ T ($T \to 0$) for B along the tetragonal axis) extrapolates to the antiferromagnetic phase boundary for $T \to 0$.

In order to investigate whether a close connection between B^* and B_c exists or whether the coincidence of B^* and B_c for $T = 0$ K is accidental, high-field high-pressure magnetoresistance experiments have been performed [28]. The experimental results are shown in fig.9. The anomalous behaviour at low fields ($B < 5$ T) is related to the field orientation of magnetic domains. The most important result, that is inferred from fig.9, is that the antiferromagnetic phase boundary and the metamagnetic-like transition, that merge at zero pressure, are separated under pressure. The pressure variation of B_c and B^* is also plotted in fig.9. B^* increases at a constant rate of 0.81 T/kbar, while the suppression of B_c takes place non-monotonously, so that $B_c = 11.3$ T at 4.9 kbar (the Néel temperature is suppressed with pressure at a rate $dT_N/dp = -0.3$ K/kbar in zero field [7]).

From the data in fig.9 it is concluded that the coincidence of B_c and B^* for $T \to 0$ at zero pressure is accidental (note that the measurements have been performed at $T = 2.0$ K, but one would have to assume an exceptionally strong temperature variation of B^* and B_c under pressure in order to arrive at $B_c = B^*$ for $T \to 0$). Under pressure $\Delta\rho$ at B^* is nearly constant, suggesting that the metamagnetic-like transition also takes place at zero pressure. At present a method for the separation of the contibutions from the long-range antiferromagnetic order and the metamagnetic-like transition to

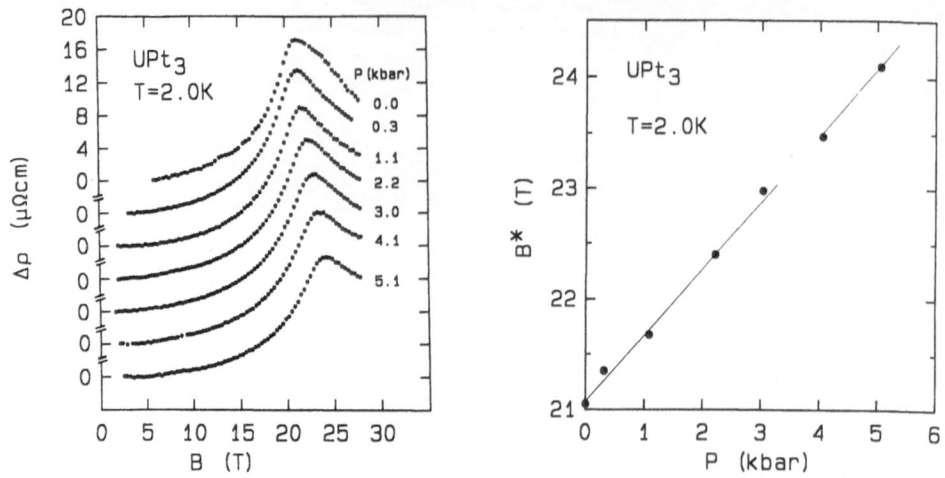

Figure 8. The magnetic phase diagram of U(Pt$_{0.95}$Pd$_{0.05}$)$_3$ for $B\|$a (after Ref.28). Data are taken from specific heat ($\square$), magnetization ($\triangle$) and magnetoresistance measurements (o). LRAO = long-range antiferromagnetic order, AFSF = antiferromagnetic spin fluctuations and PP = polarized paramagnetic phase. The dashed line represents the metamagnetic-like transition, i.e. quenching of the antiferromagnetic spin fluctuations. The solid line is the antiferromagnetic phase boundary.

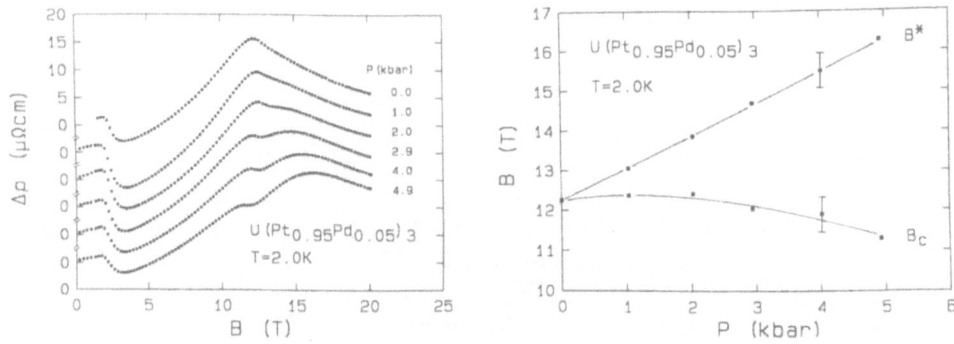

Figure 9. Left frame: High-field magnetoresistance of U(Pt$_{0.95}$Pd$_{0.05}$)$_3$ ($B\|I\|$a) at $T = 2.0$ K under hydrostatic pressures up to 4.9 kbar as indicated (after Ref.28). Right frame: Pressure variation of the metamagnetic threshold field (B^*) and the antiferromagnetic phase boundary (B_c) of U(Pt$_{0.95}$Pd$_{0.05}$)$_3$ for $B\|$a at $T = 2.0$ K.

$\Delta\rho$ is not at hand. Therefore, other types of experiments are necessary to elucidate the magnetoresistance data. In particular magnetization measurements under pressure might enable one to follow the pressure variation of the "steps" in the magnetization associated with B_c and B^*.

5. PSEUDO-METAMAGNETISM IN (Ce,La)Ru$_2$Si$_2$

Comprehensive transport, magnetic and thermal studies [32-34] have shown that heavy-fermion CeRu$_2$Si$_2$ is another exemplary system that may serve to investigate the approach of the antiferromagnetic instability. Although CeRu$_2$Si$_2$ remains a Pauli paramagnet down to the lowest temperatures investigated (20 mK) [33], long-range antiferromagnetic order is readily induced by alloying with La (the antiferromagnetic instability appears for 7 at% La) [32]. At low temperatures, $CeRu_2Si_2$ bears a close resemblance to UPt_3, in particular as far as the magnetic properties are concerned. A maximum is observed in the magnetic susceptibility at $T_{max} = 10$ K, and a metamagnetic-like transition at $B^* = 8$ T [33]. However, tetragonal $CeRu_2Si_2$ is an uniaxial system, i.e. these phenomena are only observed for a field along the tetragonal axis. Note that also in the case of $CeRu_2Si_2$ the characteristic low-temperature thermal energy ($k_B T_{max}$) is almost equal to the characteristic magnetic energy ($\mu_B B^*$) and that an one-parameter scaling law was verified by extensive experimental work [34]. In the case of $CeRu_2Si_2$, inelastic neutron-scattering experiments (in an external field) [35] have been very elucidating. Two contributions to the magnetic fluctuation spectrum have been observed: i) an on-site contribution, attributed to Kondo-type fluctuations, that is almost field independent and ii) antiferromagnetic intersite-interactions that are strongly suppressed at a threshold field B^*. Hence, the neutron-scattering data provide direct evidence that the pseudometamagnetism arises from the suppression of the intersite interactions. For an elaborate study of the low-temperature properties of the $(Ce,La)Ru_2Si_2$ system the reader is referred to Ref.34.

6. THE ANTIFERROMAGNETIC PHASE BOUNDARY OF UPd_2Al_3

Recently, the study of the interplay of antiferromagnetism and superconductivity in heavy-fermion systems received a considerable impetus by the discovery of the novel antiferromagnetic superconductors UPd_2Al_3 ($T_N = 14$ K, $T_c = 2$ K) [36] and UNi_2Al_3 ($T_N = 4.6$ K, $T_c = 1$ K) [37]. The formation of heavy quasiparticles in both compounds is evidenced by the moderately enhanced linear coefficient of the normal-state specific heat ($\gamma = 150$ mJ/molK^2 for UPd_2Al_3 [36] and $\gamma = 120$ mJ/molK^2 for UNi_2Al_3 [37]). The heavy quasiparticles take part in the superconducting condensate as follows from the large jumps in the specific heat at T_c. The analysis of the upper critical field (B_{c2}) [36,37] yields in both cases a quasiparticle mass (m^*) of ~ 70 times the free electron mass. Both uranium compounds crystallize in the hexagonal $PrNi_2Al_3$ structure. In the following we concentrate on UPd_2Al_3, as for this compound good-quality single-crystalline samples have been prepared.

The antiferromagnetic phase transition of UPd_2Al_3 at a Néel temperature $T_N = 14$ K is evidenced by a pronounced λ-type anomaly in the specific heat $c(T)$ [36,38], a kink in the magnetic susceptibility $\chi(T)$ [36,38] and a kink in the electrical resistivity $\rho(T)$ [36]. Neutron-diffraction experiments [39] on polycrystalline material revealed a magnetic structure consisting of ferromagnetic sheets that are coupled antiferromagnetically along the hexagonal axis (c-axis), i.e. a doubling of the nuclear unit cell with an ordering vector k = [0,0,1/2]. The ordered uranium moment amounts to 0.85 ± 0.03 μ_B/U-atom.

Magnetic-susceptibility data taken on single-crystalline samples revealed that the magnetic properties are strongly anisotropic [38,40]. The basal plane is the easy direction for magnetization. For $B\|a$ deviations from a Curie-Weiss behaviour ($\mu_{eff} = 3.6$ μ_B/U-atom) appear below 150 K, leading to a broad maximum centered at 40 K, while a kink is found at $T_N = 14$ K. For $B\|c$ $\chi(T)$ is only weakly temperature dependent. The magnetic anisotropy is also reflected in the high-field magnetization data $M(B)$ (see fig.10) [37]. At 4.2 K $M(B)$ is close to linear for $B\|c$, while for $B\|a$ the magnetization increases faster than linear and shows a sharp jump at $B_c = 18.0$ T. The size of the jump amounts to $\Delta M = 0.92$ μ_B/U-atom, which is almost equal to the size of the ordered moment (0.85 μ_B/U-atom). Therefore, it is likely that the anomaly at B_c reflects the antiferromagnetic phase boundary. In the high-field magnetoresistance measurements [41], the antiferromagnetic phase boundary appears as a sharp maximum in the longitudinal configuration ($B\|I\|$a) and as a sudden drop in the transverse configuration ($B\perp$b, I$\|$c) (see fig.10).

Comparing the $M(B)$ and the $\Delta\rho$ data in fig.10, it is observed that for both field directions in the basal plane (a$\perp$b) the transition occurs at almost the same field. This indicates that the basal-plane anisotropy is negligible or at least very small. The high-field data indicate that the magnetization process is rather complex, although it is of the single-step type. The large jump in the magnetization at B_c, with a size equal to the ordered moment, suggests that a spin- flip takes place. However, the absence of a

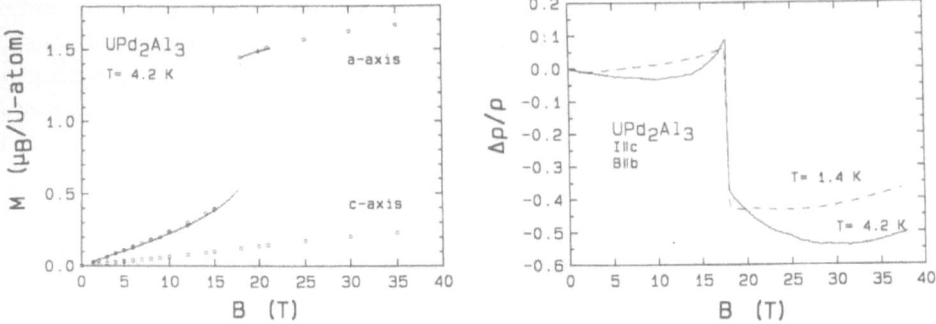

Figure 10. Left frame: Magnetization versus field of UPd$_2$Al$_3$ along the a- and c-axis at $T = 4.2$ K (after Ref.38). Right frame: Transverse magnetoresistance versus field of UPd$_2$Al$_3$ ($B\|$b, I$\|$c) at temperatures as indicated (after Ref.41).

marked basal-plane anisotropy implies that at low fields the sublattice magnetization is free to rotate towards a direction perpendicular to the field. Therefore, one expects (in a simple two-sublattice picture) that the alignment of the sublattice magnetizations with increasing field would occur gradually, which is contradicted by the experiment. However, as UPd$_2$Al$_3$ is a heavy-fermion compound, the Kondo-effect might complicate the magnetization process and transitions between different magnetic structures might only take place after the suppression of the Kondo-screening in field.

7. GAP-FORMATION IN CeNiSn

Recently, the ternary compound CeNiSn has attracted much attention because of the discovery [42] of the low-temperature formation of a pseudo-gap in the quasiparticle density of states. This novel ground state in correlated electron systems was first detected in the transport properties. Resistivity measurements revealed a logarithmic increase below ~ 7 K, that was taken as evidence for a simple activation law: $\rho(T)= \rho_0\exp(E_g/2k_BT)$ [42]. The $\rho(T)$-data, taken on a single-crystalline sample (CeNiSn has an orthorhombic structure) [43], indicated a strongly anisotropic structure of the gap, as $T_g \equiv E_g/k_B$ is estimated at 1.0, 4.8 and 8.0 K for the a, b and c axis, respectively. Magnetoresistance measurements show that the gap is suppressed for a field of ~ 14 T along the a-axis [44]. In the electronic specific heat [43] no distinct anomaly related to the opening of the gap is signalized. Below ~ 25 K, the c/T-value increases monotonously up to a maximum value of 0.18 J/molK^2 at ~ 7 K. This gradual increase of the c/T-value is attributed to the formation of a Kondo-lattice state. Below $T = 7$ K, c/T decreases again, due to the opening of the gap. Hence it appears that the gap opens in the quasiparticle density of states. Extrapolating the c/T-data measured for $T > 7$ K to $T = 0$ K, yields a γ-value of 0.2 J/molK^2.

In order to investigate the anomalies in the volume that are associated with this novel type of ground state, accurate measurements of the coefficients of thermal expansion (α_a, α_b and α_c) have been performed in the temperature interval 0.3 K < T < 12 K. Because of the strongly anisotropic gap, the measurements have been performed on a single-crystalline sample. The experimental results, taken in zero magnetic field and in an applied field of 8 T along the a-axis, are shown in fig.11. A large anisotropy is observed between α_a and α_b on the hand and α_c on the other hand. The coefficient of volume expansion, $\alpha_v = \alpha_a + \alpha_b + \alpha_c$, shows no specific structure when the gap opens. However, a broad anomalous contribution is visible in the high-temperature part of the data, just as is found in the specific heat. At low temperatures a second anomaly appears (below $\sim$ 1.5 K), the most clearly in α_c. In the volume expansion

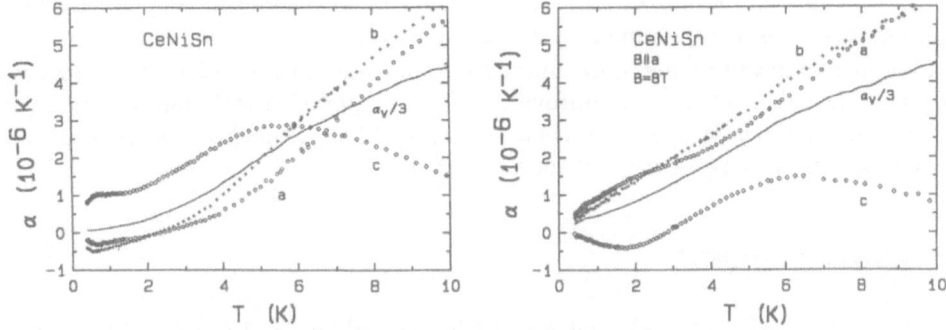

Figure 11. Left frame: Coeffient of linear thermal expansion versus temperature for CeNiSn along the a (○), b (+) and c (◇) axis. The solid line represents $\alpha_v/3$. Right frame: As in the left frame, but in a magnetic field of 8 T directed along the a-axis.

this low-temperature anomaly hardly shows up. In a magnetic field of 8 T along the a-axis (the easy direction for magnetization) a very different behaviour is observed. The anisotropy in the thermal expansion coeffients is reversed, and the low-temperature anomaly is pushed upwards considerably. The field effect on α_v is positive, but small. Recently, two other data sets of the thermal expansion of CeNiSn have been reported. Oomi *et al.* [45] have measured the *same* single-crystalline specimen for 4.2 K < T < 300 K. The pronounced anomalies detected by these authors at 7 K, when the gap opens, are not confirmed by the data in fig.11, and therefore should be discarded. Aliev *et al.* [46] investigated $\alpha_v(T)$ for a polycrystalline sample (0.3 K < T < 10 K). The pronounced negative contribution, observed below 1 K by these authors, is probably due to a second phase (or a non-isotropic distribution of the crystallites), as it is neither confirmed by the data in fig.11, nor by measurements on another single-crystalline sample [40].

A quantitative analyis of the electronic contribution to the thermal expansion of CeNiSn awaits the determination of the phonon part as may be estimated by measurements on LaNiSn. The additional anomaly (T < 1.5 K at B = 0 T) indicates the presence of a second low-temperature energy scale in CeNiSn and is likely related with the stabilization of antiferromagnetic intersite correlations. The presence of intersite correlations was recently put on firm footing by μSR experiments [47]. A full account of the magnetovolume experiments on CeNiSn will be published elsewhere [48].

8. CONCLUDING REMARKS

In the persevering attempts to unravel the physics of correlated electrons and the related electronic instability, careful experimental investigations of the transport and thermal properties play a major role. The profound studies on exemplary systems, like $U(Pt,Pd)_3$ and $(Ce,La)Ru_2Si_2$, reveal that a wealth of phenomena may occur in a *single* series of compounds: Kondo-behaviour, RKKY interactions, reduced moment magnetism, long-range antiferromagnetism, pseudo-metamagnetism and unconventional superconductivity. This indisputably deserves a large attention of solid state theorists. As far as the interpretation of the bulk of these phenomena is concerned, only a more or less satisfactory qualititative picture has been offered. Detailed theoretical descriptions are still lacking. Hopefully, the exemplary experimental studies reviewed here will trigger further theoretical work. As to the experimental side of the physics of heavy-fermion materials, research techniques in very strong magnetic fields *and* at very low temperatures ($T < 1$ K) should be developed in order to detail the suppression of the heavy-fermion state. Besides, the metallurgical aspects of heavy-fermion metals have become more and more important and, therefore, future research programs should also focus on improving sample preparation techniques. Finally, it is stressed that the rich and exotic physics of correlated electrons justifies the continuous effort to search for novel heavy-fermion materials.

ACKNOWLEDGEMENTS

The author gratefully acknowledges all his co-authors on the various topics, and is especially indebted to J.J.M. Franse and J. Flouquet for their continuous support. This work was made possible by a fellowship of the "Koninklijke Nederlandse Akademie van Wetenschappen" (The Royal Netherlands Academy of Arts and Sciences). Support from the Grant-in-Aid for the International Joint Research Program "A searching Study of New Heavy-Fermion Systems" sponsored by the Ministry of Education, Science and Culture of Japan is highly appreciated.

REFERENCES

[1] A. de Visser, J.C.P. Klaasse, M. van Sprang, J.J.M. Franse, A. Menovsky, T.T.M. Palstra and A.J. Dirkmaat, *Phys. Lett.* A 113 (1986) 489.

[2] A. de Visser, A. Menovsky and J.J.M. Franse, *Physica* 147B (1987) 81.

[3] G. Aeppli, E. Bucher, A.I. Goldman, G. Shirane, C. Broholm and J.K. Kjems, *J. Magn. Magn. Mater.* 76&77 (1988) 385.

[4] G.R. Stewart, Z. Fisk, J.O. Willis and J.L. Smith, *Phys. Rev. Lett.* 52 (1984) 679.

[5] R. Verhoef, A. de Visser, A. Menovsky, A.J. Riemersma and J.J.M. Franse, *Physica* 142B (1986) 11.

[6] P.H. Frings, B. Renker and C. Vettier, *J. Magn. Magn. Mater.* 63-64 (1987) 202.

[7] J.J.M. Franse, K. Kadowaki, A. Menovksy, M. van Sprang and A. de Visser, *J. Appl. Phys.* 61 (1987) 3380.

[8] B. Batlogg, D.J. Bishop, E. Bucher, B. Golding Jr., A.P. Ramirez, Z. Fisk and J.L. Smith, *J. Magn. Magn. Mater.* 63-64 (1987) 441.

[9] K. Kadowaki, M. van Sprang, A.A. Menovsky and J.J.M. Franse, *Jpn. J. Appl. Phys.* 26 (Suppl. 26-3) (1987) 1243.

[10] H.P. van der Meulen, J.J.M. Franse, A. de Visser, J.A.A.J. Perenboom and H. van Kempen, *Physica* B165&166 (1990) 441.

[11] J.O. Willis, J. D. Thompson, Z. Fisk, A. de Visser, J.J.M. Franse and A. Menovsky, *Phys. Rev.* B 31 (1985) 1654.

[12] B.S. Shivaram, Y.H. Jeong, T.F. Rosenbaum, D.G. Hinks and S. Schmitt- Rink, *Phys. Rev.* B35 (1987) 5372.

[13] C. Broholm, G. Aeppli, R.N. Kleiman, D.R. Harshman, D.J. Bishop, E. Bucher, D.L. Williams, E.J. Ansalo and R.H. Heffner, *Phys. Rev. Lett.* 65 (1990) 2062.

[14] R.A. Fisher, S. Kim, B.F. Woodfield, N.E. Phillips, L. Taillefer, K. Hasselbach. J. Flouquet, A.L. Giorgi and J.L. Smith, *Phys. Rev. Lett.* 62 (1989) 1411.

[15] K. Hasselbach, L. Taillefer and J. Flouquet, *Phys. Rev. Lett.* 63 (1989) 93.

[16] T. Vorenkamp, Z. Tarnawski, H.P. van der Meulen, K. Kadowaki, V.J.M. Meulenbroek, A.A. Menovsky and J.J.M. Franse, *Physica* B163 (1990) 564.

[17] K. Hasselbach, A. Lacerda, A. de Visser, K. Behnia, L. Taillefer and J.Flouquet, *J. Low Temp. Phys.* 81 (1990) 299.

[18] N.H. van Dijk, A. de Visser, J.J.M. Franse, S. Holtmeier, L. Taillefer and J. Flouquet, to be published.

[19] G. Bruls, D. Weber, B. Wolf, P. Thalmeier, B. Lüthi, A. de Visser and A. Menovsky, *Phys. Rev. Lett.* 65 (1990) 2294.

[20] R. Joynt, *Superc. Sci. Techn.* 1 (1988) 210; K. Machida, M. Ozaki and T. Ohmi, *J. Phys. Soc. Jpn.* 58 (1989) 4116; D.W. Hess, T. Tokuyasu and J.A. Sauls, *J. Phys. Cond. Matt.* 1 (1989) 8135.

[21] R. Joynt, *J. Magn. Magn. Mater.* 108 (1992) 31.

[22] A. de Visser, H.P. van der Meulen, B.J. Kors and J.J.M. Franse, *J. Magn. Magn. Mater.* 108 (1992) 61.

[23] J.J.M. Franse, H.P. van der Meulen and A. de Visser, *Physica* B165&166 (1990) 383.

[24] P.H. Frings, J.J.M. Franse, F.R. de Boer and A. Menovsky, *J. Magn. Magn. Mater.* 31-34 (1983) 240.

[25] A. de Visser, R. Gersdorf, J.J.M. Franse and A. Menovsky, *J. Magn. Magn. Mater.* 54-57 (1986) 383.

[26] H.P. van der Meulen, Z. Tarnawski, A. de Visser, J.J.M. Franse, J.A.A.J. Perenboom, D. Althof and H. van Kempen, *Phys. Rev.* B 41 (1990) 9352.

[27] K. Bakker, A. de Visser, A.A. Menovsky and J.J.M. Franse, *Phys. Rev.* B46 (1992) 544.

[28] K. Bakker, A. de Visser, A.A. Menovsky and J.J.M. Franse, to be published in: The Proceedings of the International Conference on Strongly Correlated Electron Systems (Sendai, 7-11 September, 1992).

[29] A. de Visser, J.J.M. Franse and J. Flouquet, *Physica* B 161 (1989) 324.

[30] B. Lüthi, *J. Magn. Magn. Mater.* 52 (1985) 70.

[31] A. de Visser, M. van Sprang, A.A. Menovsky and J.J.M. Franse, *J. de Physique Coll.* C8-49 (1988) 761.

[32] P. Haen, J. Voiron, F. Lapierre, J. Flouquet and P. Lejay, *Physica* B 163 (1990) 519.

[33] P. Haen, J. Flouquet, F. Lapiere, P. Lejay and G. Remenyi, *J. Low Temp. Phys.* 67 (1987) 391.

[34] C. Paulsen, A. Lacerda, L. Puech, P. Haen, P. Lejay, J.L. Tholence, J. Flouquet and A. de Visser, *J. Low Temp. Phys.* 81 (1990) 317.

[35] J. Rossat-Mignod, L.P. Regnault, J.L. Jacoud, C.˙Vettier, P. Lejay, J. Flouquet, E. Walker, D. Jaccard and A. Amato, *J. Magn. Magn. Mater.* 76&77 (1988) 376.

[36] C. Geibel, C. Shank, S. Thies, H. Kitazawa, C.D. Bredl, A. Böhm, M. Rau, A. Grauel, R. Caspary, R. Helfrich, U. Ahlheim, G. Weber and F. Steglich, *Z. Phys.* B85 (1991) 1.

[37] C. Geibel, S. Thies, D. Kaczorowski, A. Mehner, A. Grauel, B. Seidel, U. Ahlheim, R. Helfrich, K. Petersen, C.D. Bredl and F. Steglich, *Z. Phys.* B83 (1991) 305.

[38] A. de Visser, H. Nakotte, L.T. Tai, A.A. Menovsky, S.A.M. Mentink, G.J. Nieuwenhuys and J.A. Mydosh, *Physica* B179 (1992) 84.

[39] A. Krimmel, P. Fischer, B. Roessli, H. Maletta, C. Geibel, C. Schank, A. Grauel, A. Loidl and F. Steglich, *Z. Phys.* B86 (1992) 161.

[40] C. Geibel, U. Ahlheim, C.D. Bredl, J. Diehl, A. Grauel, R. Helfrich, H. Kitazawa, R. Köhler, R. Modler, M. Lang, C. Schank, S. Thies, F. Steglich, N. Sato and T. Komatsubara, *Physica* C185-189 (1991) 2651.

[41] A. de Visser, K. Bakker, L.T. Tai, A.A. Menovsky, S.A.M. Mentink, G.J. Nieuwenhuys and J.A. Mydosh, to be published in: The Proceedings of the International Conference on Strongly Correlated Electron Systems (Sendai, 7-11 September, 1992).

[42] T. Takabatake, Y. Nakazawa and M. Ishikawa, *Jpn. J. Appl. Phys.* 26 (Suppl. 26-3) (1987) 547.

[43] T. Takabatake, F. Teshima, H. Fujii, S. Nishigori, T. Suzuki, T. Fujita, Y. Yamaguchi, J. Sakurai and D. Jaccard, *Phys. Rev.* B41 (1990) 9607.

[44] T. Takabatake, M. Nagasawa, H. Fujii, M. Nohara, T. Suzuki, T. Fujita, G. Kido and T. Hiraoka, *J. Magn. Magn. Mater.* 108 (1992) 155.

[45] Y. Uwatoko, G. Oomi, T. Takabatake, F. Teshima and H. Fujii, to be published.

[46] F.G. Aliev, R. Villar, S. Vieira, M.A. Lopez de la Torre, R.V. Scolozdra and M.B. Maple, to be published.

[47] A. Kratzer, G.M. Kalvius, T. Takabatake, G. Nakamoto, H. Fuji and S.R. Kreitzman, to be published.

[48] A. de Visser, K. Bakker and T. Takabatake, to be published.

THERMAL PROPERTIES OF HEAVY FERMION SUPERCONDUCTORS

J.-P. Brison and J. Flouquet
CRTBT and DRFMC-Grenoble

K. Behnia and D. Jaccard
DRMC-Geneve

Recent thermal conductivity experiments on heavy fermion superconductors with special focus on magnetic field effects are reviewed. Comparisons are made with the theory. Our attempt to realize reliable specific heat experiments below 50 mK are discussed. Finally, we discuss the need for experimental progresses in materials and physical measurements, notably thermometry.

(DCNQI)$_2$Cu: A LUTTINGER-PEIERLS SYSTEM

Hidetoshi Fukuyama

Department of Physics, Faculty of Science, University of Tokyo
7-3-1 Hongo, Bunkyo-ku, Tokyo 113, Japan

INTRODUCTION

Among various one-dimensional organic conductors, $(R_1, R_2$-DCNQI)$_2$Cu is unique by its intriquing structural and transport properties associated with drastic changes with magnetic properties through the metal-insulator transition.[1] The phase diagram in the plane of preseure (p) and temperature (T) is shown in Fig.1(a), where the high-pressure insulating phase (I) is separated from the low-pressure metallic phase (M) by the critical line of the metal-insulator transition, T_{MI}. This indicates the existence of three characteristic regions of p denoted as I, II and III. In each region the temperature dependence of the resistivity, ρ, has a characteristic feature as shown schematically in Fig.1(b). Depending on the choice of R_1 and R_2, e.g. DMe, DMeO, MeBr etc., each member of these salts shows either one of behaviors shown in Fig.1(b) at ambient pressure and the application of the external pressure makes the system move to the right in the horizontal coordinates in Fig.1(a).

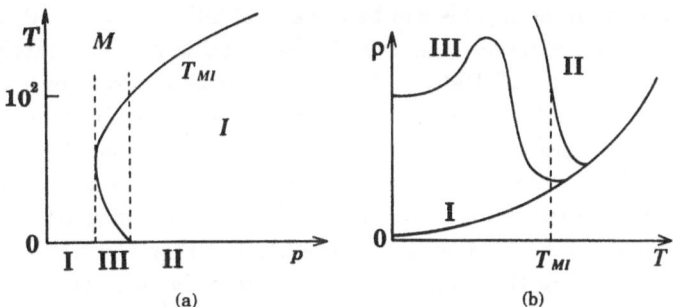

Figure 1. (a) A shematic representation of the phase diagram of (DCNQI)$_2$Cu family in the plane of pressure (p) and temperature (T); the metallic state (M) and the insulating state (I) are separated by the meal-insulator transition temperature (T_{MI}). There are three different regions, I, II and III. (b) A shematic representation of the temperature dependences of the resistivity in each region, I, II and III. There exists a reentrant metallic state in the region III.

In accordance with the classification, I, II and III, based on the temperature dependence of the resistivity, there exist following magnetic and structural characteristics in each region:[2-6]

I. Metallic down to the lowest temperatures without lattice distortions. The Pauli-like paramagnetism independent of temperature.

II. Three-fold static lattice distortions and the Curie-Weiß like paramagnetism strongly dependent on temperature below the metal-insulator transition temperature (T_{MI}). The onset of the magnetic phase transition, possibly antiferromagnetic, at around $10K$.

III. Similar features to those in the heavy electrons in the electronic specific heat, spin susceptibility and the coefficient of T^2 in the resistivity in the reentrant metallic state.

Besides these interesting behaviors in each region, the most noteworthy is the fact that the average valence of Cu is always equal or close to +4/3, i.e. $Cu^{+4/3}$, independent of pressure and temperature.[2,6,7]

A THEORETICAL MODEL

In order to study this system[8] we will employ the periodic Anderson model for the three dimensional array of chains associated with stacked DCNQI molecules and Cu ions located in between as in ref.9. Each unit cell in the plane perpendicular to the stacking axis has two DCNQI chains and one Cu chain. We first ignore the three-dimensionality and focus on the unit cell in the plane perpendicular to the chain. In this case it is sufficient to consider doubly degenerate one-dimensional π-bands of DCNQI and one flat band associated with the d-orbitals of Cu, which are to be coupled together by the mixing integral. These Bloch bands are filled as the amount of the charge transfer is varied. If $(DCNQI)_2^{-1/2}Cu^{+1}$ is stable, the band scheme will be as shown in Fig.2 based on the hole picture i.e. the Cu-d hole level, ε_d, should be located above the Fermi level, ε_F, and doubly degenerate DCNQI-π bands are 3/4 filled. In the case of $(DCNQI)_2^{-2/3}Cu^{+4/3}$ on the other hand the lower mixed band denoted as the ε_1-band will be 5/6 filled whereas the ε_2-band is 2/3 filled as shown in Fig.2(b), where the region around $k/c^* = 1$ in Fig.2(a) is enlarged together with the redefinition of the origin of k. This will be the actual filling of the band, i.e. the Fermi surface satisfying the Luttinger sum rule, if the metallic state is stabilized down to low temperature. The special feature of this filling is understood by noting that the stability of the amount of charge transfer is in general determined by the energetics involving the whole energy scheme. If the Coulomb interaction at Cu-site is strong, the spectrum weight at each momentum of the one-particle Green function associated with the ε_1-band will essentially be reduced by half in the region where the d-orbital component dominates (shaded region in Fig.2(b)) and then the whole shaded region is essentially occupied. In these approximations the ε_1-band in the shaded region in Fig.2b can be considered as the Hubbard lower band. Consequently, apart from the subtle singularity associated with the true Fermi surface, the ε_1-band is completely filled from the energetical point of view and hence the charge transfer resulting in $Cu^{+4/3}$ as an average will particularly be stabilized.

The argument given above implies that there exist fluctuations of valence of Cu as $Cu^+:Cu^{++} = 2:1$ as an average even in the metallic state.

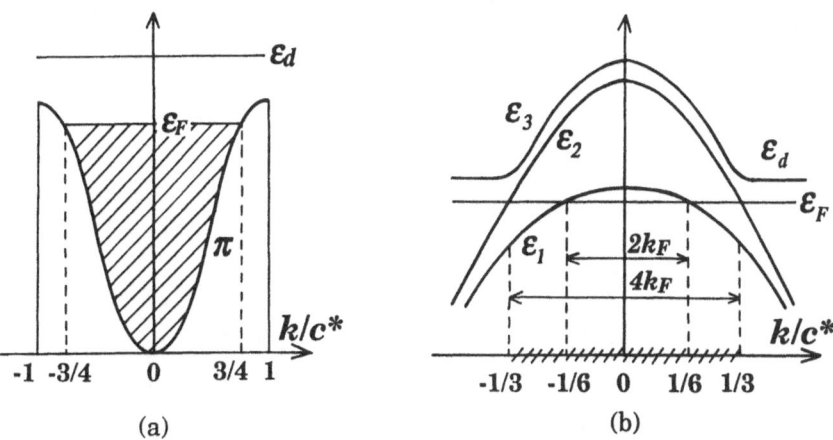

Figure 2. (a) A hole band scheme if $(\text{DCNQI})_2^{-1/2}\text{Cu}^+$ is stable; ε_d and ε_F are the atomic d-level and the Fermi energy, respectively, and $c^* \equiv \pi/c$, c being the lattice spacing in the direction of the stacking. The π-bands associated with DCNQI are shown as if completely degenerate. (b) A hole band scheme in the case of $(\text{DCNQI})_2^{-2/3}\text{Cu}^{+4/3}$, where the origin of k is shifted to $k/c^* = 1$ in Fig.2(a) and the splitting of the π-bands is explicitly shown. The "$2k_F$" and "$4k_F$" are referred to the ε_1-band, but note that this "$4k_F$" corresponds to "$2k_F$" of the ε_2-band.

METAL-INSULATOR TRANSITION

Next we will study the charge ordering process, which can be viewed as Peierls transition driven by the $4k_F$ charge fluctuations in the d-band (ε_1-band) together with $2k_F$ charge fluctuations of π-band (ε_2-band). It is to be noted that both fluctuations can contribute to the softening of the same phonon mode. The possible candidate of this phonon is the one which involves the change of N-Cu-N angle, α, since α is known experimentally to change at $T = T_{MI}$. The critical temperature, T_p, of this transition, which may be called the Tomonaga-Luttinger-Peierls transition, or simply the Luttinger-Peierls transition, is given by the solution of the following equation

$$\omega_0^2 = g_1^2 N_1(4k_F, T) + g_2^2 N_2(2k_F, T) \ , \tag{1}$$

where ω_0 is the unrenormalized frequency of the phonon mode. In this equation g_i and $N_i(q, T)$ are the coupling constant and the density-density correlation function associated with each band, i. Since the contributions from $N_1(4k_F, T)$ will be appreciable in a quantitative sense but is finite as $T \to 0$, these can be taken into account in the large renormalization of the phonon frequency. The existence of such appreciable contributions of $4k_F$ fluctuations to the lattice instability can be deduced from the fact that T_{MI} in $(\text{DCNQI})_2\text{Cu}$ family is higher than those in $(\text{DCNQI})_2\text{Ag}$.[4] On the other hand the $2k_F$-charge fluctuations represented by $N_2(2k_F, T)$ is divergent as $T \to 0$ in purely one-dimensional systems logarithmically for non-interacting electrons and with power-law in the presence of mutual interactions (Luttinger liquid) and is expected to be very sensitive to the interchain transfer integral, $t_\perp$, i.e. the three-dimensionality. Actually it has been demonstrated[10-12] that the solution, T_p, of eq.(1) for a fixed value of $\omega^2(T) \equiv \omega_0^2 - g_1^2 N_1(4k_F, T)$ with $N_2(2k_F, T)$ for non-interacting electrons has a critical dependence on $t_\perp$ as schematically shown in Fig.3. In $(\text{DCNQI})_2\text{Cu}$ the effective value of $t_\perp$ is tightly related with the N-Cu-N angle (α) which is inferred from the fact that α increases as the temperature is lowered in accordance with the expan-

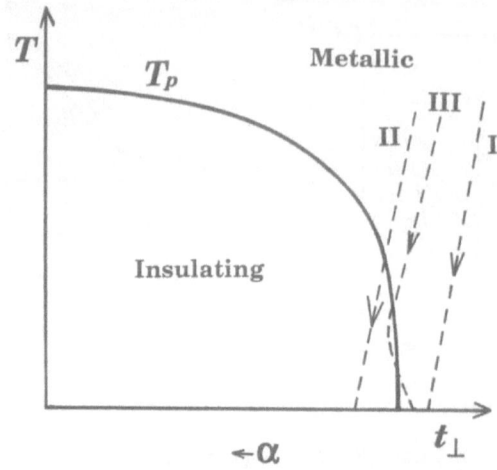

Figure 3. The shematic dependence on the interchain transfer integral, $t_{\perp}$, of the critical temperature, T_p after refs. 10–12. As the N-Cu-N angle, α, increases, $t_{\perp}$ decreases. (Since T_p is of first order transition in the present case, there exists a discontinuous change of α at T_p.)

$\oslash : Cu^{++}$ $\bigcirc : Cu^{+}$

Figure 4. The shematic representation of the charge ordering below T_p. Depending on the magnitude of the $4k_F$ periodic lattice distortion, the spatial extents of the variation of the valence and the local magnetic moment (Cu^{++}) will be different and are not necessarily so localized as in this figure.

sion in the direction perpendicular to the chains[4] and that the polarized reflectance spectra in the perpendicular direction loses the Drude tail in the insulating phase[13] where α takes larger values. Hence the increase of $t_{\perp}$ in Fig.3 will correspond to the decrease of α, and each member of $(DCNQI)_2Cu$ family follow the dotted lines in the figure corresponding to the classification of groups I, II and III in Fig.1(a). The fact that the reentrant behavior seen in resistivity goes together with that of α has been experimentally indicated by the X-ray diffraction on $(DMeDCNQI)_2Cu$ carried out by Kagoshima *et al.*[5]

In the present view the spin degree of freedom of the ε_1-band, *i.e.* the d-spin of Cu^{++}, is not directly coupled to the transition and the spin fluctuations survive even $T < T_P$ in accordance with the experimental findings of the Curie-Weiß like paramagnetism in the insulating state. This is in sharp contrast of the ordinary $2k_F$-Peierls transition. In the latter case both spin and charge fluctuations are suppressed below the transition. Actually the ε_1-band is half-filled in the presence of the three fold lattice distortion and then result in the Mott insulator with a local magnetic moment associated with Cu^{++} in the presence of strong correlation as shown schematically in Fig.4.

SUMMARY

In summary it is indicated that the stability of the average valence of $Cu^{+4/3}$ in $(DCNQI)_2Cu$ is due to the strong correlations in Cu sites and that the metal-insulator transition of this family can be understood as the Peierls transition driven by the $2k_F$-charge fluctuations of the nearly free DCNQI-π band (ε_2-band) together with the $4k_F$-charge fluctuations of the narrow d band (ε_1-band), which behaves like a Tomonaga-Luttinger liquid. It is also argued that there exists a competition between the Peierls transition in the Luttinger liquid and the three-dimensionality resulting from the interchain transfer integral, which would stabilize the (heavy) Fermi liquid.

ACKNOWLEDGEMENT

The author thanks R. Kato, S. Kagoshima and T. Takahashi for informative discussions on the experimental facts and Y. Suzumura for many useful discussions on the subject. He also thanks H. Kobayashi for his sending him an illuminating review article. This work is financilally supported by Grant-in-Aid for Science Research from the Ministry of Education, Science and Culture No.3303017 and Monbusho International Scientific Research Program: Joint Research "Magnetism and Superconductivity in Highly Correlated Systems" (03044037).

REFERENCES

1. Various papers in the Proceedings of the International Conference on Science and Technology of Synthetic Metals (ICSM'90), Synth. Metals 41-43 (1991).
2. A. Kobayashi, R. Kato, H. Kobayashi, T. Mori and H. Inokuchi, Solid State Commun. 64 : 45 (1987).
3. T. Mori, H. Inokuchi, A. Kobayashi, R. Kato and H. Kobayashi: Phys. Rev. B38 : 5913 (1988).
4. R. Kato, H. Kobayashi and A. Kobayashi, J. Am. Chem. Soc. 111 : 5224 (1989).
5. S. Kagoshima, N. Sugimoto, T. Osada, A. Kobayashi, R. Kato and H. Kobayashi, J. Phys. Soc. Jpn. 60 : 4222 (1991).
6. H. Kobayashi, A. Miyamoto, R. Kato, F. Sakai, A. Kobayashi, Y. Yamakita, Y. Furukawa, M. Tasumi and T. Watanabe, preprint.
7. I.H. Inoue, A. Kakizaki, H. Namatame, A. Fujimori, A. Kobayashi R. Kato and H. Kobayashi, Phys. Rev. B 45 : 5828 (1992).
8. H. Fukuyama, submitted to J. Phys. Soc. Jpn.
9. Y. Suzumura and H. Fukuyama, J. Phys. Soc. Jpn. 61 no.9 (1992) .
10. B. Horovitz, H. Gutfreud and M. Weger, Phys. Rev. B 12 : 3174 (1975).
11. K. Yamaji: J. Phys. Soc. Jpn. 51 : 2787 (1982).
12. Y. Hasegawa and H. Fukuyama, J. Phys. Soc. Jpn. 55 : 3978 (1986).
13. K. Yakushi, A. Ugawa, G. Ojima, T. Ida, H. Tajima, H. Kuroda, R. Kato and H. Kobayashi, Mol. Ceyst. Liq. Cryst. 181 : 217 (1990).

QUASIPARTICLES IN HEAVY FERMION SYSTEMS BELOW AND ABOVE THE COHERENCE TEMPERATURE

Peter Fulde, Uwe Pulst and Gertrud Zwicknagl
Max-Planck-Institut für Festkörperforschung
D-7000 Stuttgart 80, Germany

Abstract

The quasiparticle concept of Landau is based on the existence of a Fermi surface. Arguments are given why in heavy fermion systems below the coherence temperature, the f electrons should be treated as itinerant when the Fermi surface is determined while above the Kondo temperature they are to be considered as part of the core. Possible experimental deviations from Luttinger's theorem are pointed out. For temperatures above the coherence temperature the Anderson impurity model within the Noncrossing Approximation (NCA) may be applied. It is shown that a simple approximation scheme for solving the coupled integral equations of the NCA is very useful for the interpretation of a number of experiments. Thereby the crystal field splittings of the $4f$ ($5f$) ions. can be taken into consideration. Examples are the temperature dependence of the quadrupole moment of $YbCu_2Si_2$ and the inelastic neutron scattering experiments on $YbPd_2Si_2$. The theory gives simple results for the transport coefficients.

I. Introduction

We know from experiments that at low temperatures heavy fermion systems behave like Landau liquids [1-3]. This implies that there is a one to one correspondence between the excitations of a heavy fermion system and those of a noninteracting electron gas. With this in mind several questions arise which we will try to answer. How accurately can the Fermi surface of a system of heavy fermions be calculated and how large is the volume in momentum space enclosed by it? Does Luttinger's theorem hold and under which circumstances might it be violated?

When the temperature is sufficiently high, quasiparticle collisions dramatically reduce the quasiparticle lifetimes [4]. As a consequence coherence is lost in the lattice of f sites (formed e.g. by Ce ions) and the problem reduces to the Anderson impurity model. The latter can be handled in the Noncrossing Approximation (NCA) [5]. The resulting coupled integral equations are solved with the help of a simple approximation scheme [6]. The solutions may be used in order to calculate the temperature dependence of a number of physical quantities like the magnetic response. The effects of the crystalline electric field (CEF) are thereby included. A comparison with Bethe *ansatz* results can be made as well as with exact numerical solutions of the NCA equations. We show that the method gives surprisingly good results. The experimentally observed temperature dependence of the quadrupolar susceptibility of $YbCu_2Si_2$ can be explained [6] as well as the inelastic neutron scattering results on related materials [7]. The approximation allows to calculate also transport properties like the resistivity, thermal conductivity or thermopower.

II. Quasiparticles and the Fermi surface

The assumption that heavy fermion systems are Fermi liquids as defined by Landau implies, that there must exist a Fermi surface. Indeed, by de Haas-van Alphen measurements e.g. on UPt_3 the Fermi surface was determined and it was shown that the quasiparticle excitations have the large relative mass required for the explanation of the large γ-value in the low temperature specific heat $C = \gamma T$. Furthermore, it was demonstrated by means of renormalized band-structure calculations that not only the topology of the Fermi surface can be calculated correctly, but that also the large relative mass anisotropies e.g. in $CeRu_2Si_2$ can be explained without any fit parameter [8]. But repeatedly the question has been raised under which circumstances the f-electrons must be included in a band calculation and when they have to be treated as part of the core [9]. Since there seems to be a lack of mutual consent we want to address this problem here.

Without doubt, the low lying excitations which are characteristic for heavy fermion systems involve predominantly spin degrees of freedom. Direct evidence is the magnitude of the entropy which is associated with the excess specific heat. It is of order $S \simeq k_B ln N_f$ per f site where N_f is the degeneracy of the ground-state of the atomic f shell which depends on the CEF. Like in the one-impurity Kondo problem a singlet-triplet excitation has to be associated with each f site (e.g. Ce). The excitation energy is of order $k_B T^*$ and defines a characteristic low-energy scale of the system (Kondo temperature). The lower T^* is, the smaller is the change in the f charge associated with the excitation. In the case of Ce, it is of order $(1 - < n_f >)$ and for $T^* \to 0$ the f count approaches $< n_f > \to 1$. The simplest model for this excitation is found in Ref. [10]. Because the f-sites form a lattice, the excitations are coupled with each other and at sufficiently low temperatures, i.e., below $T_{coh} < T^*$ coherent states are formed. The details of this coupling are not yet well understood, but from experiments we know that the density of states of the excitations is like the one in a noninteracting electron gas, though with renormalized parameters. Let us first assume that de Haas-van Alphen experiments are performed at a temperature $T << T_{coh}$. In that case the f electrons have to be included in the renormalized band calculations. In many Ce (or Yb) compounds with one $4f$ electron (hole) the topological shape of the Fermi surface is essentially determined by the conduction electrons [4]. The $"f - band"$ is situated by an energy of order of $k_B T^*$ above the Fermi energy ϵ_F (Kondo resonance). The situation is different in systems like UPt_3, where the magnetic ion, i.e., U has more than one f electron. In that case the Fermi surface is intersecting with the f bands and its shape is strongly influenced by the presence of f electrons.

Returning to the case of Ce compounds, we note that the electrons are scattered by the Ce sites, resulting in energy dependent phase shifts $\eta_\nu^{Ce}(\epsilon - \epsilon_F)$. Through a generalized Friedel sum rule the f phase shift at ϵ_F is fixed by the requirement that there is slightly less than one f electron contained in the CEF ground-state of a Ce site. The slope of the f phase shift at ϵ_F is related to the high effective mass m^*. The volume enclosed by the Fermi surface includes the $4f$ electrons in the counting. Clearly, Luttinger's theorem is fulfilled.

Next we assume that a de Haas-van Alphen experiment on a Ce compound is done at a temperature $T > T^*$. In that case the low lying excitations of the f electrons are saturated and they can be treated as part of the core. The (small) charge fluctuations associated with their hybridizations are eliminated by a Schrieffer-Wolff transformation and replaced by an effective, spin dependent interaction between f and conduction electrons (Coqblin-Schrieffer Hamiltonian). A mass enhancement of the conduction electrons can be derived from this interaction. The Fermi surface has shrunk in its electronic part and expanded its hole pockets as compared with the previous case ($T < T_{coh}$). The enclosed volume should contain one electron less than before, because the $4f$ electron is part of the core. By excluding the f electron from counting, Luttinger's theorem is again satisfied. This is precisely the situation found when de Haas-van Alphen measurements on $CeRu_2Si_2$ and $CeRu_2Ge_2$ are compared with each other. Because of the large atomic distances the hybridization of the $4f$ electron with its surroundings is much less in $CeRu_2Ge_2$ than it is in $CeRu_2Si_2$. In a series of beautiful experiments [11, 12] it was demonstrated that the Fermi surface looks very much alike in both materials but with the hole parts extended and the electronic parts reduced in $CeRu_2Ge_2$. The difference in volume of phase space is roughly one electron per unit cell. Because in $CeRu_2Ge_2$ the characteristic energy scale $k_B T^*$ and hence also T_{coh} are very small (in fact they have not been measured) parts of the Fermi surface of $CeRu_2Si_2$ have disappeared. These are the parts on which the lifetime of the quasiparticles has become too short in order to retain a coherent motion. It is worth pointing out that the calculated band masses of $CeRu_2Ge_2$ are by a factor of 1.3-5.4 smaller than the measured ones. This is what one might expect. The virtual excitation of higher CEF levels of the $J = 5/2$ ground-state multiplet of Ce^{3+} as well as the excitation of spin waves ($CeRu_2Ge_2$ is a ferromagnet with a Curie temperature of $T_0 \simeq 8$ K) are not included in the calculation. They will certainly lead to an enhancement of the band masses [12]. The measured γ coefficient in the specific heat agrees well with the one determined from the measured effective masses. This suggests that there are no other but the quasiparticle contributions to the linear specific heat term.

From the above considerations the following scenario can be deduced. For $T << T_{coh}$ heavy, coherent quasiparticles exist and the large specific heat coefficient γ is directly related to the large quasiparticle mass. When T increases above T_{coh}, the mean-free path of the (spin dominated) excitations of the f electron system becomes so short that coherence can no longer be maintained. The heavy quasiparticles gradually disappear in de Haas-van Alphen measurements and the volume enclosed by the measured Fermi surface does no longer correspond to an integer electron number per unit cell. Luttinger's theorem is violated in that temperature regime. As the temperature increases further, the Fermi surface reduces to the one with the f electrons taken as part of the cores. But as long as $T < T^*$ there is still a large specific heat present. For $T_{coh} \leq T \leq T^*$ this specific heat can no longer be calculated from the measured quasiparticle dispersions alone. Instead, it contains large contributions from the incoherent f electron excitations. Only when $T > T^*$ does one have again a correspondence between the quasiparticle dispersion and the specific heat. This is the regime to which $CeRu_2Ge_2$ belongs, because T^* is apparently extremely low in that system. King and Lonzarich [12] have suggested that by applying pressure to $CeRu_2Ge_2$ one should be able to reach a situation encounted in $CeRu_2Si_2$. This is a fascinating perspective. According to what has been suggested above one should first observe a strong increase in the specific heat without a corresponding enhancement of the quasiparticle masses. At higher pressures the Fermi surface should extend so that it incorporates the $4f$ electron and the large measured quasiparticle masses should match with the specific heat coefficient γ.

III. Low frequency response

At temperatures $T > T_{coh}$ the problem of the Kondo lattice reduces to that of independent impurities. Although the heavy quasiparticles have lost their coherence properties in that temperature regime, their dynamics must still be properly accounted for. Therefore the starting point is the Anderson impurity Hamiltonian and we limit ourselves to the case of one electron (Ce^{3+}) or one hole (Yb^{3+}):

$$H = \sum_{\underline{k}\sigma} \epsilon(\underline{k}) c_{\underline{k}\sigma}^{\dagger} c_{\underline{k}\sigma} + \epsilon_f \sum_m n_m^f \tag{1}$$

$$+ U/2 \sum_{m \neq m'} n_m^f n_{m'}^f + \tag{2}$$

$$+ \sum_{\underline{k}m\sigma} (V_{m\sigma}(\underline{k}) f_m^{\dagger} c_{\underline{k}\sigma} + V_{m\sigma}^*(\underline{k}) c_{\underline{k}\sigma}^{\dagger} f_m). \tag{3}$$

The operators $c_{\underline{k}\sigma}^{\dagger}(c_{\underline{k}\sigma})$ create (annihilate) conduction electrons with momentum $\underline{k}$, and spin σ The $f_{km}^{\dagger}(f_{km})$ are the creation (annihilation) operators for f-electrons on the impurity site. They are chracterized by the total angular momentum J and a quantum number m which denotes the different states within the J multiplet. Only the lowest spin-orbit multiplet is considered which is $J = 5/2$ for Ce^{3+} and $J = 7/2$ for Yb^{3+}. The orbital energies are

$$\epsilon_{fm} = \epsilon_f + \Delta_m \tag{4}$$

where Δ_m are the CEF excitation energies. The hybridization matrix element is denoted by $V_{m\sigma}(\underline{k})$. The Coulomb repulsion between two f-electrons is given by U. It is assumed that U is much larger than the other energy scales and therefore we may let $U \to \infty$. The conduction electron degrees of freedom which do not couple to the impurity are omitted here. The interconfiguration energy ϵ_f is negative. The energies Δ_m and the corresponding wave functions depend on the symmetry and the strength of the CEF.

The properties of the single impurity model can be calculated from two many-body spectra $\rho_f(\omega, T)$ and $\chi''(\omega, T)$ which are referred to as the f-density of states and the f-moment spectrum, respectively. These spectra can be determined experimentally, i.e. $\rho_f(\omega, T)$ from photoelectron and inverse photoelectron spectroscopy and $\chi''(\omega, T)$ from inelastic neuton scattering, for example. The spectra are calculated from the resolvents of the empty state $|0 >$ (i.e. the f^0 or f^{14} configuration) and the occupied f states $|m >$, respectively, denoted by $R_0(z)$ and $R_m(z)$

$$R_0(z) = \frac{1}{z - \Sigma_0(z)} \tag{5}$$

$$R_m(z) = \frac{1}{z - \Sigma_m(z)}. \tag{6}$$

The self-energies Σ_0 and Σ_m are coupled. In the Noncrossing Approximation (see Ref. [5, 13, 14]) the coupling is given through the integral equations

$$\Sigma_0(z) = \frac{\Gamma}{\pi} \sum_m \int_{-\infty}^{+\infty} d\xi \rho_m(\xi) K(z - \xi) \tag{7}$$

$$\Sigma_m(z) = \frac{\Gamma}{\pi} \int_{-\infty}^{+\infty} d\xi \rho_0(\xi) K(z - \xi) \tag{8}$$

with $z = \omega + i\delta$ and Γ denoting the width of the $4f$ level due to hybridization. In the absence of hybridization, the spectral densities $\rho_\alpha(\xi)(\alpha = 0, m)$ of the resolvents R_α

are δ-functions $\delta(\xi - \omega_\alpha)$ centered at $\omega_0 = 0$ and $\omega_m = \epsilon_{fm}$. The kernel $K(z)$ is defined through

$$K(z) = \frac{1}{N(0)} \int_{-\infty}^{+\infty} d\epsilon N(\epsilon) \frac{f(\epsilon)}{z + \epsilon} \qquad (9)$$

where $N(\epsilon)$ is the density of states of the conduction electrons and $f(\epsilon)$ is the Fermi distribution function. The energy cut-off is provided by the conduction electron band width.

Our *ansatz* exploits the information about the self-consistent solution for $\rho_0(\omega)$ (see Ref. [5] and references therein). For temperatures $T \to 0$ the function consists of three rather different parts. The most prominent feature is a narrow peak of weight $(1 - n_f)$ centered at $\omega_0 = -|\epsilon_f| - T^*$. It reflects an admixture of states with no f-electron/hole to the many-body ground- state and the low-lying excitations. It is this feature which gives rise to the anomalous low-temperature behavior of Kondo alloys. The residual weight is distributed among a structureless background and a broad peak describing "charge fluctuations".

The spectral function $\rho_0(\omega)$ is approximated by

$$\rho_0(\omega, T) \simeq (1 - n_f)\delta(\omega - \omega_0) \qquad (10)$$

where $n_f = n_f(T = 0)$ is the low-temperature f-valence. This *ansatz* is inserted into the self-consistency equations to obtain the spectral function of the occupied f-states, $\rho_m(z)$ as

$$\rho_m(\omega, T) = \frac{1}{\pi} \frac{(1 - n_f)\Gamma f(\omega_0 - \omega)}{(\omega + |\epsilon_f| - \Delta_m)^2 + [(1 - n_f)\Gamma f(\omega_0 - \omega)]^2} . \qquad (11)$$

The simplified forms (8,9) yield explicit expressions for experimental quantities which can easily be evaluated. In particular, CEF effects can be incorporated in a straight-forward way.

For example, the temperature dependence of the f-electron occupational numbers $n_{fm}(T) = < f_m^+ f_m >$ is given by

$$n_{fm}(T) = \frac{1}{Z_f} \int_{-\infty}^{+\infty} d\omega \rho_m(\omega) e^{-\beta(\omega - \mu)} \qquad (12)$$

where μ is the chemical potential and Z_f denotes the partition function of the f electron system (i.e., after the conduction electron degrees of freedom have been integrated out)

$$Z_f = \int_{-\infty}^{+\infty} d\omega \rho_0(\omega) e^{-\beta\omega} + \sum_{m=1}^{2J+1} \int_{-\infty}^{+\infty} d\omega \rho_m(\omega) e^{-\beta\omega} . \qquad (13)$$

A detailed discussion of the method is found in Ref. [6].

The simplified solutions were used to calculated the temperature dependence of the $4f$-quadrupole moment $Q(T)$ in $YbCu_2Si_2$. The latter is given by

$$Q(T) = \sum_{m=1}^{8} < m \mid (3J_z^2 - J^2) \mid m > n_{fm}(T) . \qquad (14)$$

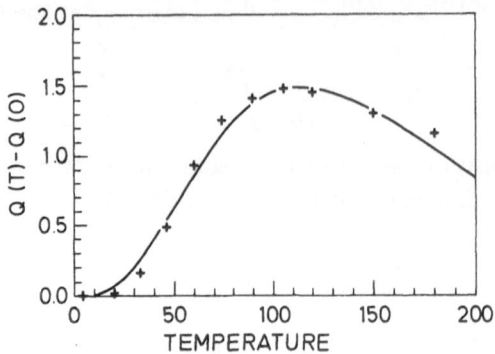

Figure 1. Temperature-dependent quadrupole moment $Q(T)$ for the $4f$ shell of $YbCu_2Si_2$: theory [6] (solid line) and experiments [15] (crosses). The following parameters were used: $T^* = 200$ K, $\Gamma = 47.4$ meV and $W = -1.67$ meV.

Here $| \, m >$ denotes the different CEF eigenstates of the $J = 7/2$ multiplet of the Yb^{3+} ions with population $n_{fm}(T)$. Results of the calculation are shown in Fig. 1 which also contains the measurements of Tomala et al. [15]. Thereby the input parameters $n_f(T = 0), T^*$ and the CEF parameter $W = 3B_2^0$ had to be chosen properly. They are listed in the figure caption. The agreement of the theoretical curve with the experimental data is very good. Another test of the simplified forms (8,9) was made by calculating the T dependence of static magnetic susceptibility and comparing it with the Bethe *ansatz* result. After adjustment of the characteristic temperature T^* the computations based on (8,9) reproduce well the Bethe *ansatz* data, even in the case of two-fold degeneracy $N_f = 2$.

The imaginary part of the diagonal elements of the susceptibility tensor is proportional to

$$\chi''_{\alpha\alpha}(\omega, T) \sim \frac{1}{Z_f} \sum_{mm'} |< m \, | \, J_\alpha \, | \, m' >|^2 \int_{-\infty}^{+\infty} d\omega' e^{-\beta\omega'} \rho_m(\omega') \rho_{m'}(\omega' + \omega) \ . \qquad (15)$$

For $T = 0$ this expression can be evaluated analytically when the approximation (9) is used. For details see Ref. [6]. A comparison of $\chi''(\omega, T)$ has been made for $T << T^*$ in the absence of CEF splittings, when on one hand Eq. (9) is used and when in distinction the NCA equations are solved numerically. It was found that the two results agree very well with each other [6].

The theory was recently used by Polatsek and Bonville [7] to interpret inelastic neutron scattering data on $YbPd_2Si_2$ [16] and $YbAgCu_4$ [17]. The inelastic neutron scattering cross-section is

$$\frac{d^2\sigma}{d\Omega d\omega} \sim \frac{1}{1 - e^{-\omega/(k_B T)}} \, \chi''(\omega, T) \ . \qquad (16)$$

Regarding $YbPd_2Si_2$ it was found that all available experimental data can be best explained by assuming $T^* = 60 K, \Gamma = 40 meV$ and the following CEF parameters: $B_2^0 = -50 \pm 5 \ cm^{-1}, B_4^0 = 15 \pm 5 \ cm^{-1}, B_4^4 = 250 \pm 30 \ cm^{-1}, B_6^0 = -25 \pm 5 \ cm^{-1}$ and $B_6^4 = -140 \pm 40 \ cm^{-1}$. The corresponding splitting energies of the four Kramer doublets are 0, 27, 199 and 209 K, respectively.

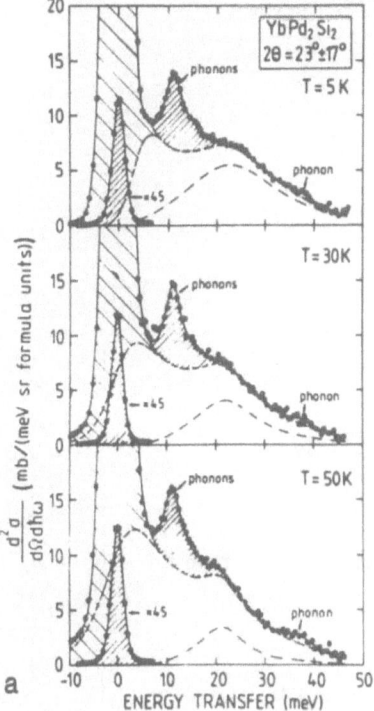
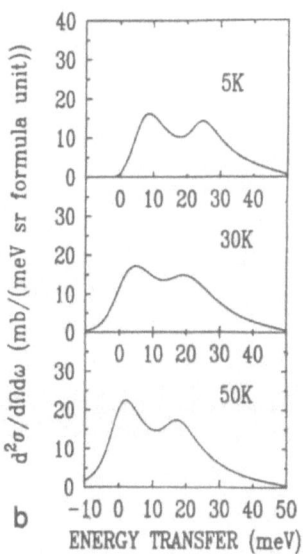

Figure 2. Temperature dependence of magnetic scattering of $YbPd_2Si_2$ as measured by inelastic neutron scattering [16] and as calculated from theory [7] by using Eqs. (13, 9).

A comparison between the measured cross-section and the computed ones for different temperatures is shown in Fig 2. The incident neutron energy is $\omega_{in} = 50.8$ meV. One notices that the theory can well describe the experimental data, even in details. In passing we note that also for $YbAgCu_4$ and $YbCu_2Si_2$ [18] good agreement is found between inelastic neutron scattering measurements and theory. The above examples demonstrate convincingly, that the NCA and the simplified solutions (8,9) provide a successful and easy to handle scheme for interpreting the low-energy response of heavy fermion or intermediate-valency systems.

IV. Transport properties

The absence of lattice coherence in heavy fermion compounds for temperatures $T > T_{coh}$ is also reflected in the transport properties. Their variation with temperature resembles those of dilute magnetic alloys. In the linear response regime, it is determined by a scattering rate $1/\tau(\epsilon)$ per impurity which is related to the f density of states $\rho_f(\epsilon)$

$$1/\tau(\epsilon) \propto \rho_f(\epsilon) \tag{17}$$

where

$$\rho_f(\omega) = \frac{1}{Z_f} \left(1 + e^{-\beta\omega}\right) \int_{-\infty}^{+\infty} d\epsilon \, e^{-\beta\epsilon} \rho_0(\epsilon)\rho_m(\epsilon + \omega) \tag{18}$$

is measuring the weight for transitions between f^0 and f^1 configurations. The simplified approximation scheme (8,9) which focusses on the universal low-energy features and yields the explicit expression

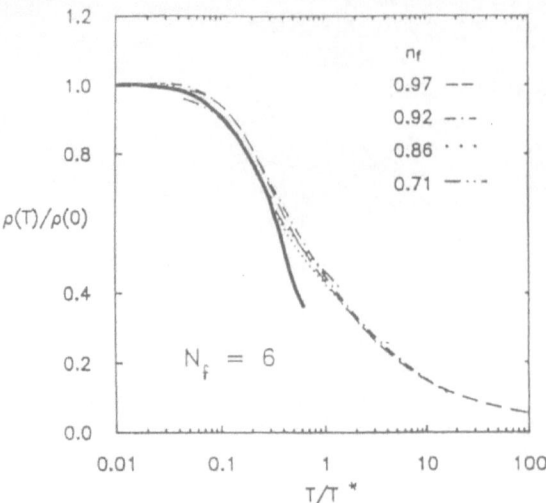

Figure 3. Temperature dependence of the resistivity as calculated from Eq. (20) for $n_f = 0.9$ (full line) and by using the complete NCA solution [5].

$$\rho_f(\omega) = (1 - n_f(T)) \sum_m \frac{1}{\pi} \frac{\Gamma(1 - n_f)}{(\omega - T^* - \Delta_m)^2 + (\Gamma(1 - n_f)f(-\omega))^2} .$$ (19)

The transport properties such as the resistivity ρ, the thermopower S and the thermal conductivity κ are expressed in terms of this quantity through the transport integrals $L_n(T)$ [19]

$$L_n(T) \propto \int_{-\infty}^{+\infty} d\epsilon \left[-\frac{df}{d\epsilon}\right] \epsilon^n \tau(\epsilon)$$ (20)

which are moments of the f density of states. Within the present approximation scheme they can be directly evaluated. The result for the resistivity ratio

$$\rho(T)/\rho(0) = (L_0(T)/L_0(0))^{-1}$$ (21)

is particularly simple,

$$\frac{\rho(T)}{\rho(0)} = \frac{1 - n_f(T)}{1 - n_f} \frac{1}{1 + \pi^2 \left(\frac{T}{T^*}\right)^2 \left[3 + \pi^2 \frac{n_f^2}{N_f^2}\right]^{-1}} .$$ (22)

The material-dependent prefactors are eliminated by dividing through the corresponding zero-temperature values.

Transport properties test energy moments of the f density of states and hence the shape of the Kondo resonance. Figure 3 compares the results of the full NCA and of the approximation scheme for the resistivity ratio $\rho(T)/\rho(0)$. The NCA results are reasonably well reproduced in the temperature regime where the behavior of the dilute magnetic alloy is universal. This rule of thumb is also valid for the other transport properties.

The deviations in the transport data may be partially due to the fact that the *ansatz* focusses exclusively on the universal features but entirely neglects the influence of charge fluctuations. This approximation certainly represents an oversimplification which affects the higher energy moments of the conduction-electron life times entering the thermopower, the thermal conductivity and the Lorentz ratio. The approximation scheme is extended in a straight forward way to approximately account for the charge fluctuations. A detailed discussion of the results will be reported elsewhere.

Acknowledgement

We want to thank Prof. V. Zevin for numerous discussions. Chapter II is a review of work done with him. We also thank Dr. G. Polatsek for making available to us his data.

References

[1] G.R. Stewart. Rev. Mod. Phys. 56,755 (1984)

[2] H. R. Ott, Progr. Low. Temp. Phys. 11, 215 (1987)

[3] N. Grewe and F. Steglich in "Handbook on the Physics and Chemistry of Rare Earths" Vol. 14, ed. by K.A. Gscheidner, Jr. and L. Eyring (North-Holland, Amsterdam 1991)

[4] G. Zwicknagl, Adv. Phys. 41, 203 (1992)

[5] For a review see N.E. Bickers, Rev. Mod. Phys. 59, 845 (1987)

[6] G. Zwicknagl, V. Zevin and P. Fulde, Z. Phys. B79, 365 (1990); see also V. Zevin, G. Zwicknagl and P. Fulde, Phys. Rev. Lett. 60, 2331 (1988)

[7] G. Polatsek and Bonville, Z. Phys. B88, 189 (1992)

[8] G. Zwicknagl, E. Runge and N.E. Christensen, Physica B163, 97 (1990)

[9] see e.g. Progr. Theor. Phys. Supplement 108 (1992) " Physics of f-electrons and related phenomena in strongly correlated systems", ed. by O. Sakai and T. Saso

[10] P. Fulde, J. Keller and G. Zwicknagl, Solid State Physics, Vol. 41, ed. by H. Ehrenreich and D. Turnbull (Academic, San Diego 1988) p.1

[11] B.K. Howard and G.G. Lonzarich, (private communication)

[12] C.A. King and G.G. Lonzarich, Physica B171, 161 (1991)

[13] L. Kuramoto, Z. Phys. B53, 37 (1983)

[14] M. Keiter and G. Czycholl, J. Magn. Magn. Mater. 31, 477 (1983)

[15] K. Tomala, D. Weschenfelder, G. Czjzek and E. Holland-Moritz, J. Magn. Magn. Mater. 89, 143 (1990)

[16] W. Weber, E. Holland-Moritz and A.P. Murani, Z. Phys. B76, 229 (1989)

[17] A. Severing, A.P. Murani, J.D. Thompson, Z. Fisk, C.K. Loong, Phys. Rev. B41, 1739 (1990)

[18] E. Holland-Moritz, D. Wohlleben and M. Loewenhaupt, Phys. Rev. B25, 7482 (1982)

[19] N.W. Ashcroft and N.D. Mermin, "Solid State Physics" (Holt, Rinehart and Winston, Philadelphia, 1976)

COMPETITION BETWEEN CEF SINGLET AND KONDO SINGLET AS ORIGIN OF WEAK ANTIFERROMAGNETISM AND RESISTIVITY ANOMALY

Y. Kuramoto

Department of Physics
Tohoku University
Sendai 980, Japan

INTRODUCTION

In some uranium compounds with $5f^2$ configuration (U^{4+}) the ground state in the crystalline electric field (CEF) is a nonmagnetic singlet. The singlet CEF state is also realized in the case of praseodymium compounds with $4f^2$ configuration (Pr^{3+}). In such cases the spin entropy of the system goes to zero as temperature decreases even though interactions with conduction electrons or with f electrons at other sites are absent. This is in striking contrast with the case of cerium compounds with $4f^1$ configuration (Ce^{3+}) where the Kramers degeneracy associated with the $4f^1$ configuration should lead to nonvanishing entropy if a Ce ion were isolated. It has been considered from this fact that the case of the singlet CEF ground state is much simpler because one need not consider the Kondo effect and related infrared effects which originate from interaction with conduction electrons. However, URu_2Si_2 for which the singlet CEF model accounts for the gross feature of the highly anisotropic susceptibility [1], gives clear indication of the Kondo effect in the resistivity[2] at temperatures higher than 100 K. The Kondo-type behavior of the resistivity has also been found in other U systems with a singlet CEF ground state such as UPt_2Si_2 [3], UPd_3 [4], UNi_2Al_3 [5], and UPd_2Al_3 [6].

The purpose of this paper is to elucidate some novel features caused by competition between the CEF effect and the Kondo effect for a singlet CEF ground state. To our surprise we have not found any theoretical work on the Kondo effect in the singlet CEF system. As the first step to explore characteristic properties of the system, we mainly discuss features which already appear in the single impurity system with the singlet CEF state. We take the simplest possible approach and apply the scaling method[7] with account of leading logarithmic terms. It is shown that the exchange interaction between f and conduction electrons does give rise to the Kondo effect at finite temperatures. Furthermore, provided that the interaction is strong enough, the ground state of the system can be a Kondo singlet instead of the CEF singlet. We discuss possibility of controlling the competition between the Kondo and CEF effects by diluting the system or by applying pressure.

In order to understand the low-temperature properties of URu$_2$Si$_2$ such as the tiny antiferromagnetic moment and the superconductivity, which coexists with the antiferromagnetism, we obviously need analysis of intersite interactions. In this paper we merely suggest that a cooperative collapse of the CEF levels may be a driving force of the Néel-type transition, but wait for a more detailed study to prove (or disprove) the suggestion.

CONSTRUCTION OF AN EXCHANGE MODEL

We take URu$_2$Si$_2$ as an exemplary target of our analysis. According to ref.[1], magnetic susceptibility of URu$_2$Si$_2$ can roughly be understood in terms of CEF levels of three singlets referred to as $|a\rangle$, $|b\rangle$ and $|a'\rangle$ with energies 0, 46 K and 170 K, respectively. The lowest doublet $|c_\pm\rangle$ lies at 550 K. In terms of the basis set $|J_z\rangle$ with $|J_z| \leq 4$ in the $^3H_4$ configuration, these wave functions are given by

$$
\begin{aligned}
|a\rangle &= \epsilon[|4\rangle + |-4\rangle] + \gamma|0\rangle, \\
|b\rangle &= 2^{-1/2}[|4\rangle - |-4\rangle], \\
|a'\rangle &= 2^{-1/2}\gamma[|4\rangle + |-4\rangle] - 2^{1/2}\epsilon|0\rangle, \\
|c_\pm\rangle &= \alpha|\mp 3\rangle + \beta|\pm 1\rangle,
\end{aligned}
$$

where α, β, γ and ϵ are numerical constants. From these equations the matrix elements of the magnetic moment operator M_i with $i = x, y, z$ can easily be derived. For example $\langle a|M_i|b\rangle$ is nonzero only for $i = z$. This leads to the van Vleck susceptibility along the z direction. On the other hand, the x and y components of the moment have nonzero elements between $|a\rangle$ and $|c_\pm\rangle$ with energy much higher than that of $|b\rangle$, or between $|b\rangle$ and $|c_\pm\rangle$. Thus the CEF scheme is consistent with the strong Ising-type anisotropy of the susceptibility.

To construct the simplest model for studying the competition between the CEF and Kondo effects, we neglect $|a'\rangle$ and other levels higher than $|c_\pm\rangle$. Then the 5f-electron part of the Hamiltonian with a single U site is given by

$$
\mathcal{H}_f = \Delta_0|b\rangle\langle b| + \Delta_1 \sum_\pm |c_\pm\rangle\langle c_\pm|, \tag{1}
$$

where the state $|a\rangle$ does not appear because its energy is taken to be 0. In order to introduce the exchange interaction we define the following operators which constitute the magnetic moment by a suitable linear combination:

$$
\begin{aligned}
m_z &= |b\rangle\langle a| + |a\rangle\langle b| \\
m_+ &= |c_+\rangle\langle a| + |a\rangle\langle c_-| \\
m_- &= |c_-\rangle\langle a| + |a\rangle\langle c_+| \\
\tilde{m}_z &= |c_+\rangle\langle c_+| - |c_-\rangle\langle c_-| \\
\tilde{m}_+ &= |c_+\rangle\langle b| - |b\rangle\langle c_-| \\
\tilde{m}_- &= |b\rangle\langle c_+| - |c_-\rangle\langle b|.
\end{aligned}
$$

The x and y components are derived from the spin flip components in the standard way:

$$
m_\pm = m_x \pm im_y, \qquad \tilde{m}_\pm = \tilde{m}_x \pm i\tilde{m}_y.
$$

For conduction electrons we take a single band and write the creation operator of a conduction electron using the Wannier basis at the f-electron site as $d_\sigma^\dagger$ with spin σ. Then the exchange interaction is given by

$$\mathcal{H}_{ex} = \sum_i (I_i m_i + K_i \tilde{m}_i) \sum_{\sigma\rho} (\tau_i)_{\sigma\rho} d_\sigma^\dagger d_\rho, \qquad (2)$$

where τ_i is the Pauli matrix with $i = x, y, z$ and the exchange constants I_i and K_i have a tetragonal symmetry: $I_x = I_y \equiv I_\perp, K_x = K_y \equiv K_\perp$. Figure 1 illustrates the CEF energy levels and the nonzero matrix moments of the magnetic moment. Thus our model is characterized by eqs.(1),(2) and the spectrum of the conduction band to be specified in the next section.

A consequence of the competition between the CEF and Kondo effects is understood by the following qualitative argument. Let us for simplicity assume $\Delta_0 = \Delta_1 = \Delta$ and introduce the Kondo temperature T_K which would result in the case of $\Delta = 0$. Obviously the Kondo effect is of no importance in the limit of $\Delta/T_K \to \infty$. In the opposite limit of $\Delta/T_K \to 0$, the CEF splitting is of no importance and the ground state should be a Kondo singlet. The crucial point here is that the two kinds of singlets have different number of screening conduction electrons, and therefore the wave functions are orthogonal to each other. If one can increase Δ/T_K from zero to infinity, there must be a change of the ground state. At zero temperature this change accompanies abrupt decrease in the number of screening electrons. Near the critical ratio of Δ/T_K the characteristic energy of the system is smaller than either Δ or T_K.

At finite temperature the decrease in the number of screening electrons should be continuous, but the effect of competition appears, for example, as the anomalous increase of the van Vleck susceptibility at temperatures lower than Δ.

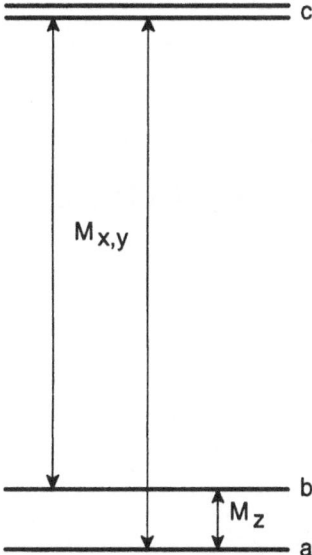

Figure 1. The CEF level scheme in the present model. The vertical lines with arrows indicate the presence of nonzero matrix elements of the moment M_i with $i = x, y$ or z.

SCALING THEORY FOR EFFECTIVE EXCHANGE INTERACTIONS

Let us consider the lowest-order correction to the exchange scattering as shown in Fig.2. Figure 2(a) shows a potential scattering generated from the exchange interaction. As in the case of the standard Kondo model there is no logarithmic correction because of cancellation of the two processes in Fig.2(a). On the other hand the processes in Fig.2(b) give rise to a logarithmic correction to the exchange interaction I_z. In contrast to the Kondo model, the present model has a low-energy cut-off Δ_0 so that the logarithm does not diverge even at $T = 0$. To be specific we consider a conduction band with constant density of states ρ_c between the band edges $-D_0$ and D_0. Following the method of "poor man's scaling" in ref.[7] we pursue the change of the coupling constants upon shifting the cut-off energy from D_0 to a smaller value D. We introduce the resolvents as follows:

$$G_a(z) = \frac{1}{z - D}, \qquad G_b(z) = \frac{1}{z - \Delta_0 - D}, \qquad G_c(z) = \frac{1}{z - \Delta_1 - D}.$$

The variable z represents the excitation energy measured from energy of the Fermi sea plus the lowest CEF state $|a\rangle$. Let us introduce a dimensionless vertex part Γ_z which reduces to $I_z\rho_c$ in the lowest order. Similarly we introduce $\Gamma_\perp$, Λ_z, $\Lambda_\perp$ which reduce to $I_\perp\rho_c$, $K_z\rho_c$ and $K_\perp\rho_c$, respectively. Then we obtain the following scaling equations:

$$\frac{\partial\Gamma_z}{\partial D} = \Gamma_\perp\Lambda_\perp G_c,$$

$$\frac{\partial\Gamma_\perp}{\partial D} = \Gamma_z\Lambda_\perp G_b + \Gamma_\perp\Lambda_z G_c,$$

$$\frac{\partial\Lambda_z}{\partial D} = \frac{1}{2}(\Gamma_\perp^2 G_a + \Lambda_\perp^2 G_b),$$

$$\frac{\partial\Lambda_\perp}{\partial D} = \Gamma_z\Gamma_\perp G_a.$$

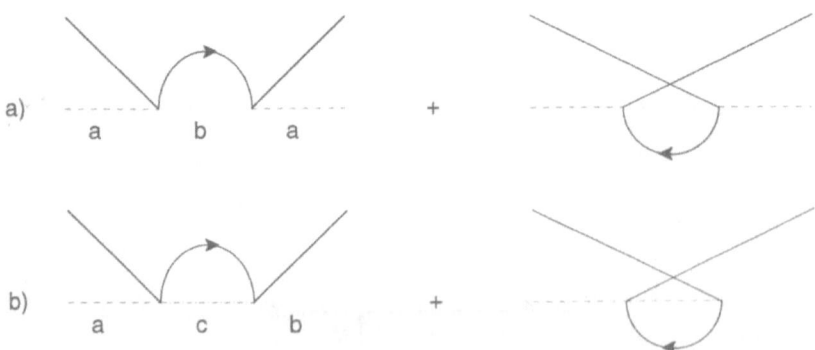

Figure 2. Examples of second-order contributions to the t-matrix: (a) potential scattering with the same initial and final f-states, (b) exchange scattering with different initial and final f states.

We have integrated the scaling equation numerically putting $z = 0$. Figure 3(a) shows the results in the case of

$$I_z \rho_c = I_\perp \rho_c = K_z \rho_c = K_\perp \rho_c = 1/4.$$

In the course of scaling the vertex parts grow and finally diverge simultaneously at $\log(D_0/D) \simeq 3.5$ or $D/D_0 \simeq 0.03$. We interpret this divergence in the same way as in the case of the standard Kondo model. Namely the ground state of the system is the Kondo singlet which is not accessible by perturbation theory with respect to the exchange interactions. In the case of an exchange model with a degenerate CEF ground state, the characteristic energy corresponding to the Kondo temperature has

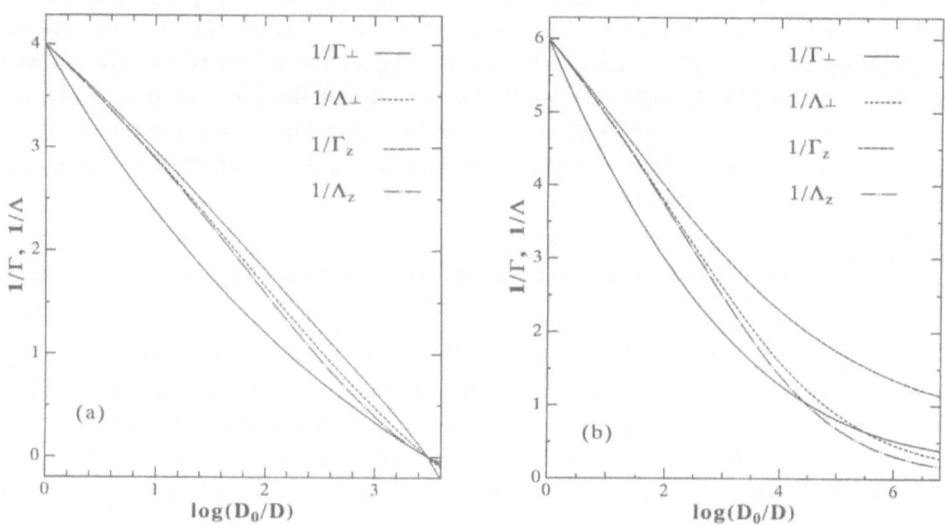

Figure 3. Renormalization of vertex parts against the change of the band cut-off: (a) a case where the vertex parts diverge and the Kondo singlet is realized, (b) another case where the vertex parts tend to saturate to finite values. In this case the singlet CEF ground state is realized with a renormalized CEF splitting.

been obtained as the value of $z \, (< 0)$ which gives divergence of vertex parts at $D = 0$ [8]. In the present singlet CEF system the number of screening electrons is likely to be 2. Then derivation of the energy scale requires more elaborate procedure [9].

In contrast to the standard Kondo model, there is a finite threshold in the coupling constant for the divergence to occur. For example if we start from smaller coupling constants such as

$$I_z \rho_c = I_\perp \rho_c = K_z \rho_c = K_\perp \rho_c = 1/6,$$

we obtain the results shown in Fig.3(b). In this case the vertex parts grow but tend to saturate to finite values. This means that the ground state of the system is connected continuously with the unperturbed state. The CEF singlet without screening electrons remains the ground state.

RESISTIVITY AND MAGNETIC SUSCEPTIBILITY

The electrical resistivity due to exchange scattering with the singlet CEF ground state has a characteristic dependence on temperature. In terms of the t-matrix $t(\epsilon)$ of conduction electrons the conductivity $\sigma(T)$ at temperature T is given by

$$\sigma(T) = C \int d\epsilon \left(-\frac{\partial f}{\partial \epsilon}\right) \frac{1}{|\operatorname{Im} t(\epsilon)|},$$

where C is a constant and f is the Fermi distribution function. Then the resistivity $\rho(T)$ is given by $1/\sigma(T)$. In contrast to the degenerate CEF ground state, the lowest-order scattering of a conduction electron has an activation energy. As a result the exchange scattering becomes exponentially small at temperatures lower than the CEF splitting. This situation is most clearly seen in a simplified treatment of the exchange scattering where only the singlet levels $|a\rangle$ and $|b\rangle$ are kept. In the lowest-order scattering, the initial f state $|a\rangle$ goes to $|b\rangle$ in the final state. The reverse process can also occur with the thermally excited level $|b\rangle$ as the initial state. We use the resolvent method[10] to calculate the t-matrix. In this method one first evaluates diagrams with a loop of f-state lines and the incoming Matsubara frequency $i\epsilon_n$ of a conduction electron. After analytic continuation $i\epsilon_n \to \epsilon + i\delta$ with ϵ real and δ positive infinitesimal we obtain

$$\frac{|\operatorname{Im} t(\epsilon)|}{\pi I_z^2 \rho_c} = f(-\Delta_0)[f(-\epsilon+\Delta_0) + f(\epsilon+\Delta_0)] + f(\Delta_0)[f(\epsilon-\Delta_0) + f(-\epsilon-\Delta_0)] \equiv g(\epsilon).$$

$$(3)$$

We note that $g(\epsilon)$ in eq.(3) tends to 1 in the limit $\Delta_0 \ll T$. Thus the resistivity tends to a constant value as the temperature increases. In the opposite limit of low T, the resistivity is determined by $\operatorname{Im} t(\epsilon)$ at $\epsilon = 0$. It is easily checked that $g(\epsilon)$ at $\epsilon = 0$ is $\cosh^{-2}[\Delta_0/(2T)]$. Hence the resistivity becomes exponentially small. Figure 4 shows the numerical result for $\rho(T)/\rho(\infty)$ in the lowest-order theory.

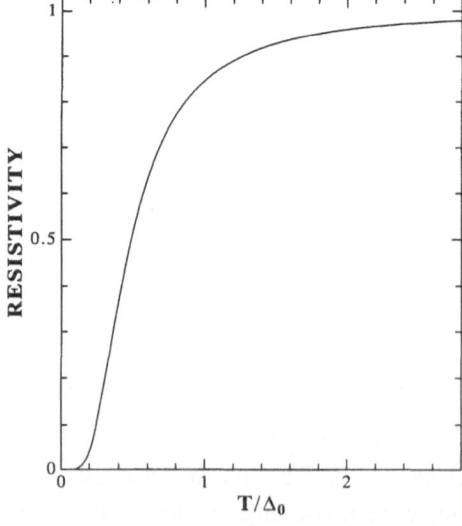

Figure 4. Temperature dependence of the resistivity in the lowest-order theory with the singlet CEF ground state.

Inclusion of leading logarithmic terms in the t-matrix proceeds just as in the case of the Kondo model [11]. Within the logarithmic accuracy we replace the bare coupling constant $I_z \rho_c$ by $\Gamma_z(z = T)$ and obtain

$$\mathrm{Im}\, t(\epsilon) = -\pi |\Gamma_z|^2 \rho_c g(\epsilon),$$

where $g(\epsilon)$ has been defined in eq.(3). If the temperature is much higher than the CEF splitting the vertex part Γ_z depends logarithmically on T as in the case of the Kondo model. On the other hand if the bare coupling constants are below the threshold for the Kondo singlet, the vertex part at $T \ll \Delta_0$ is almost independent of T, and $\rho(T)$ becomes exponentially small as in the case of the lowest-order theory. Thus $\rho(T)$ has a peak at $T \simeq \Delta_0$. This behavior is qualitatively the same as the one found experimentally in URu_2Si_2. We thus suggest a possibility that the decrease of the resistivity below 100 K is *not* due to the lattice effect, but due essentially to the single-site effect with the singlet CEF ground state. This is in striking contrast with the case of Ce heavy fermion systems.

Let us now consider the magnetic susceptibility. In zero-th order the van Vleck susceptibility is given by

$$\chi_{vv} = |\langle b|M_z|a\rangle|^2 \frac{2}{\Delta_0} \tanh\left(\frac{\Delta_0}{2T}\right).$$

At temperatures much higher than Δ_0, $\chi_{vv}(T)$ follows the Curie law. As T decreases $\chi_{vv}(T)$ continues to increase but finally saturates to $2|\langle b|M_z|a\rangle|^2/\Delta_0$. This type of behavior should appear in the singlet CEF system if interactions with conduction electrons and with other f sites are weak. In the presence of the exchange interaction we have seen two possible cases for the ground state: the CEF singlet and the Kondo singlet. If the coupling constants are near the threshold the characteristic energy of the system can be very small. Intuitively one may interpret this small scale as the renormalized CEF splitting $\tilde{\Delta}_0$. If $\tilde{\Delta}_0$ is much smaller than Δ_0 the Curie-type behavior of χ_{vv} should persist down to temperatures much lower than Δ_0.

DISCUSSIONS

Relevance to Experimental Results

We first discuss possible relevance of the results obtained to actual systems, in particular URu_2Si_2 and its dilute counterparts like $U_xTh_{1-x}Ru_2Si_2$. Since the intersite interaction between U ions is neglected in the present treatment, the dilute U system is the best candidate for comparison between theory and experiment. In this connection we mention recent experimental work[12] on $U_xTh_{1-x}Ru_2Si_2$. A remarkable feature found in $U_xTh_{1-x}Ru_2Si_2$ is that the characteristic temperature which corresponds to the peak in the magnetic susceptibility $\chi(T)$ shifts to lower temperatures as x becomes smaller than 0.07. In the case of $x = 0.01$, $\chi(T)$ continues to increase down to 0.01K which is the lowest temperature measured. Correspondingly $\rho(T)$ and the specific heat also indicate the presence of a very small energy scale. We note that $\chi(T)$ in pure URu_2Si_2 has a peak at $T = 55$ K, which we ascribe to the presence of antiferromagnetic intersite interactions. It has been suggested in ref.[12] that the CEF ground state in $U_xTh_{1-x}Ru_2Si_2$ may in fact be a doublet which accompanies the Kondo effect. An alternative possibility is that the effective CEF splitting between the two singlets becomes extremely small by the competition described in this paper. If this is indeed the case one may cross between the Kondo fixed point and the CEF singlet one by applying pressure or by changing the alloying composition.

Comparison with the Two-Impurity Kondo System

The existence of two kinds of singlet in the system is analogous to the two-center Kondo system with antiferromagnetic RKKY interaction. It has been shown that the competition of the single-site Kondo effect and the pair singlet caused by the RKKY interaction can lead to a very small energy scale[14]. The staggered susceptibility and the specific heat coefficient γ become subsequently large. If one neglects charge fluctuation of f electrons the characteristic energy can become zero, and then γ diverges.

In actual systems there should always be the charge fluctuation, and then the f-electron levels of the two-center system undergo splitting into bonding and antibonding levels[15, 17]. As a result at $T = 0$ the two Kondo singlets in the limit of small RKKY interaction change continuously to the pair singlet as the RKKY interaction increases. If the charge fluctuation is small the transient region becomes narrow.

In the present model for the singlet CEF system, the charge fluctuation of f electrons has been neglected. It remains to be seen whether the slight amount of charge fluctuation makes smooth the change from the CEF singlet to the Kondo singlet even at $T = 0$.

Possibility of More Quantitative Theory

In this paper, as the first step to study the competition between the Kondo singlet and the CEF singlet, we have used the poor man's scaling with account of leading logarithmic terms only. For more quantitative study we mention two possible approaches: the numerical renormalization-group (NRG) method[16] and the quantum Monte Carlo (QMC) method[17] which can be combined with the expansion in terms of reciprocal of interacting f-electron sites[18, 19, 20]. The NRG has been successfully applied to the single-impurity and two-center Anderson models[15]. It seems straightforward to apply the NRG method to the present CEF model. The QMC method has also been applied to both models[17]. This method has an advantage of being able to treat finite temperatures.

Recently the QMC has been applied to the infinite dimensional Hubbard model[18, 19, 20]. For infinite dimensional hypercubic lattice the problem is reduced to the effective single-site model in a medium which is to be self-consistently determined. It should be possible to apply the same idea to the lattice of singlet CEF sites plus conduction electrons. Then the intersite interaction may lead to a phase transition between the CEF singlet phase and the Kondo singlet phase both of which are affected by the presence of many f sites.

Weak Antiferromagnetism and the Lattice Anomaly

The origin of weak antiferromagnetism in URu_2Si_2 is a long-standing mystery. In a previous paper[21] we assumed the presence of nesting in the Fermi surface of conduction electrons, and considered that the resultant enhancement of the polarization function plays an important role in stabilizing the small moment. Since the anomaly in the specific heat is large in spite of the tiny moment, it is natural to suspect the presence of a hidden order parameter other than the Néel order. A possible driving force is the cooperative collapse of the CEF splittings by the Kondo effect as the temperature is increased. This interpretation is consistent with the disappearance of a clear dispersive mode, which is due to the singlet-singlet CEF excitation with intersite interactions, above the Néel temperature[22].

Recently similar weak antiferromagnetism has been found[23] in UPd_3 which also has the CEF singlet as the ground state. UPd_3 undergoes a structural transition at $T_1 = 6.5$ K, and then Néel-like transition at 4.5 K. Dispersive modes have been found below T_1 by neutron scattering[24, 25]. The temperature dependence of the resistivity in UPd_3 is characterized by a peak at around 150 K and a Kondo-type behavior above 150 K[4]. Further similarity to URu_2Si_2 is the sensitivity of the phase transition to impurities. Namely inclusion of Th in place of U suppresses the large peak in specific heat at T_1 drastically[26]. For comparison we note that the analogous peak in URu_2Si_2 is much suppressed in $U_{1-y}La_yRu_2Si_2$ [27].

Presence of the large anomaly[28] in the lattice parameter in URu_2Si_2 suggests a strong coupling to the lattice degrees of freedom. In this connection we notice that tilting of the quadrupole moment associated with each 5f wave function has been argued[25] to be the origin of the weak antiferromagnetism observed in UPd_3. It should be interesting to check whether such tilting is also relevant to URu_2Si_2.

SUMMARY

In summary, we have shown that the single-site f-electron system with a singlet CEF ground state has highly nontrivial physics when the Kondo singlet is competing for the stability. Of particular interest is the emergence of a new energy scale which can be much smaller either than the CEF splitting or the hypothetical Kondo temperature without the CEF splitting. The temperature dependence of the resistivity in URu_2Si_2 with the Kondo effect is naturally explained by the present theory. We have proposed an interpretation of the unusual temperature dependence of $\rho(T)$, $\chi(T)$ and specific heat in $U_xTh_{1-x}Ru_2Si_2$ in terms of the accidental tuning to the critical region of the competition upon alloying. In order to make the theory more quantitative we have proposed application of the NRG and QMC methods.

I would like to thank N. Sato, T. Komatsubara, F. Steglich and H. Amitsuka for informative discussions on experimental results. This work was supported by a Grant-in-Aid for Scientific Research from the Ministry of Education, Science and Culture of Japan.

REFERENCES

[1] G.J. Nieuwenhuys, Phys. Rev. **B35**, 5260 (1987).

[2] T.T.M. Palstra, A.A. Menovsky, and J.A. Mydosh, Phys. Rev. **B33**, 6527 (1986).

[3] R.A. Steeman et al., J. Phys. Condens. Matter **2**, 4059 (1990).

[4] M.-T. Beal-Monod, D. Davidov and R. Orbach, Phys. Rev. **B14**, 1189 (1976).

[5] C. Geibel et al., Z. Phys. **B83**, 305 (1991).

[6] N. Sato et al., J. Phys. Soc. Jpn. **61**, 32 (1992).

[7] P.W. Anderson, J. Phys. **C3**, 2436 (1970).

[8] K. Yamada, K. Yosida, and K. Hanzawa, Prog. Theor. Phys. **71**, 84 (1984).

[9] T. Saso, Prog. Theor. Phys. Suppl. No.108, 89 (1992).

[10] Y. Kuramoto, Z. Phys. **B53**, 57 (1983).

[11] A.A. Abrikosov, Physics **2**, 5 (1965).

[12] H. Amitsuka *et al.,* to be published in Physica B.

[13] T. Sakakibara *et al.,* to be published in Physica B.

[14] B.A. Jones, C.M. Varma and J.W. Wilkins, Phys. Rev. Lett. **61**, 125 (1988).

[15] O. Sakai, Y. Shimizu and T. Kasuya, J. Phys. Soc. Jpn. **58**, 3666 (1989); Prog. Theor. Phys. Suppl. No.108, 73 (1992).

[16] K.G. Wilson, Rev. Mod. Phys. **47**, 773 (1975).

[17] R.M. Fye and J.E. Hirsch, Phys. Rev. **B40**, 4780 (1989).

[18] M. Jarrel, Phys. Rev. Lett. **69**, 168 (1992).

[19] M.J. Rozenberg, X.Y. Zhang, and G. Kotliar, Phys. Rev. Lett. **69**, 1236 (1992).

[20] A. Georges and W. Krauth, Phys. Rev. Lett. **69**, 1240 (1992).

[21] Y. Kuramoto and K. Miyake, Prog. Theor. Phys. Suppl. No.108, 199 (1992).

[22] C. Broholm *et al.,* Phys. Rev. **B43**, 12809 (1991).

[23] U. Steigenberger *et al.,* J. Magn. & Magn. Mater. **108**, 163 (1992).

[24] W.J.L. Buyers *et al.,* Physica **102B+C**, 291 (1980).

[25] K.A. McEwen, U. Steigenberger, and J.L. Martinez, to be published in Physica B.

[26] K. Andres *et al.,* Solid State Commun. **28**, 405 (1978).

[27] H. Amitsuka *et al.,* J. Magn. & Magn. Mater. **104-107**, 60 (1992).

[28] A. de Visser *et al.,* Phys. Rev. **B34**, 8168 (1986).

THE GROUND STATE OF THE ONE DIMENSIONAL
KONDO LATTICE MODEL

M. Sigrist[a,b], H. Tsunetsugu[a,c], K. Ueda[b],
Y. Hatsugai[d], and T.M.Rice[a]

a) Theoretische Physik,ETH-Hönggerberg, 8093 Zürich, Switzerland
b) Paul Scherrer Institut, 5232 Villigen PSI, Switzerland
c) Interdisziplinäres Projektzentrum für Supercomputing
   ETH-Zentrum, 8092 Zürich, Switzerland
d) Institute for Solid State Physics, University of Tokyo 106, Japan

INTRODUCTION

The rich variety of phenomena observed in the class of heavy fermion materials has attracted attention from experimental and theoretical side. It is believed that the generic features of these materials can be understood from the subtle interaction between localized f-orbitals forming local magnetic moments and extended conduction electron orbitals. The typical theoretical models to describe this are the periodic Anderson model (PAM) and the Kondo lattice model (KLM).[1] In spite of big efforts in theoretical study it is still an open question whether these models contain the right ingredients to account for all the essential properties of the heavy fermion systems, the transport properties as well as the ordered phases, like antiferromagnetism or exotic superconductivity.[2]

Our study concentrates on the ground state of the KLM. This model is given by the following Hamiltonian

$$\mathcal{H}_{KLM} = -t \sum_{i,j} \sum_{s} c_{is}^{\dagger} c_{js} - J \sum_{i} \mathbf{S}_{i} \cdot \left(\sum_{s,s'} c_{is}^{\dagger} \frac{\sigma_{ss'}}{2} c_{is'} \right) \tag{1}$$

The conduction electrons denoted by the operators $c_{is}^{(\dagger)}$ move via nearest neighbor hopping with a matrix element $-t$ and have a local exchange interaction with the localized spin S, on each site. No exact solutions are known for this model in the sense as one obtains by Bethe ansatz for the 1D Hubbard model or the single impurity Kondo model. Furthermore, the fact, that the Hilbert space of the KLM is considerably larger than, for example, that of the Hubbard model, makes numerical investigations more difficult.

On the other hand, a few exact results are known for some limiting cases of the KLM, which provide a good test for approximate treatments. Let us here briefly review two results which are valid in any spatial dimension. The first is concerning the ground state of the KLM for very small conduction electron concentrations. Restricting on the case of just one electron in a finite lattice (L lattice sites) with periodic boundary conditions, it is found that the ground state is ferromagnetic for all finite values of $J < 0$. with a total spin quantum number $S_{tot} = (L - 1)/2$. The ground state whose wave function is obtained exactly, is characterized by the formation of a spin polaron dressing the conduction electron.[3]

The second case is the KLM with a half-filled conduction band. It can be shown rigorously that the ground state of the symmetric PAM is a total spin singlet in the case of half filling for any choice of the parameters in the model.[4] This result can be transferred partially to the KLM in its weak coupling limit as this represents an effective model for the strong coupling PAM. On the other side, the infinite coupling KLM ($J = -\infty$) is trivially a spin singlet since each site is occupied by one conduction electron locked into an on site spin singlet pair with the localized spin. In some range of the coupling constant J the ground state has probably antiferromagnetic long range order, at least for three dimensions. However, it has not been possible to give a proof for this yet.

In this paper we would like to review our recent work on the 1D KLM in order to clarify at least partially the phase diagram for the ground states. Our main attention is devoted to the strong coupling limit. Since this fix point is rather easy to understand it is a useful starting point for the perturbative study of the 1D KLM. It provides also the starting point for the understanding of the phase diagram of this system.

EFFECTIVE STRONG COUPLING MODEL

In the strong coupling limit the low energy states are contained in a very restricted Hilbert space. All electrons form onsite singlet pairs (OSSP) with the localized spins. For concentrations less than half filling the charge transfer in the KLM Hamiltonian leads to the motion of the OSSP with the nearest neighbor hopping matrix element $-t^* = -t/2$ accompanied by a spin back flow, exactly like the hole motion in the $U = \infty$ Hubbard model. The breaking of an OSSP is connected with an energy cost of the order $|J|$. In the restricted Hilbert space consisting of configurations which contain only OSSPs and unpaired localized spins, these high energy states can be included via virtual process leading in the lowest non-trivial order to an new Hamiltonian of the strong coupling limit. In one dimension it has the form

$$\tilde{\mathcal{H}} = \mathcal{H}_1 + \mathcal{H}_2 + \mathcal{H}_3 + \mathcal{H}_4 + \mathcal{H}_5 \tag{2}$$

with

$$\mathcal{H}_1 = -\frac{t}{2}\sum_{i,s}(f^{\dagger}_{i+1,s}f_{is} + h.c.) + \frac{3J}{4}\sum_{i}(1-n_i)$$

$$\mathcal{H}_2 = -\frac{t^2}{2J}\sum_{i,s}(f^{\dagger}_{i+1,s}n_i f_{i-1,s} + h.c.)$$

$$\mathcal{H}_3 = \frac{t^2}{4J}\sum_{i,s,s'}(f^{\dagger}_{i+1,s'}f^{\dagger}_{is}f_{is'}f_{i-1,s} + h.c.)$$

$$\mathcal{H}_4 = -\frac{t^2}{2J}\sum_{i,s}(f^{\dagger}_{i+1,s}(1-n_i)f_{i-1,s} + h.c.)$$

$$\mathcal{H}_5 = -\frac{5t^2}{6J}\sum_{i}n_{i+1}n_i + \frac{t^2}{6J}\sum_{i}(n_i+4)$$

We introduced here hard core Fermi operators $f^{(\dagger)}_{is}$ for the unpaired localized spins considering the OSSP as holes, so that the Hamiltonian is expressed entirely in an f-electron basis ($n_i = \sum_s f^{\dagger}_{is}f_{is}$).

In the case of infinite J this 1D system exhibits a complete spin degeneracy like the 1D $U = \infty$ Hubbard model. Thus the charge and spin degrees of freedom are separated so that the charge is described by the picture of non interacting spinless fermions with a wave function given by a Slater determinant. The complete wave function can therefore be written in product form:
$|\Psi\rangle = \sum_{\{r_i\}}\sum_{\{s_i\}}\phi_c(r_1,...,r_N)\phi_s(s_1,...,s_N)|r_1,...,r_n\rangle \otimes |s_1,...,s_N\rangle$ where N is the number of f-electrons ($L-N$ is the number of conduction electrons in the KLM).

How is the spin degeneracy lifted when J is turned to finite values? This question can be answered by the lowest order perturbation theory in $|t/J|$ for the degenerate ground state of $\mathcal{H}_0$. The only process which connects different spin configurations is contained in $\mathcal{H}_2$ which denotes assisted next nearest neighbor hopping. All other terms of the Hamiltonian describe hopping processes or interactions which do not change the spin sequence. Thus, considering the Hamiltonian matrix of the degenerate ground state the spin space basis $\{|s_1,...,s_N\rangle\}$, $\langle s'_1,...,s'_N|\mathcal{H}_2|s_1,...,s_N\rangle$, it is easy to prove that the lowest eigenvalue belongs to the state which is fully spin polarized, $S_{tot} = N/2$ [5]. This result is valid for all f-electron numbers N, except $N = L$ which corresponds to the KLM without conduction electrons. Especially, for the half-filled KLM, $N = 0$ the singlet ground state is recovered trivially.

ELEMENTARY EXCITATIONS IN THE STRONG COUPLING LIMIT

The elementary excitations of the effective model in Eq.(2) are given essentially by the collective modes of the symmetry broken state, the spin waves, and the particle-hole excitations. The former can easily be described by deriving a spin Hamiltonian describing the spin degrees of freedom effectively on a squeezed spin chain (L sites $\rightarrow$ N sites).

$$\mathcal{H}_{spin} = \sum_{i=1}^{N} J_{eff} \mathbf{S}_{i+1} \cdot \mathbf{S}_i \tag{3}$$

with $J_{eff}(\rho) = (t^2/\pi J)[(2/\pi\rho)\sin^2(\pi\rho) - \sin(2\pi\rho)]$ ($J_{eff} \leq 0$ for all $\rho = N/L$). This leads to the textbook spin wave spectrum $\omega_q = J_{eff}(1 - \cos q)$, $q = 2\pi m/N$. In the thermodynamic limit the momentum q of the spin wave on the squeezed chain is related to the momentum $\tilde{q}$ of the complete system by $\tilde{q} = \rho q$ [5].

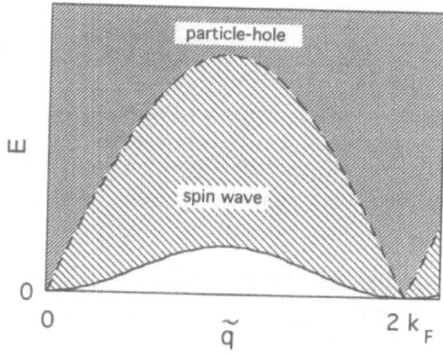

Figure 1. Spectrum of elementary excitations. The full line is the one magnon excitation and the dashed line is the particle-hole excitation. The shaded areas denote the excitation continuum for both types of excitations.

The particle-hole excitations are essentially those of spinless fermions. This picture is justified because in the ground state all f-electrons are spin polarized yielding this behavior even for finite values $|t/J|$ where the separation of spin and charge is not allowed anymore in a strict sense.

In Fig.1 the spectrum of both excitations is shown. Both are gapless at $\tilde{q} = 0$ and $2k_F'$, where k_F' is the Fermi momentum for the f-electrons). [1] The shaded areas in Fig.1

---

[1] Note, that $2k_F'$ corresponds to $4k_F$ with k_F the Fermi momentum for the conduction electrons.

denote the excitation continua. Since the energy scale of the two types of excitations is different by a factor $|t/J|$, the two spectra do not intersect so that there is no damping of the single magnon modes through particle-hole excitations in this 1D system. This is different for ferromagnetic states in higher dimensions (see for example Ref.6).

THE HALF-FILLED 1D KLM

Let us now briefly turn to the case of the half-filled 1D KLM. From our previous description it is clear that this system has an excitation gap in the strong coupling limit. The lowest excitation of the singlet ground state is produced by breaking one OSSP and transforming it into a triplet at the energy cost of $|J|$. In a numerical study we have investigated the evolution of this gap as we turn the exchange coupling towards

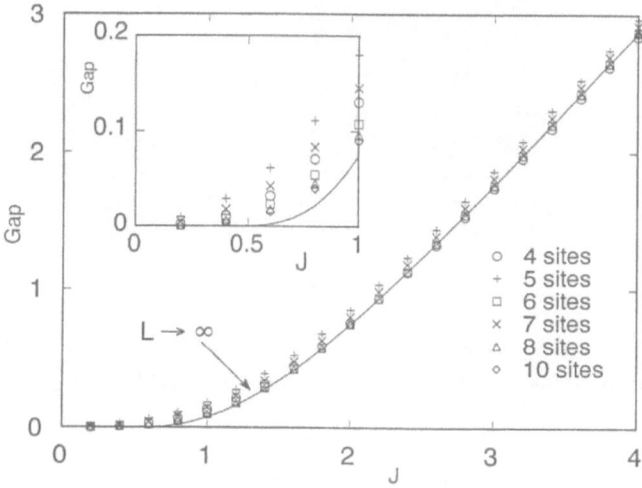

Figure 2. Spin excitation gap of the half-filled 1D Kondo lattice. The numerical results for systems from 4 to 10 sites are given. The line in the figure denotes the finite size extrapolation for the spin gap.

the weak coupling limit.[7] In Fig.2 the excitation gap as a function of J is shown for various system sizes. A careful finite size analysis is necessary to extract a meaningful result from this numerical data in the weak coupling region. The comparison of the numerical result with a mean field BCS-like description leads to a clear scaling for all computationally accessible finite J-values. This scaling allows to conclude that the gap is finite for all $J < 0$. Furthermore, it is found from this scaling that the functional dependence of the gap has an essential singularity at $J = 0$ of the form $G(J) \propto \exp(-1/bN(0)|J|)$ ($N(0)$: the density of states at the Fermi level of the conduction electrons). A comparison with the energy scale of the single impurity Kondo problem given by the Kondo temperature, $T_K \propto \exp(-1/N(0)|J|)$ shows that there is a slight lattice enhancement found in the KLM, $1 \leq b \leq 5/4$. The lowest

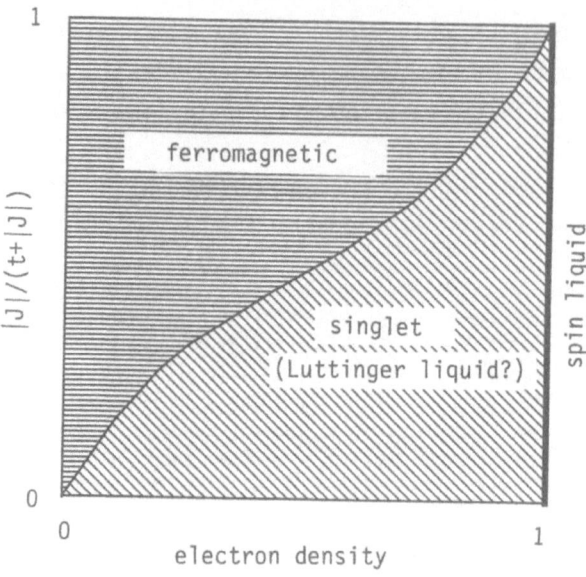

Figure 3. Schematic phase diagram of the 1D KLM, electron density versus coupling constant The emphasized line at half-filling denotes the spin liquid ground state.

spin excitation (singlet $\rightarrow$ triplet) is lower than the charge excitations for all finite J. Consequently, the 1D KLM is an *incompressible spin liquid* at half filling.[7] There is no indication of a transition to a ground state with antiferromagnetic quasi long range order which would give rise to gapless spin excitations.

PHASE DIAGRAM

For the construction of a phase diagram of the 1D KLM we can rely now on several rigorous results for limiting cases. One important limit, however, is still missing, the weak coupling limit At present it is not clear how the lift of the spin degeneracy at $J = 0$ can be described in one dimension. The concept of an effective RKKY-interaction among the localized spins is not applicable in one dimension. The reason is that this interaction produces a logarithmic singularity for the $2k_F$-spin correlation which leads to the unphysical result that the ground state energy had no lower bound. This means that for one dimension, opposite to higher dimensions, the Fermi sea of conduction electrons and the localized spins cannot be treated in a separate way as soon as J is finite. The system may show Luttinger liquid behavior for all fillings away from half-filling in the weak coupling regime. Results from numerical diagonalization of finite systems show that the spin correlation in the ground state are strongly peaked at $2k_F$ which distinguishes this phase clearly from the ferromagnetic phase in the strong coupling regime. This phase might have a strong spiral spin correlation with $Q = 2k_F$ as suggested by Fazekas and Müller-Hartmann.[8]

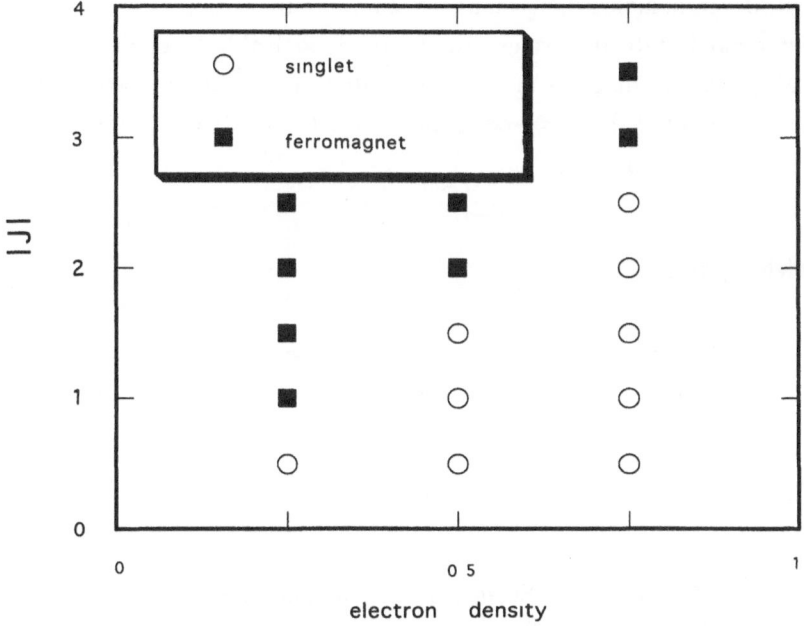

Figure 4. Phase diagram obtained from numerical diagonalization of the 8 sites Kondo lattice model.

On this basis the schematic phase diagram shown in Fig.3 may be appropriate for the 1D KLM. It shows a wide region of ferromagnetism in the low conduction electron density and the strong coupling region. This is in contrast to phase diagrams found on the basis of mean field and variational calculations.[8] Especially the "Kondo singlet" phase usually suggested for the strong exchange coupling exists only for half-filling in the 1D KLM. This schematic phase diagram is supported by numerical calculations on finite systems, exact diagonalization (Fig.4) and Quantum Monte Carlo studies [9] of a finite Kondo chain. However, in these calculations the exact location of the phase boundary lines is difficult because of various finite size effects. From these results it is also not clear whether the transition between these two phases is continuous or not.

FINAL REMARKS

We have studied the KLM in one dimension starting from the strong coupling limit and found that the ground state is ferromagnetic (spin polarized) in this regime for all conduction band fillings away from half-filling (by particle-hole symmetry also for band fillings larger than half-filled). This result supports the idea that the ferromagnetism established for only one electron in a finite system would be possibly stable for a small but finite electron concentration in the thermodynamic limit of the higher dimensional KLM.[6] However, in higher dimensions the infinite coupling regime yields for most fillings a singlet ground state (equivalent to the $U = \infty$ Hubbard model). The extension of the Nagaoka ferromagnetism - this is related here to the low conduction electron concentration limit of the $J = -\infty$ KLM - in the thermodynamic limit is still a matter of intense investigation.

Acknowledgment

We are grateful to the Swiss National Science Foundation and the Ministry of Education, Science and Culture of Japan (monbusho) for financial support of our studies.

References

1. P.A. Lee, T.M. Rice, J.W. Serene, L.J. Sham, and J.W. Wilkins, Comments Cond. Mat. Phys. **12**, 99 (1986).

2. For an experimental review see H.R. Ott, Helv. Phys. Acta **60**, 62 (1987).

3. M. Sigrist, H. Tsunetsugu, and K. Ueda, Phys.Rev.Lett. **67**, 2211 (1991).

4. K. Ueda, H. Tsunetsugu, and M. Sigrist, Phys.Rev.Lett. **68**, 1030 (1992).

5. M. Sigrist, H. Tsunetsugu, K. Ueda, and T.M. Rice, preprint.

6. M. Sigrist, K. Ueda, and H. Tsunetsugu, Phys. Rev. **B46**, 175 (1992).

7. H. Tsunetsugu, Y. Hatsugai, K. Ueda, and M. Sigrist, Phys. Rev. **B46**, 3175 (1992).

8. P. Fazekas, and E. Müller-Hartmann, Z. Phys. B- Condensed Matter **85**, 285 (1991).

9. M. Troyer, and D. Würtz, preprint.

ANISOTROPIC TRANSPORT PROPERTIES OF CERIUM KONDO COMPOUNDS

A.K. Bhattacharjee[1], B. Coqblin[1], S.M.M. Evans[2], C. Ayache[3], P. Haen[4] and F. Lapierre[4]

[1]Laboratoire de Physique des Solides, Bât. 510, Université Paris-Sud, 91405 Orsay, France
[2]Department of Mathematics, Imperial College, 180 Queens Gate, London SW7 2BZ, U.K.
[3]DRFMC, C.E.N.G., BP 85 X, 38041 Grenoble-Cédex, France
[4]CRTBT, C.N.R.S., BP 166, 38042 Grenoble-Cédex 9, France

A theoretical calculation of the anisotropy in the transport properties of cerium Kondo compounds is presented. It uses the third-order perturbation treatment of the Coqblin-Schrieffer hamiltonian at high temperatures (for temperatures T larger than T_K) and the slave-boson technique for the Anderson lattice hamiltonian at low temperatures (T smaller than T_K). It is shown that this model accounts for the main features of the anisotropic transport properties of single crystals of Ce Kondo compounds actually available. However the need for more experiments is emphasized.

1. INTRODUCTION

Anomalous cerium compounds have been extensively studied from both an experimental and theoretical point of view since more than two decades, i.e. since the date when a theoretical computation for the so-called "high temperature" (HT) regime of the resistivity (ρ) of Ce Kondo lattice compounds was first reported.[1] The same technique, i.e. third order perturbation calculation (for T higher than the Kondo Temperature T_K) with the Coqblin-Schrieffer hamiltonian including crystal field effects, was successively applied to the thermoelectric power[2] (S) and to the thermal conductivity[3] (K). Other developments have been made during the same period,[4,5] but in each of them only the transport properties of polycrystalline compounds were considered. This was because at this time, almost no data on transport properties of anisotropic single crystals were reported. This was particularly true for intermetallic compounds, single crystals of which are quite difficult to grow.

A real need and a great incitation to perform further theoretical work were raised only a few years ago with the first reports of anisotropies in the transport properties of some Ce Kondo compounds such as the orthorhombic compound $CeCu_6$,[6,7] or the tetragonal compound $CePt_2Si_2$.[8] For the latter, a complete study of the anisotropy of the three transport properties ρ, K and S in the HT regime has been performed. The experimental data have been accounted for within a theoretical model that extended the method of refs.1-3 by introducing an anisotropic relaxation time. Similar calculations[9] also lead to a reasonable agreement with the experimental HT anisotropy of ρ and S in $CeCu_6$, in particular the three positive maxima occuring[7] in S around 50 K. An anisotropic behaviour is also observed in the "low temperature" (LT) Fermi liquid regime of the resistivity of heavy fermion compounds, i.e. in both the AT^2 and the BT terms, as well as in the residual resistivity, ρ_0. Experimental values for these parameters have been reported for $CeCu_6$ [6,7] and, more recently, for the tetragonal compound $CeRu_2Si_2$.[10] The anisotropy of ρ_0 and of the A coefficients has been calculated[11] using the slave boson technique on the periodic Anderson hamiltonian and compared to the experimental values. The anisotropy of the magnetic susceptibility and of the relaxation time T_1 has been also theoretically studied[12,13] by the latter approach and applied to recent results on cerium Kondo compounds such as $CeCu_6$ or $CeRu_2Si_2$.

We shall present here only the key points of the theoretical calculations of ref. 11. Then, we shall compare the predictions with some experimental results, limiting ourselves to the resistivity. For this pupose, a short review of available data concerning the anisotropy of ρ will be made and discussed.

2. THEORETICAL MODEL

2.1 "High temperature" regime $(T > T_K)$

The model we present here is based on a third-order perturbation theory calculation on the Coqblin-Schrieffer hamiltonian including crystal-field effects, like those previously performed for of ρ,[1] S,[2] and K,[3] for polycrystals. In the latter, the relaxation time of a conduction electron was approximated by an isotropic average over the different $\vec{k}$ directions of the conduction electrons. On the contrary, in the case of single crystals, we compute the transport properties along the principal axes i and we must calculate the relaxation time $\tau_{\vec{k}\sigma}$ for a conduction electron plane wave of wavevector $\vec{k}$ and spin σ, which turns out to be highly anisotropic.

The electrical resistivity ρ is given by :

$$\frac{1}{\rho} = e^2 K_0 ,$$ (1)

the thermoelectric power by :

$$S = \frac{K_1}{eTK_0} ,$$ (2)

and the thermal conductivity κ by :

$$\kappa = \frac{1}{T}\left[K_2 - \frac{(K_1)^2}{K_0} \right] .$$ (3)

In the case of a single crystal, the integrals K_n (written in the following K_n^i for each direction i) are given by :

$$K_n^i = \frac{1}{8\pi^3} \int \left(\frac{\partial \varepsilon_k}{\hbar \partial k_i}\right)^2 \left(-\frac{\partial f_k}{\partial \varepsilon_k}\right) \varepsilon_k^n (\tau_{\vec{k}\uparrow} + \tau_{\vec{k}\downarrow}) \, dk \tag{4}$$

ε_k is the conduction-electron energy and f_k the Fermi-Dirac distribution. The relaxation time $\tau_{k\sigma}$ is given here by :

$$\frac{1}{\tau_{k\sigma}} = \sum_\mu |\langle \vec{k}\sigma|k\mu\rangle|^2 \frac{1}{\tau_{k\mu}}, \tag{5}$$

as a function of the partial wave relaxation time $\tau_{k\mu}$, where the partial wave $|k\mu\rangle$ corresponds to one of the eigenfunctions in presence of crystalline-field effects. The partial wave relaxation times $\tau_{k\mu}$ are computed by the usual third-order perturbation theory and are given by the equations (8), (9) and (10) of ref. 11.

In general, the partial wave function $|k\mu\rangle$ (or the corresponding 4f eigenfunction $|\mu\rangle$ in the presence of crystal-field effects) is a linear combination :

$$|k\mu\rangle = \sum_M a_{M\mu} |kM\rangle, \tag{6}$$

of the elementary wavefunctions $|kM\rangle$ with $M = \pm 1/2, \pm 3/2, \pm 5/2$ in the case of cerium compounds and with $M = \pm 1/2, \pm 3/2, \pm 5/2, \pm 7/2$ in the case of ytterbium compounds. Then we insert (6) in the expression (5) and the weight of the partial wave $|kM\rangle$ inside the plane wave $|k\sigma\rangle$ is given by (with $\sigma = \pm 1/2$) :

$$|\langle \vec{k}\sigma|kM\rangle|^2 = 4\pi \left(\frac{7 - 4\sigma M}{14}\right) |Y_3^{M-\sigma} (\Omega_k)|^2 . \tag{7}$$

The key point of the model yielding anisotropy in the transport properties is the $\vec{k}$-dependence of the relaxation time $\tau_{\vec{k}\sigma}$, because the angular integration over Ω_k in (4) gives different results for the different directions. Integration of the products of spherical harmonics leads to three values of the integrals K_n^x, K_n^y and K_n^z. We then proceed, as usual, for the calculation of the transport properties by using the third-order perturbation approximation for the inversion of the relaxation times.

Then, the electrical resistivity ρ_i, the thermoelectric power S_i and the thermal conductivity K_i along the i-direction have been computed for cerium or ytterbium compounds and we do not present here the detailed formulae which are given by the equations (13)-(19) of ref. 11. Let us remark that the anisotropy disappears in cubic crystals and that the basal plane anisotropy disappears in hexagonal or tetragonal crystal structures.

2.2 "Low temperature" regime $(T < T_K)$

By low temperature regime we understand the temperatures below T_K, where Ce Kondo compounds are characterized by a Fermi liquid behaviour. The calculation as reported in ref. 11, starts from the Anderson lattice hamiltonian in the Kondo limit and in the $U \rightarrow \infty$ and large spin-orbit coupling limit, yielding $N = (2j+1)$ 4f states with different $M = j_z$ values. Then, we include crystalline field effects for non cubic cerium Kondo compounds, thus splitting the 4f level into three doublets of energy E_μ. The hybridization matrix element between a 4f wavefunction of M value on site i and a plane wave of wavevector $\vec{k}$ and spin σ equal to :

$$V_{M\sigma}(\vec{k}) = V \langle \vec{k}\sigma|kM\rangle , \tag{8}$$

where V is a constant and $\langle \vec{k}\sigma|kM\rangle$ is given by the equation (7).

Then we use the "slave boson" technique which consists in rewriting the hybridization term in the $U \rightarrow \infty$ limit, in order to avoid double occupancy for 4f electrons. In the "mean field" approximation, we obtain renormalized parameters for the hybridization and the different energies of the three doublets split by the crystalline field effect. Then, we include fluctuations to go beyond the mean field solution, in order to compute the self-energy and consequently the transport properties.

In the low temperatures ($T < T_K$) Fermi liquid regime, we get a T^2 dependence of the resistivity, but different coefficients along the different principal directions. The reason of this anisotropy lies here, as in high temperature treatment, in the $\vec{k}$-dependence of the hybridization term as resulting from the $\ell = 3$ spherical harmonic entering the formulae (7) and (8).

These calculations have been extended in several directions. First of all, it is well known that the residual resistivity ρ_0 is large in cerium Kondo compounds and we need to introduce the effect of disorder to account for it. In a first step, we take an isotropic and temperature independent scattering rate $1/\tau_i$ resulting from disorder purely on the conduction electron sites. Including such an effect yields a smaller anisotropy of the resistivity. But moreover, as has been noted experimentally[14] the coefficient A of the T^2 law and ρ_0 might be not independent from each other, which suggests a temperature dependent impurity scattering. Thus, in a second step, we include f electron disorder in which ρ_0 can be thought as arising from Kondo hole scattering.[15]

The thermoelectric power computed within the same model, is linear in temperature at low temperatures in the case of a pure lattice, while in the case where impurity scattering is important, S behaves as T^3 ; the anisotropy of this T^3 term is approximately the same as that for the T^2 term in the resistivity. Finally, the thermal conductivity diverges as $1/T$ for a pure lattice; the addition of impurities suppresses the divergence and instead κ goes to 0 linearly with decreasing temperature, but there is no longer any anisotropy in this case.

3. EXPERIMENTAL REVIEW AND COMPARISION WITH THE MODEL

There are not many available data on the anisotropy of transport properties on Ce Kondo single crystals, even in the "high temperature" domain. The three anisotropic transport properties ρ, S and K were studied in both the HT and LT regimes[6,7,16] only for the case of the orthorhombic compound $CeCu_6$. As mentioned before, the studies[8] performed on the tetragonal compound $CePt_2Si_2$ were principally devoted to the HT domain, although preliminary measurements of ρ down to 80 mK have been also reported.[17] However, only few compounds crystallize in the same non-symmetric $CaBe_2Ge_2$-type structure as $CePt_2Si_2$ (let us mention some resitsivity measurements performed[18] on $CeNi_2Sn_2$). Thus, one could expect to find more data for compounds which crystallize in the symmetric tetragonal $ThCr_2Si_2$-type structure. Except for a study[19] of the anisotropy of S in $CeRu_2Si_2$, most of the works reported for such compounds concern the anisotropy of ρ. We have already mentioned a study[10] performed down to 20 mK on $CeRu_2Si_2$, while other measurements have been recently reported.[20] In the other cases, the anisotropy of ρ has been measured only in the HT domain. Measurements on $CeCu_2Si_2$ were reported in several papers.[21-24] One of them[21] deals also whith $CeNi_2Ge_2$ while interesting data have been reported[25] for $CePd_2Si_2$. Works on compounds with other crystallographic structure can be mentioned, such as the recent ones[26,27] performed on $CeAl_3$ in the entire range of temperature and the HT measurements performed on $CeSi_x$ compounds[28] and on compounds of the Ce-Sn system.[29] In the following paragraphs, we will reproduce and discuss some of the above results, limiting ourselves to the electrical resitivity anisotropy.

3.1 "High temperature" domain

As mentioned in introduction, the preceding model has been successfully applied to the three anisotropic transport properties of $CePt_2Si_2$ and to the electrical resistivity and the thermoelectric power of $CeCu_6$ in the HT domain. These results have been discussed in ref. 8 for $CePt_2Si_2$ and ref. 9 for $CeCu_6$.

Let us however emphasize some points concerning the resistivity anisotropy in $CePt_2Si_2$. Figure 1 represents the experimental results down to 80 mK of ref. 17 and the calculated curves for the HT regime, showing good agreement between the model and experiment in this temperature range. However, the experimental curves should be corrected for the respective phonon contributions, in order to deal only with the Ce contributions. Since there are no data for a reference compound such as $LaPt_2Si_2$ available yet, we have estimated the Ce contributions $\Delta\rho//c$ and $\Delta\rho//a$ by subtracting the resistivity of a $LaRu_2Si_2$ polycrystal[30], which is plotted in fig. 1 (although the crystallographic structures are not the same). The latter, assumed to be isotropic, shows a reasonable value of 30 μΩcm at room temperature and a linear decrease down to about 100 K. Fig. 2 represents the resulting experimental anisotropy ratio $\Delta\rho_{///}\Delta\rho_{\perp}$, i.e. $(\Delta\rho//c)/(\Delta\rho//a)$ together with the anisotropy (ρ_z/ρ_{xy}) calculated by the model. The general features of the two curves are the same, i.e. an increase of the anisotropy as T decreases, especially below about 100 K (the minimum at high T in the experimental curve can be an artefact due to the subtraction of the non-magnetic reference).

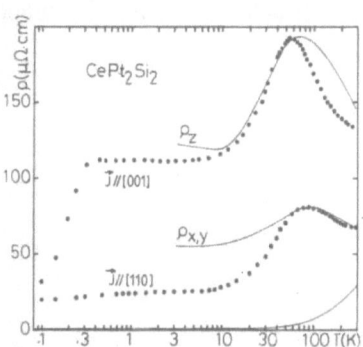 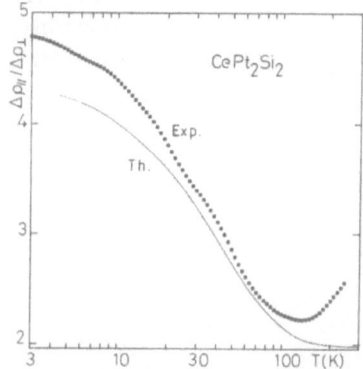

Figure 1. (Left frame) : Variation of the experimental resistivity for i//a and i//c (after ref. 17) and calculated curves for the HT regime in $CePt_2Si_2$, as a function of logT. Continuous line at right bottom represents the expected phonon contribution (see text).

Figure 2. (Right frame) : Variation of the anisotropy ratio $\Delta\rho_{///}\Delta\rho_{\perp}$ as derived from the experimental curves of fig. 1 after subtraction of the phonon contribution and the theoretical anisotropy curve (full curve).

An important feature observed in fig. 1 is the decrease of ρ//c below ≈ 0.4 K, from a quite high value of ≈100 μΩcm to a residual one of the same order as ρ//a (≈ 20 μΩcm). As a consequence, the experimental anisotropy is expected to decrease below this temperature. Notice that this decrease is not well understood yet and will be further studied. Besides, the residual values are quite high for both ρ//c and ρ//a, which is not so surprising, since $CePt_2Si_2$ crystallizes in the non-symmetric $CaBe_2Ge_2$ structure, in which site inversions can easily occur.

The data recently reported[18] for the compound $CeNi_2Sn_2$ show also $\rho//c > \rho//a$, in agreement with the classification reported in ref. 31. It would be interesting to check also the model on these data.

Let us consider now the tetragonal compounds which crystallize in the symmetric $ThCr_2Si_2$ structure. In the case of $CeRu_2Si_2$, an anisotropy $\rho//c < \rho//a$ is observed, at least at room temperature, also in agreement with the classification reported in ref. 31. Fig. 3 represents the variations of the Ce contributions $\Delta\rho//c$ and $\Delta\rho//a$ in this compound, as reported by two of us in ref 10. For the HT regime, similar variations can be deduced from the data of ref. 20, in spite of small differences in the absolute values of ρ at room temperature, and of the fact that the resistivity of $LaRu_2Si_2$ is reported as slightly anisotropic in ref. 10 ($\rho//c < \rho//a$) while it is not in ref. 20. These small differences, as well as the introduction of uncertainties in the absolute values of ρ do not greatly affect the general variation of the anisotropy ratio $\Delta\rho_{//}/\Delta\rho_\perp$ which is plotted in fig. 4. The main feature of this ratio is a maximum around 14 K, which might be related to T_K and/or to the onset of coherence.

In the case of $CePd_2Si_2$, actually reported data[25] show $\rho//a$ about three times larger than $\rho//c$ down to 4.2 K, i.e. even below the ordering temperature $T_N = 8.5$ K. In fig. 5, we have plotted the HT part (T > 10 K) of these results vs. logT. Since no data for a reference compound (such as $LaPd_2Si_2$) were reported, we have calculated the Ce contributions, $\Delta\rho$, by assuming simply reference resitivity variations identical to those reported[10] for $LaRu_2Si_2$. The $\Delta\rho$ curves are plotted also in fig. 5. They are typical for a Kondo lattice compound with a crystal field shoulder separating two lnT variations. The resulting "experimental" anisotropy is plotted in fig. 6, down to 30 K. It shows a shallow maximum around 80-100 K, i.e. in the vicinity of the crystal field structure. However, the total variation is small. More precise measurements are needed to estimate the anisotropy of ρ below 30 K in this compound, especially below T_N and down to the residual value.

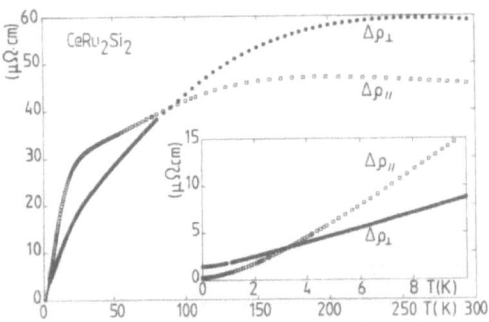
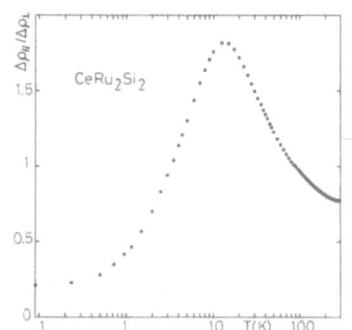

Figure 3. (Left frame) : Temperature dependence of the Ce contributions $\Delta\rho//c$ and $\Delta\rho\perp c$ to the resistivity of $CeRu_2Si_2$, after subtraction of the respective resistivities $\rho//c$ and $\rho\perp c$ of $LaRu_2Si_2$ (after ref. 10).
Figure 4. (Right frame) : Variation of the anisotropy $\Delta\rho_{//}/\Delta\rho_\perp$ vs. logT in $CeRu_2Si_2$.

Although several measurements on $CeCu_2Si_2$ have been reported,[21-24] it is not possible to determine precisely the thermal variation of the anisotropy of ρ in this compound. Quite large anisotropies ($\rho//c < \rho//a$) have been first reported[21] for a non-superconducting crystal. In another study,[22] large differences in absolute values are reported for crystals grown by the Czochralski method, but issued from different batches. In the two other reports,[23,24] $\rho//c$ and $\rho//a$ are normalized to the same value at ambient temperature. One can simply observe in these data the occurence of an anisotropy below about 200 K, which increases on cooling, at least in the HT regime.

For a long time, only polycrystals of the hexagonal compound CeAl₃ could be prepared, since this compound is formed in a peritectoidal reaction. But some years ago, small single crystals were extracted from large polycrystals and their resistivity anisotropy measured.[26] Very recently, new measurements have been reported.[27] In the HT domain, $\rho//c < \rho\perp c$ is observed. We have subtracted from the latest $\rho//c$ and $\rho\perp c$ results[27] the resistivity of LaAl₃ reported in ref. 32. This lead to the differences $\Delta\rho_{//}$ and $\Delta\rho_\perp$ and to their ratio $\Delta\rho_{//}/\Delta\rho_\perp$

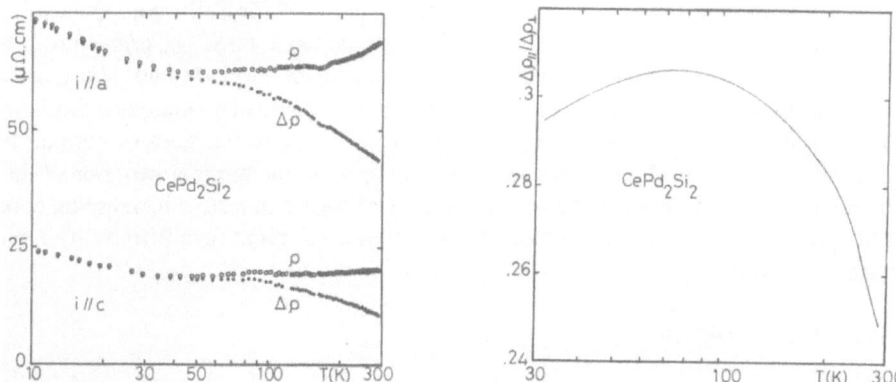

Figure 5. (Left frame) : Variation vs. logT of $\rho//c$ and $\rho\perp c$ of CePd₂Si₂ (after ref. 25) and of the expected Ce contributions $\Delta\rho//c$ and $\Delta\rho\perp c$, obtained after subtraction of phonon contributions evaluated from LaRu₂Si₂.[10]

Figure 6. (Right frame) : Resulting variation vs. logT of the anisotropy $\Delta\rho_{//}/\Delta\rho_\perp$ in CePd₂Si₂.

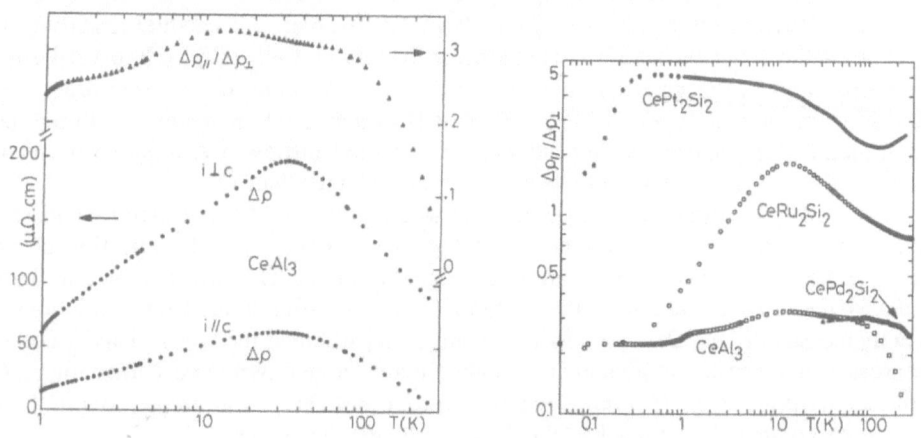

Figure 7. (Left frame) : Variation of Ce contributions $\Delta\rho//c$ and $\Delta\rho\perp c$ in CeAl₃ and the resulting anisotropy versus logT. These curves are deduced from the results of ref. 27, by subtraction of the contribution of LaAl₃ (after ref. 32).

Figure 8. (Right frame) : Comparison of the anisotropy curves of figs. 2, 4, 6, and 7, in log-log scales.

which are plotted in fig. 7. Both $\Delta\rho_{//}$ and $\Delta\rho_\perp$ show a maximum near 30-40 K, certainly due to crystal field effect, as in polycrystals.[1,32] This crystal field effect gives raise to the change of slope observed in $\Delta\rho_{//}/\Delta\rho_\perp$ around 80 K. (Small changes in these characteristic temperatures would be obtained by another choice for the resistivity of LaAl₃. The latter was measured[32] on a polycrystal and it is of course isotropic, but it seems to be of a good order of magnitude). A maximum is observed in $\Delta\rho_{//}/\Delta\rho_\perp$ near 14 K and some kind of change of slope near 4 K in the decrease below this maximum. The origin of these structures is not quite clear. The first one seems to be related in some way to the onset of coherence and the lower one to T_K. It is noticeable that a kind of plateau or slow increase has been observed[30,33] in the Hall coefficient R_H of CeAl$_3$ between about these two temperatures. The magnetic order occuring[26,27] in this compound lead to another anomaly in the anisotropy ratio near 1 K. Resistivity measurements on LaAl$_3$ single crystals would be useful in order to confirm the existence of the above anomalies in the anisotropy ratio. Actually they are small compared to those occuring in CeRu$_2$Si$_2$ and CePt$_2$Si$_2$, as seen in fig. 8 where all curves are plotted together in log scales.

Finally, it seems worthwhile to mention the data reported[28] for CeSi$_x$ compounds. In the tetragonal (ThSi$_2$ structure) non-ordered Kondo lattice compound CeSi$_{1.86}$, $\rho//c$ is lower than $\rho//a$ above about 100 K and larger below. The crossing of the $\rho//a$ and $\rho//c$ curves is reminiscent of that observed[10] in CeRu$_2$Si$_2$. In the ferromagnetically ordered orthorhombic compound CeSi$_{1.71}$, $\rho//c$ is larger than $\rho//a$ down to 1.5 K. No resistivities of normal metal references are reported in ref. 28. However, one can expect in the thermal variation of the Ce contribution ratios, the occurence of a maximum for CeSi$_{1.86}$, and a slow increase on cooling for CeSi$_{1.71}$. As discussed by the authors, these differences might result from different T_K and crystal field values.

3.2 Fermi liquid regime

The resistivity of cerium Kondo compounds has been measured at low temperatures with respect to T_K along the different principal axes i, only in a few cases, namely CeCu$_6$,[6,7] CeRu$_2$Si$_2$,[10] and CeAl$_3$.[26,27] The residual resistivities ρ_{0i} and the coefficients A_i of the quadratic term $\rho_i = A_i T^2$ depend generally on the axis i considered. We have shown in ref. 11 that for CeCu$_6$ the agreement between calculated and experimental values of ref. 6 is satisfying for the ratios between the A_i. The calculation also predicts anisotropy ratios between the ρ_{0i} which agree with those reported in refs. 6 and 7.

For CeRu$_2$Si$_2$, the experimental results of ref. 10 lead to a residual resistivity ratio $\rho_{0\perp}/\rho_{0//}$ of the order of 3 after correction for the ρ_{0i} of LaRu$_2$Si$_2$. (Here we write the anisotropy ratio in the way they are used in ref. 11). A value of the same order can be deduced from the data reported in ref. 20. This is in reasonable agreement with the model which predicts $\rho_{0\perp}/\rho_{0//} = 5.8$. But, the experimental anisotropy of A reported in ref. 10 is small : $A_\perp/A_{//} \approx 0.7$, while the model predicts values larger than 1.

For CeAl$_3$, the first measurements[26] showed a crossing of the resistivity curves near 0.6 K, leading to ratios $\rho_{0\perp}/\rho_{0//} \approx 0.7$ and $A_\perp/A_{//} \approx 2.5$ (for the AT^2 variation observed below 0.35 K, i.e. well below the antiferromagnetic order). This was compatible with the model for the A_i ratio, but not for the ρ_{0i} ratio. However, different resistivity variations were recently measured[27] for CeAl$_3$ at low temperature. In particular, the new resistivity curves do not cross each other any more and anisotropy remains large down to very low temperature : both the $\rho_{0\perp}/\rho_{0//}$ and $A_\perp/A_{//}$ ratios are of the order of 4. These values are now in the right sense, eventhough twice as much as the theoretically predicted ones.

4. CONCLUDING REMARKS

There are certainly several origins for the anisotropy, but we have shown in the model[11] summarized above that the Coqblin-Schrieffer or Anderson lattice hamiltonians, which describe both the Kondo effect and crystalline field effects, can yield a large anisotropy in transport properties and provide usefull informations for the study of the Kondo effet in Ce compounds. For the HT regime, up to now the model has been quantitatively applied to CePt$_2$Si$_2$ and CeCu$_6$ yielding good agreement.

The experimental resistivity anisotropies we have reviewed here show interesting features. In particular, the temperature dependence of the anisotropy ratio reveals structures which are not obvious in the resistivity curves. Whatever be situation at room temperature ($\rho_{//}$ higher or lower than $\rho_\perp$), the anisotropy ratio, $\rho_{//}/\rho_\perp$, increases upon cooling, then it is eventually followed by a maximum and a decrease at lower temperature. The observed structures can be generally related to characteristic effects and/or temperatures of the systems, i.e. crystal field, T_K, coherence, magnetic order. However, the amplitude of the anisotropy ratio variation is very different in different systems (see fig. 8). Large variations are observed in $CePt_2Si_2$ and $CeRu_2Si_2$. The effect is relatively small for $CePd_2Si_2$ and $CeAl_3$. However, it should be checked whether a reduction of the anisotropy occurs far below T_N in $CePd_2Si_2$, and whether this could be a general behaviour in ordered compounds. Also, the case of $CePt_2Si_2$ needs to be further studied below 1 K. On the other hand, it would be interesting to extend the theoretical model to cover the "intermediate" temperature range.

Focussing now on the very low temperature T^2 variations, we must say that experimental data in this temperature domain must be well confirmed before any quantitative comparison with theoretical results can be made. Nevertheless, reasonable agreement has been found in the cases of $CeCu_6$ and $CeAl_3$.

ACKNOWLEDGEMENTS

We are indebted to Dr. D. Jaccard for supplying us the recent resistivity data of ref. 27 for $CeAl_3$ and for interesting discussion about the resistivity anisotropy in this compound.

REFERENCES

1. B. Cornut and B. Coqblin, *Phys. Rev. B* 5:4541 (1972).
2. A.K. Bhattacharjee and B. Coqblin, *Phys. Rev. B* 13:3441 (1976).
3. A.K. Bhattacharjee and B. Coqblin, *Phys. Rev. B* 38:338 (1988).
4. S. Maekawa, S. Kashiba, M. Tachiki, and S. Takahashi, *J. Phys. Soc. Jpn* 55:3194 (1986); and in: "Theory of Heavy Fermions and Valence Fluctuations," T. Kasuya and T. Saso, eds., Springer Verlag, New York (1985) p. 90.
5. A. Guessous, Thesis, Grenoble (1987, unpublished).
6. A. Sumiyama, Y. Oda, H. Nagano, Y. Onuki, K. Shibutani, and T. Komatsubara, *J. Phys. Soc. Jpn* 55:1294 (1986).
7. A. Amato, D. Jaccard, E. Walker, and J. Flouquet, *Solid State Commun.* 55:1131 (1985).
8. A.K. Bhattacharjee, B. Coqblin, M. Raki, L. Forro, C. Ayache, and D. Schmitt, *J.Physique* 50:2781 (1989).
9. B. Coqblin, A.K. Bhattacharjee, and S.M.M. Evans, *J. Mag. Mag. Mat.* 90&91:393 (1990).
10. F. Lapierre and P. Haen, *J. Mag. Mag. Mat.* 108:167 (1992).
11. S.M.M. Evans, A.K. Bhattacharjee, and B. Coqblin, *Phys. Rev. B* 45:7244 (1992).
12. S.M.M. Evans, *J. Phys. Cond. Mater* 2:9097 (1990).
13. S.M.M. Evans, and B. Coqblin, *Phys. Rev. B* 43:12790 (1991).
14. D. Jaccard, R. Cibin, and J. Sierro, *Helv. Phys. Acta* 61:530 (1988).
15. J.M. Lawrence, J.D. Thompson, and Y.Y. Chen, *Phys. Rev. Lett.* 54:2537 (1985).
16. A. Amato, D. Jaccard, J. Flouquet, F. Lapierre, J.L. Tholence, R.A. Fisher, S.E. Lacy, J.A. Olsen, and N.E. Phillips, *J. Low Temp. Phys.* 68:371 (1987).
17. R.M. Marsolais, C. Ayache, D. Schmitt, A.K. Bhattacharjee, and B. Coqblin, *J. Magn. Magn. Mat.* 76&77:269 (1988).
18. T. Takabatake, F. Teshima, H. Fujii, S. Nishigori, T. Suzuki, T. Fujita, Y. Yamaguchi, and J. Sakurai, *J. Magn. Magn. Mat.* 790&91:474 (1990).

19. A. Amato, D. Jaccard, J. Sierro, P. Haen, P. Lejay, and J. Flouquet, *J. Low Temp. Phys.* 77:195 (1989).

20. Y. Ōnuki, I. Umehara, A. K. Albessard, T. Ebihara, and K. Satoh, *J. Phys. Soc. Jpn* 61:960 (1992).

21. H. Schneider, Z. Kletowski, F. Oster, and D. Wohlleben, *Solid State Commun.* 48:1093 (1983).

22. Y. Ōnuki, Y. Furukawa, and T. Komatsubara, *J. Phys. Soc. Jpn* 53:2197 (1984).

23. W. Asmuss, M. Herrmann, U. Rauchschwalbe, S. Riegel, W. Lieke, H. Spille, S. Horn, G. Weber, F. Steglich, and G. Cordier, *Phys. Rev. Lett..* 52:469 (1984).

24. B. Batlogg, J. P. Remeika, and A.S. Cooper, *J. Appl. Phys.* 55:2001 (1984).

25. R.A. Steeman, E. Frikkee, R.B. Helmholdt, A.A. Menovsky, J. van den Berg, G.J. Nieuwenhuys, and J.A. Mydosh, *Solid State Commun.* 66:103 (1988).

26. D. Jaccard, R. Cibin, J.L. Jorda, and J. Flouquet, *Jpn J. Appl. Phys.* 26 (suppl. 26-3):517 (1987).

27. G. Lapertot, R. Calemczuk, C. Marcenat, J.H. Henry, J.X. Boucherle, J. Flouquet, J. Hammann, R. Cibin, J. Cors, D. Jaccard, and J. Sierro, SCES'92, Sendai (Japan) sept. 7-11, 1992, to be published in *Physica B*.

28. J. Pierre, O. Laborde, E. Houssay, A. Rouault, J.P. Sénateur, and R. Madar, *J. Phys.: Cond. Mater* 2:431 (1990).

29. A. Stunault, Thesis, Grenoble (1988); and to be published

30. P. Haen, J. Flouquet, F. Lapierre, P. Lejay, and G. Remenyi, *J. Low Temp. Phys.* 67:391 (1987).

31. R.A. Steeman, A.J. Dirkmaat, A.A. Menovsky, E. Frikkee, G.J. Nieuwenhuys, and J.A. Mydosh, *Physica B* 163:382 (1990).

32. A. Percheron, J.C. Achard, O. Gorochov, B. Cornut, D. Jérome, and B. Coqblin, *Solid State Commun.* 12:1289 (1973).

33. Y. Ōnuki, T. Yamazaki, T. Omi, I. Ukon, A. Kobori, and T. Komatsubara, *J. Phys. Soc. Jpn* 58:2126 (1989).

EFFECT OF PRESSURE ON THE ELECTRICAL RESISTIVITY OF
A GAP–TYPE VALENCE FLUCTUATING COMPOUND CeNiSn

Makio Kurisu,[1] Toshiro Takabatake,[2] and Hironobu Fujii[2]

[1]Faculty of Engineering, Iwate University, Morioka 020, Japan
[2]Faculty of Integrated Arts and Sciences
Hiroshima University, Nakaku Hiroshima 730, Japan

INTRODUCTION

CeNiSn is an unconventional example of a valence fluctuating cerium intermetallic compound with an energy gap at the Fermi level [1–5]. As expected from its crystal structure of the orthorhombic ε–TiNiSi–type, many physical properties are anisotropic. Transport measurements indicate that the appearance of gap is highly anisotropic below 7 K and that the anisotropic inherent single–impurity Kondo scattering dominates at high temperatures [3]. Magnetic measurements also show a strongly anisotropic behavior; susceptibility indicates a character of strongly anisotropic valence–fluctuation at high temperatures and a pronounced peak only along the a–axis at T_{coh} = 12 K at which the resistivity also takes a peak. These anomalies at T_{coh} were attributed to the development of an antiferromagnetic coherency which results in the gap formation [3]. Furthermore, the resistivity and specific–heat measurements in high magnetic field demonstrate an anisotropic suppression of the energy gap [4]. Pressure, as well as magnetic field, has been accepted to be a clean and effective tool to control the degree of hybridization of the $4f$ and conduction electrons in so–called valence–fluctuating systems or dense–Kondo regime materials. In a previous high pressure study on a polycrystal sample CeNiSn, we showed that the energy gap is continuously decreased with increasing pressure and would be closed at a pressure of ~ 30 kbar. In addition, a Kondo–like scattering in the presence of crystal field was elucidated by an application of pressure of 20 kbar [2]. This paper reports a further pressure study of these resistive anomalies of single crystal CeNiSn.

EXPERIMENTAL PROCEDURE

The samples used in this study are single crystals of CeNiSn prepared by a floating–zone method in an infrared mirror furnace. Details of the sample preparation method, the purity of starting materials and the characterization of the samples were described elsewhere [4,5].

The electrical resistivity ρ was measured with a dc four–probe method in the temperature range 1.4 K to 300 K and under hydrostatic pressures up to 20 kbar. Voltage and current leads of gold wires (0.05 mm ϕ) were soldered to rectangular single crystals, dimensions of $0.3 \times 0.5 \times 3mm^3$. Typical distance between two voltage probes was 2 mm. Hydrostatic pressure was generated by a self–clamping piston cylinder cell using a 1:1 mixture of isoamyl alcohol and n–pentane as the pressure transmitting medium. The pressure was monitored by a manganin wire gauge. On cooling, the loss of pressure was about 2 kbar independently of the clamped pressure at room temperature and the pressure at low temperatures was constant up to 150 K. The temperature of the cell was measured by Au+0.07 at.% Fe vs. chromel thermocouple in good thermal contact with the cell. The zero pressure value of ρ was redetermined after a set of high pressure measurements. The $\rho(T)$ behavior was reversible against pressure/temperature cycling.

RESULTS AND DISCUSSION

Figures 1(a)–(d) show the temperature dependence of the resistivity $\rho(T)$ for single crystal of CeNiSn at four different pressures. The zero–pressure results (Fig.1 (a)) are in good agreement with those of previous measurements [3]. The $\rho(T)$ behavior of CeNiSn at $P = 0$ kbar is characterized by the inherent and strongly anisotropic Kondo scattering and semiconductor–like dependence which is also very much anisotropic at low temperatures. As the temperature is decreased, $\rho_a(T)$, $\rho_b(T)$ and $\rho_c(T)$ show a quasilogarithmic increase for 100 K < T < 300 K. For T < 100 K, $\rho_a(T)$ and $\rho_b(T)$ reach a local maximum around $T_{max} = 100$ K and 60 K, respectively, while for the $\rho_c(T)$ there is a faint knee around 40 K. For $\rho_a(T)$, another ln T dependence is seen for 15 K < T < 40 K. Furthermore, a second well resolved maximum appears at $T_{coh} = 12$ K, which indicates the onset of some coherent scattering [3]. It is also noted that below 6 K rapid increase is seen in both $\rho_b(T)$ and $\rho_c(T)$, whereas in $\rho_a(T)$ less pronounced increase is seen.

The salient features under high pressure of the resistivity curves of single crystalline CeNiSn are the followings:

(i) The behavior of $\rho(T)$ changes continuously from that of a narrow–gap semiconductor to that of a usual dense–Kondo metal. Steep increase in $\rho_b(T)$ and $\rho_c(T)$ at low–temperatures are strongly suppressed with pressure. On the other hand, $\rho_a(T)$ is still increasing with decreasing T below 7 K even at $P = 20$ kbar (more clearly seen in Fig. 2(a)).

(ii) The resistive anomaly at T_{coh} (12 K at $P = 0$ kbar) along the a–axis is shifted to higher temperatures with increasing pressure at a rate of 1.6 K/kbar and the peak is still resolved at $P = 20$ kbar.

(iii) The T_{max} is shifted to higher temperatures with increasing pressure; T_{max} changes by a factor of 2 at a pressure of ≈ 20 kbar (see Fig. 3).

For the highest pressure of ≈ 20 kbar (Fig. 1(d)), the low–temperature $\rho(T)$ behaviors for the b and c axes indicate that the ground state is metallic. As in many unstable Ce compounds, on cooling, the $\rho(T)$ shows a logarithmic increase, reaches a maximum and then decreases rapidly at low temperatures. The low–temperature resistivity at high pressures above 12 kbar is displayed in Figs. 2(a)–(c) as a function of temperature squared. The low–temperature $\rho(T)$ data for the three principal axes show a quadratic temperature dependence between 5 K and 12 K, $\rho=\rho_0+AT^2$ with a large quadratic coefficient A. Values of the slope A determined from least squares fit to the data are plotted as a

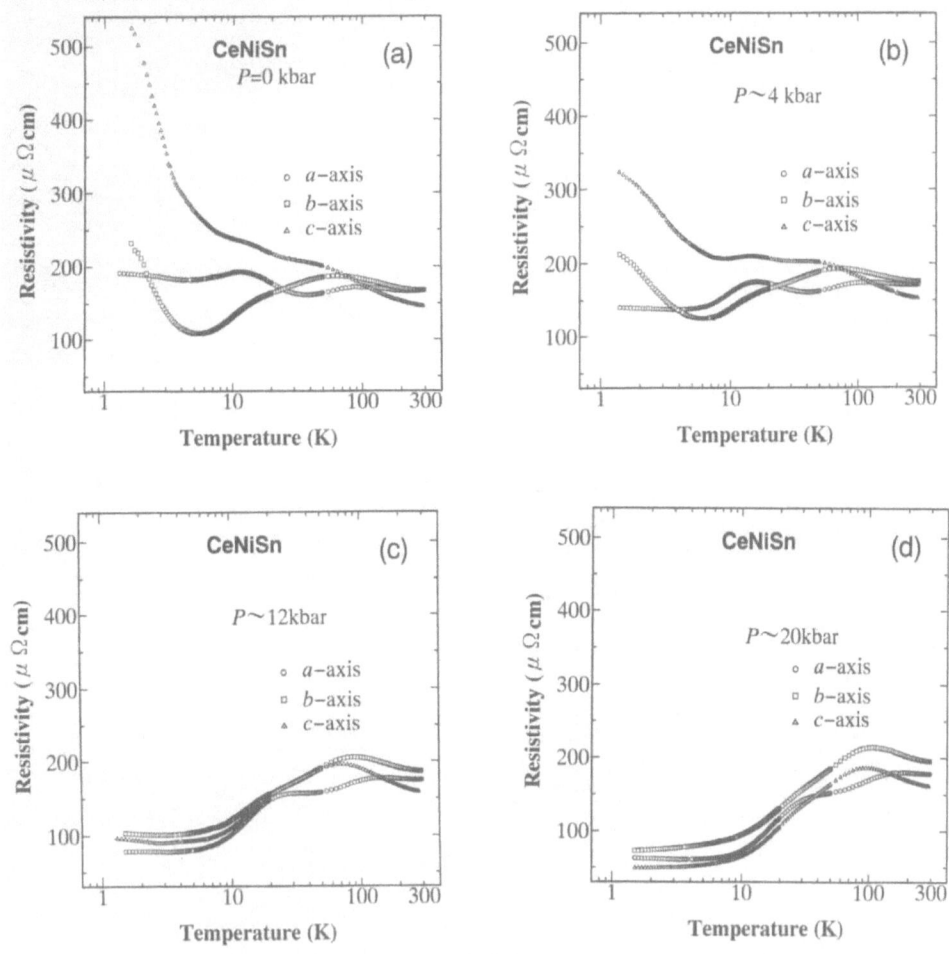

Figure 1. Resistivity of single crystal CeNiSn along the three principal axes plotted vs the logarithm of temperature for four different pressures: a) $P = 0$ kbar, b) $P \approx 4$ kbar, c) $P \approx 12$ kbar, and d) $P \approx 20$ kbar.

function of pressure in the insets of Fig. 2. A for the three axes decreases steadily with increasing pressure, to ≈ 0.2 $\mu\Omega$cm/K^2 at $P \approx 20$ kbar. When we adopt an empirical relationship between the low–temperature resistivity and the specific heat in heavy fermion compounds, i.e., $A/\gamma^2 = 1\times10^{-5}\mu\Omega$cm(mol K^2/mJ)^2 [6], we obtain $\gamma \approx 140$ mJ/mol K^2. This hypothetical metallic state γ value is comparable with both $C_m/T \approx 190$ mJ/mol K^2 at 7 K below which a gap opens at ambient pressure and $\gamma \approx 130$ mJ/mol K^2 in the magnetic field of 12 T parallel to the a–axis [4]. Thus, it is strongly suggested that a metallic state with moderately large density of states at the Fermi level is attained under pressure of ≈ 20 kbar as well as a high magnetic field of 12 T. The decrease in A with increasing pressure therefore indicates a transition from a strongly interacting Fermi liquid state towards a state with relatively weak correlations (*valence fluctuating regime*) as pressure increases. On the other hand, the specific heat measurement on single crystalline CeNiSn under high magnetic field suggests a transition to a highly correlated heavy fermion state (*Kondo regime*) as magnetic field increases [4].

For $P \approx 12$ kbar, a small upturn below 7 K which may be associated with the gap

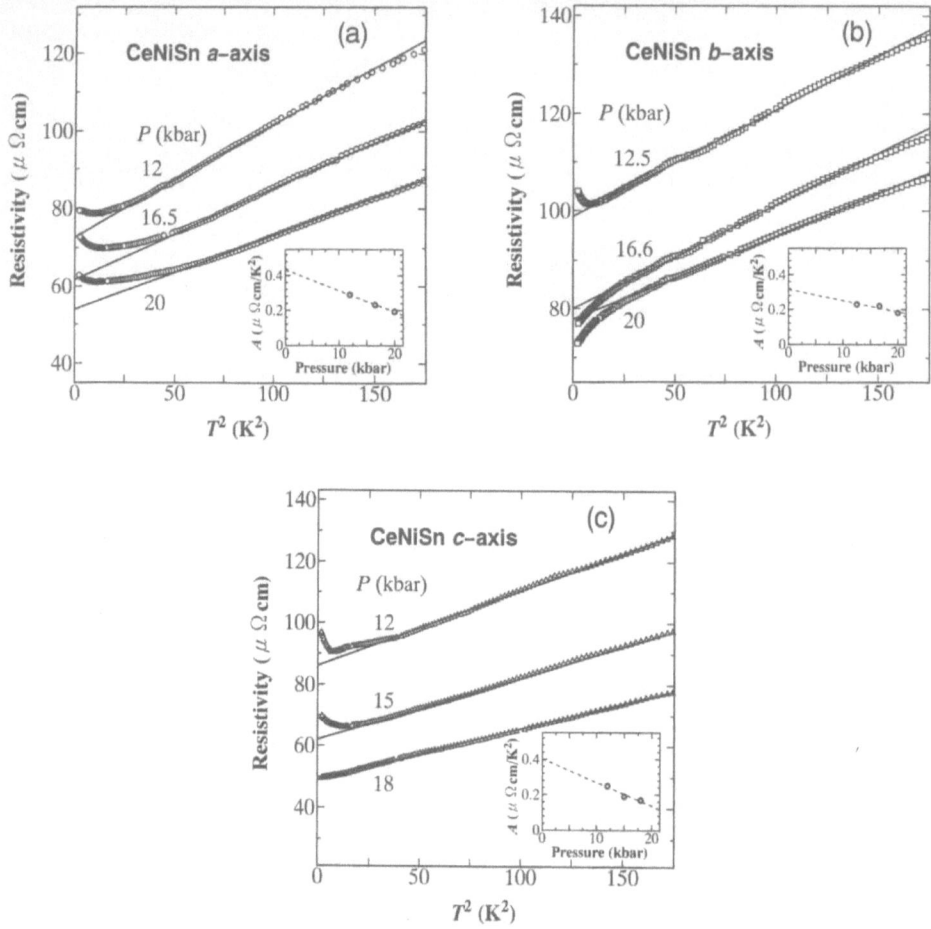

Figure 2. Resistivity of single crystal CeNiSn plotted vs the square of the temperature at pressures: (a) a–axis, (b) b–axis, and (c) c–axis. Insets: A vs pressure P, where A is the coefficient of the T^2 term in $\rho=\rho_0+AT^2$.

opening is still visible for the three axes. The more striking fact is that for the highest pressure only the $\rho_a(T)$ shows an upturn below 7 K yet, we believe that it is an intrinsic effect, suggesting that the energy gap along the a direction is not closed at 20 kbar. This is not inconsistent with the result of previous high pressure study on a polycrystal sample; the gap would be closed at 30 kbar [2].

Pressure dependence of T_{max} is shown in Fig. 3. T_{max} increases linearly with increasing pressure at rates of 3.8, 2.1 and 2.7 K/kbar for the a, b and c–axes, respectively. This behavior, as well as the increase of T_{coh} with pressure, is in agreement with the general observation that T_{max} and T_{coh} increase with pressure in usual Kondo systems, which also supports the idea that an enhancement of the degree of hybridization by an application of pressure drives CeNiSn into a weakly correlated Fermi liquid regime. Of further interest is that T_{max}'s are all different in directions. For $P = 0$ kbar, T_{max} for the a–axis is two times as large as that for other two axes. Recent inelastic neutron scattering experiment on CeNiSn has exhibited no well–defined crystal field excitations [7]. The significant difference in T_{max} for the three principal directions reflects the strongly anisotropic c–f hybridization in the orthorhombic structure. This anisotropic hybridization may play an indispensable role in the formation of a narrow–gapped insulating ground state in CeNiSn.

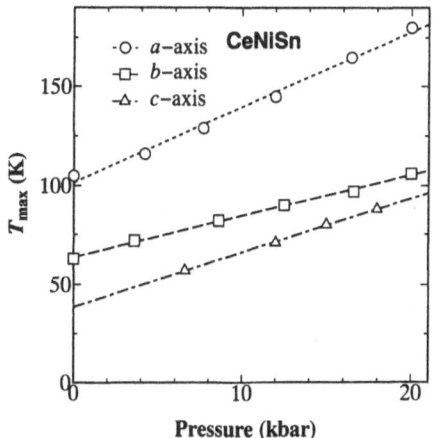

Figure 3. Pressure dependence of T_{max} for single crystal CeNiSn, where T_{max} is the temperature at which the resistivity maximum occurs.

In summary, we have shown that the behavior of $\rho(T)$ of single crystal CeNiSn changes continuously with pressure from that of a narrow–gap semiconductor to that of a usual dense Kondo metal; application of pressure drives the system into a state with relatively weak correlations from a relatively strongly interacting Fermi–liquid state.

Acknowledgements

This work was partly supported by a Grant–in–Aid for Scientific Research from the Ministry of Education, Science and Culture of Japan.

REFERENCES

[1] T.Takabatake, Y.Nakazawa, M.Ishikawa, T.Sakakibara, K.Koga, and I.Oguro, J.Magn.Magn.Mat.**76&77**:87(1988).
[2] M.Kurisu, T.Takabatake, and H.Fujiwara, Solid State Commun.**68**:595(1988).
[3] T.Takabatake, F.Teshima, H.Fujii, S.Nishigori, T.Suzuki, T.Fujita, Y.Yamaguchi, J.Sakurai, and D.Jaccard, Phys.Rev.**B41**:9607(1990).
[4] T.Takabatake, M.Nagasawa, H.Fujii, G.Kido, M.Nohara, S.Nishigori, and T.Fujita, Phys. Rev.**B45**:5740(1992).
[5] T.Takabatake, and H.Fujii, to be published in Jpn.J.Appl.Phys.Ser.8:(1992).
[6] K.Kadowaki and S.B.Woods, Solid State Commun.**58**:507(1986).
[7] M.Kohgi, T.Osakabe, K.Ohoyama, M.Kasaya, T.Takabatake and H.Fujii, to be published in Physica B.

THE HALL EFFECT IN $U_3T_3M_4$(T=Ni,Cu,Au,M=Sn,Sb)

T.Hiraoka,[1] T.Sada,[1] T.Takabatake,[2] and H.Fujii[2]

[1]Faculty of Science and Engineering, Saga University, Saga, 840 Japan
[2]Faculty of Integrated Arts and Sciences, Hiroshima University, Hiroshima, 730 Japan

INTRODUCTION

Hall effect studies in heavy fermion (HF) systems of uranium compounds[1-8] as well as in Ce ones[8-12] have recently been progressively done. The Hall effect in HF systems offers useful tool to study the HF state in the coherent region as well as in the incoherent region. In fact various behaviors are reported at low temperature region; strong temperature and magnetic field dependence including change in sign for $CeCu_6$[10] or strong enhancement at T_N for UR_2Si_2[2-5] and for CePtSi.[11]

The $U_3T_3M_4$ (T=Ni,Cu,Au,M=Sn,Sb) compound crystallizes in the cubic $Y_3Au_3Sb_4$-type structure.[13] Various electronic states appear[7]; HF states in $U_3Cu_3Sn_4$(UCS), $U_3Au_3Sn_4$(UAS) and $U_3Ni_3Sn_4$ (UNS), a semiconducting state in $U_3Ni_3Sb_4$(UNSB) and a ferromagnetic state in $U_3Cu_3Sb_4$(UCSB). UCS orders antiferromagnetically at T_N=12 K and UCSB orders ferromagnetically at T_C=91 K. UNSB shows semiconducting behavior at high temperature range with an energy gap of 0.27 eV.

This variety arises from the difference in the hybridyzation of 5f electrons with d, p and s electrons of T and M metals.[7] The present paper reports the study of this variety of states especially HF ones from measurements of the Hall effect. In the HF systems the discussion is made based on the skew scattering theory.[14] We also discuss the general aspect of the Hall effect in the present systems in comparison with other known HF systems.

SAMPLE PREPARATION

Polycrystalline samples were prepared by arc melting of constituent elements under argon atmosphere.[7] To account for the evaporation losses due to the high vapor pressure of antimony, an excess of antimony was added. The samples were subsequently annealed at about 800 ℃ for 10 days. Powder diffraction X-ray analysis showed the samples having almost single phase of the $Y_3Au_3Sb_4$-type crystal structure. The samples were spark cut into rectangular bars. For the simultaneous measurement of Hall effect and resistivity, 6 fine gold wires were spot welded to these samples using short pulse spot welder. Measurements were done by usual standard DC method in a field of 1 T.

EXPERIMENTAL RESULTS AND DISCUSSION

Figures 1-5 show the Hall effect and resistivity for UNS, UAS, UCS, UNSB and UCSB respectively. All the three HF systems UNS (Fig.1), UAS (Fig.2) and UCS (Fig.3) show positive and large Hall effect which is the characteristic of heavy fermion systems. These behaviors can be considered due to the skew scattering mechanism as discussed below. UNSB (Fig.4) and UCSB (Fig.5) also show positive Hall effect but in the former it is not due to the skew scattering. As shown in the insert, both Hall effect and resistivity show an activation-type behavior of $\exp(E_g/2kT)$ above about 230 K from which the energy gap of 0.27 eV is estimated. Band calculation[15] on UNSB confirmed the semiconducting structure. On the other hand, the Hall effect in UCSB above T_C is considered as due to the same mechanism as the above HF systems. Both the Hall effect and resistivity of UAS and UNS show maximum at low temperatures and both decrease as going into the coherent regime. However both in UCS continue to increase below T_N. The behavior of Hall effect in UCS resembles that of CePtSi[11] and URu$_2$Si$_2$[4] in which the abrupt increase below T_N is followed by a sharp decrease as the temperature approaches to zero. The Hall effect and resistivity in UCSB show very different behavior; the resistivity shows little change in the whole temperature range while the Hall effect increases below T_C and saturates as the temperature is lowered. This Hall effect is compared with that of a ferromagnet UCuAs$_2$[8] in which R_H takes maximum below Tc and decreases below it as the temperature is decreased.

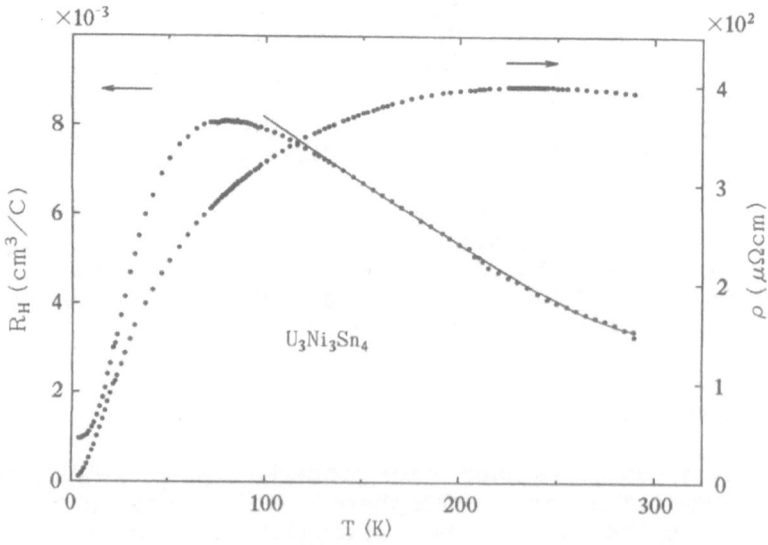

Fig.1. Hall effect and resistivity in U$_3$Ni$_3$Sn$_4$. The solid line (also in Figs. 2, 3 and 5) is the best fit using the eq.(1).

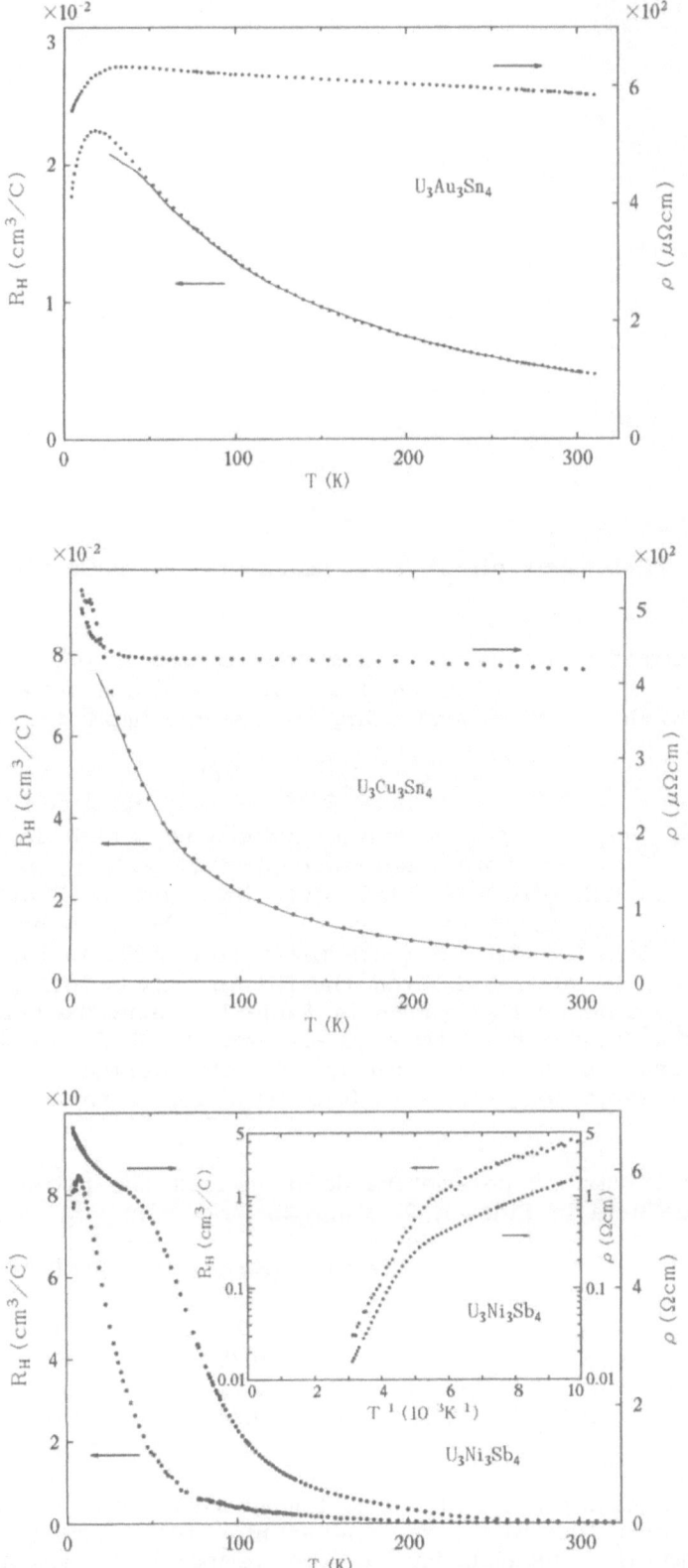

Fig.2. (upper) Hall effect and resistivity in $U_3Au_3Sn_4$.
Fig.3. (middle) Hall effect and resistivity in $U_3Cu_3Sn_4$.
Fig.4. (lower) Hall effect and resistivity in $U_3Ni_3Sb_4$.

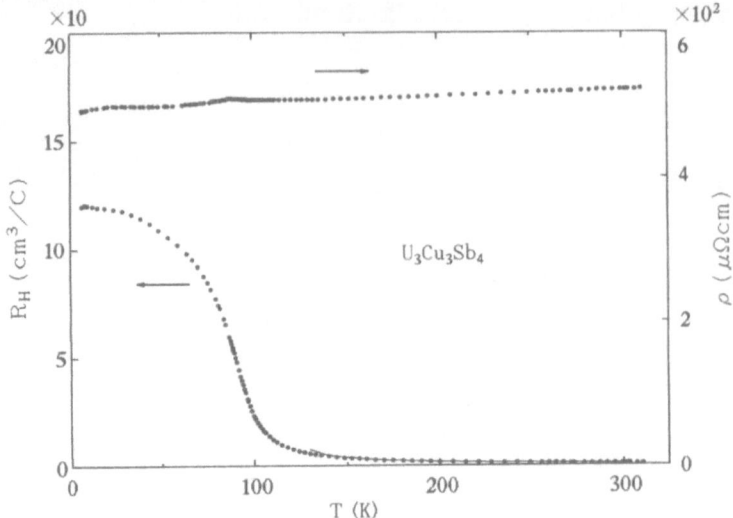

Fig.5. Hall effect and resistivity in $U_3Cu_3Sb_4$.

Hall effect of heavy fermion system in the incoherent region is expressed[14] as a sum of normal Hall effect and the extraordinary Hall effect (EHE) due to skew scattering from single impurity.

$$R_H = R_0 + \gamma_1 \tilde{\chi} \rho \qquad (1)$$

In eq.(1), ρ is the magnetic resistivity which is temperature dependent, $\tilde{\chi}$ is the normalized susceptibility and γ_1 is a constant. The observed Hall effect in UAS, UCS, UNS and UCSB are well fitted by eq.(1) at high temperatures as shown by the solid line as in Figs. 1, 2, 3, 5. Here the data of χ are taken from ref.7 in which for UNS 0.0011emu/mol is subtrated. From the fitting curves the parameters of R_0 and γ_1 are derived as shown in Table 1. n is the carrier density derived from R_0 and the lattice parameters in ref 7. Values of γ_1 are the same order of 0.226 K/T for nonmagnetic $CeRu_2Si_2$.[8] Value of n for UCSB is large compared with 0.18 e/f.u. for ferromagnet $UCuAs_2$.[8]

Table 1. Various parameters deduced from the fitting of eq.(1) to the data in Figs. 1, 2, 3 and 5.

	$R_0(10^{-3}cm^3/C)$	n(elect./U)	$\gamma_1(K/T)$
$U_3Ni_3Sn_4$	−2.12	0.20	0.49
$U_3Au_3Sn_4$	−1.92	0.25	0.48
$U_3Cu_3Sn_4$	−2.19	0.21	0.62
$U_3Cu_3Sb_4$	−0.688	0.64	0.10

In the fitting procedure, the temperature range using eq.(1) is generally restricted to narrow one at high temperature region. This comes from the dissimilarity of the temperature dependence of ρ compared with those of R_H and χ; in $CeRu_2Si_2$ it is in the range of 125<T<300 K, that R_H is well fitted by $\chi\rho$.[8] If it is fitted, instead of using eq.(1), by the usual expression of $R_H = R_0 + 4\pi R_s\chi$, the range extends down to T=86 K.[8] However, in the low temperature range about at the temperature of Hall effect maximum, T_{RHmax}, (in this case, γ_1 should be changed as γ_2 in eq.(1)) the fitting procedure using eq.(1) generally becomes well again. This is because that in this temperature range, ρ generally decreases as the temperature decreases and it compensates the increase of χ (in UCS, χ decreases instead of ρ does). Thus there appear two fitting regions of R_H between the maximum and room temperature.[5,8,14] This consideration clarifies the role of ρ in eq.(1) which is derived theoretically its presence.[14]

In low-temperature coherent region, resistivity of HF system obeys $\rho=AT^2$ law. UNS has high resistivity maximum temperature of 238 K this law is well obeyed with the coefficient $A=0.36\mu\Omega cm/K^2$ up to 10 K as shown in Fig. 6. On the other hand, in the same region, the relation $R_H(EHE)=c\rho^2$ is shown theoretically.[16] In UNS this law

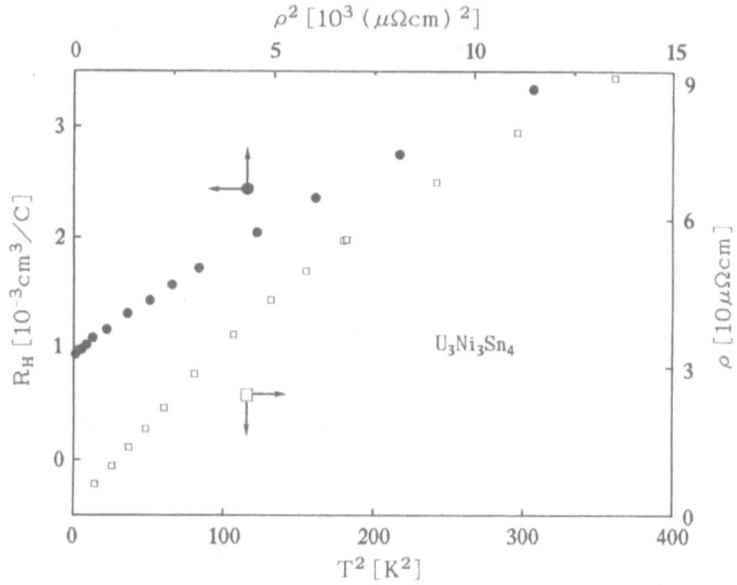

Fig.6. Hall effect vs ρ^2 and resistivity vs T^2.

is obeyed well at the low temperature region below 10 K as shown in Fig. 6. In Fig. 6 positive residual Hall effect should be considered as the normal one which would be changed from negative at high temperature region to positive one due to the change of electronic state similarly to the case in CeCu$_6$[10] thus EHE vs ρ^2 law could be obeyed strictly.

As discussed above, eq.(1) could be generally applied at R_H maximum. From this, we show in Fig.7 the Hall effect at its maximum compared with $\chi\rho$ for several U and Ce based HF systems in which ρ and χ are taken at T_{RHmax}. For the systems which order at low temperature it is taken at T_N or T_c. In Fig.7, (and also in the Figs 1, 2, 3, 5) the value of ρ is taken from the total resistivity but could be considered almost the magnetic one (for UNS and UAS) since in this temperature range phonon part could be almost negligible (in UNS, it is 3% of the total one at 300 K). The residual resistivities in UAS, UCS and UCSB can not be estimated but for UAS the measurement of ρ down to 0.3 K shows steady decrease of it. In Fig. 6, R_H is EHE except which subtraction of R_0 can not be available. It can be seen that R_H (EHE) is roughly proportional to $\chi\rho$ or γ_2 is roughly constant. It is noted that the present HF systems have larger size compared with the other HF systems. There seems to be less correspondence between γ value and R_H. The large γ of 1.6J/K^{2}mol in

CeCu$_6$ and CeAl$_3$ is known but R$_H$'s of these systems are not so large compared with the present system. The large size of R$_H$ in the present system mainly originates from that of ρ at T$_{RHmax}$ since in the present system ρ is almost an order of magnitude larger compared with the other HF systems. The residual resistivity in UCS and UCSB might be large but eq.(1) seems to be obeyed in this case also as in the case Ce$_x$La$_{1-x}$Cu$_6$ in which large residual resistivity is observed[9] but eq.(1) is well obeyed.[12]

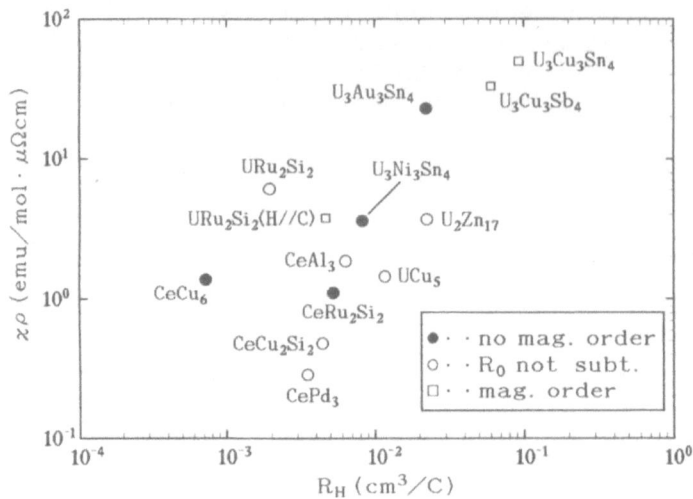

Fig.7. Hall effect vs $\chi\rho$ for HF systems including present systems.

REFERENCES

1. T.Siegrist, M.Olivier, S.P.McAlister and R.W.Cochrane, Phys.Rev.B 33: 4370(1986).
2. J.Schoenes, C.Schönenberger, J.J.M.Franse, and A.A.Menovsky, Phys.Rev. B35: 5375(1987).
3. C.Schönenberger, J.Schoenes, J.J.M.Franse, and A.A.Menovsky, Helv.Phys.Acta 60: 785(1987).
4. Y.Onuki, T.Yamazaki, I.Ukon, T.Komatsubara, A.Umezawa, W.K.Kwok, G.W.Crabtree, and D.G.Hinks, J.Phys.Soc.Japan 58: 2119(1989).
5. A.LeR.Dawson, W.R.Datars, J.D.Garrett and F.S.Razavi, J.Phys. Cond.Mat. 1: 6817(1989).
6. D.Kaczorowski and J.Schoenes, Solid State Comm 74:143(1990).
7. T.Takabatake, S.Miyata, H.Fujii, Y.Aoki, T.Suzuki, T.Fujita, J.Sakurai and T.Hiraoka, J.Phys.Soc.Japan 59: 4412(1990).
8. F.Lapierre, P.Haen, R.Briggs, A.Hamzic, A.Fert and J.P.Kappler, J.Mag.Mag.Mat 63&64: 338(1987).
9. Y.Onuki, Y.Shimizu, M.Nishihara,Y.Machii and T.Komatsubara, J.Phys. Soc.Jpn. 54: 1964(1985).
10. T.Penny,F.P.Milliken, S.von Molnar and F.Holtzberg, Phys.Rev. B34:5959(1986).
11. A.Hamzic, A.Fert, M.Miljak, and S.Horn, Phys.Rev.B38: 7141(1988).
12. Y.Onuki, T.Yamazaki, T.Omi, I.Ukon, A.Kobori, and T.Komatsubara, J.Phys.Soc.Japan 58: 2116(1989).
13. A.E.Dwight, Acta Cryst. B33: 1579(1977).
14. A.Fert, P.M.Levy, Phys.Rev. B36: 1907(1987).
15. K.Takegahara, Y.Kaneta and T.Kasuya, J.Phy.Soc.Jpn. 59: 4394(1990).
16. H.Kohno and K.Yamada, J.Magn.Magn.Mat. 90&91: 431(1990).

SOME ISSUES CONCERNING TRANSPORT AND THERMAL
BEHAVIOR OF f–ELECTRON SYSTEMS

E.V. Sampathkumaran

Tata Institute of Fundamental Research
Homi Bhabha Road, Bombay–400005
India

INTRODUCTION

During the last decade, significant progress has been made in identifying novel Ce compounds with exotic thermal and transport properties. Inspite of availability of voluminous data, complete quantitative understanding of these properties is still lacking. There are often discrepancies even in qualitative interpretation of the data of some materials. In this article, we make some remarks on certain basic issues pertinent to the interpretation of the electrical resistivity (R) and heat–capacity (C) behaviour of Ce systems.

ANTIFERROMAGNETIC ENERGY–GAP

It is quite well–known that the long–range antiferromagnetic ordering induces superzone gaps in the conduction band.[1] Such energy gaps at the Fermi surface result in an increase of R with decreasing temperature below Néel temperature (T_N). However, possible modifications of these distortions of the Fermi surface are often ignored while interpreting the experimental results of rare–earth based solid solutions. This may probably be because the existence of such effects has not been sufficiently demonstrated in the alloys containing localised moments.

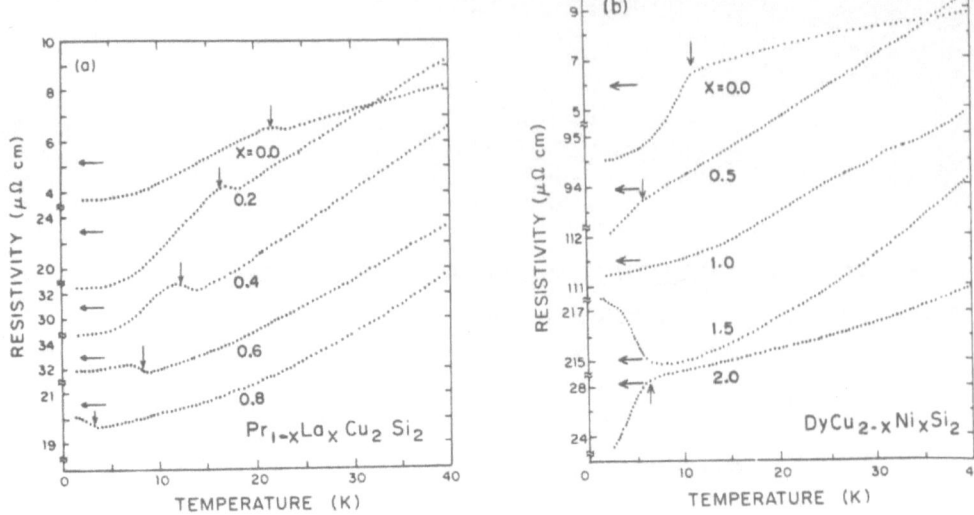

Figure 1. Electrical resistivity as a function of temperature for the alloys of the two series, (a) $Pr_{1-x}La_xCu_2Si_2$ and (b) $DyCu_{2-x}Ni_xSi_2$. Vertical arrows represent the Néel temperature determined from magnetic susceptibility measurements

To demonstrate the importance of the above phenomenon with respect to the topic of present interest we show in figure 1 the electrical–resistivity behaviour[2,3] across antiferromagnetic transition in the pseudo–ternary alloys, $Pr_{1-x}La_xCu_2Si_2$, and $DyCu_{2-x}Ni_xSi_2$. Vertical arrows represent the peak position in the magnetic susceptibility due to antiferromagnetic ordering. In many alloys, we have noticed[2,3] a peak in the temperature dependent R data near T_N, characterizing the existence of energy–gap. A careful look at the composition dependence of this peak reveals some interesting features. While there is no peak at T_N for $DyCu_2Si_2$ and $DyNi_2Si_2$, there is a dominant rise of R with decreasing temperature, surpassing the drop due to the loss of spin–disorder contribution. For $DyCuNiSi_2$, both these contributions seem to cancel each other. Likewise, in the case of Gd series, the energy–gap features appear only for the intermediate concentrations.[2] These results demonstrate the movement of the Fermi–surface across the magnetic Brillouin–zone boundary gaps with the changes in the number of conduction electrons and this finding bears significant relevance to the comparison of the properties of solid solutions of Ce–based antiferromagnets. Another finding is that the energy–gap persists even for the moderately dilute limits of magnetic ions (see the data for x=0.8 in the La series). We have also tracked the behaviour of this peak as a function of unit–cell volume.[3] The results emphasize the need to be cautious while utilizing the data in the paramagnetic state to draw any conclusion in the antiferromagnetically ordered state even for the moderately dilute concentrations of magnetic ions.

The above results may also suggest that possible existence of spin density waves[4] below 1K in the case of $CeCu_2Si_2$ can in principle result in a similar energy–gap.

ORIGIN OF THE LOW–TEMPERATURE RESISTIVITY PEAK IN Ce–SYSTEMS

It is well–known that a peak in the resistivity (in the R versus temperature plot) appears at low temperatures even in non–magnetic Ce systems. (Needless to point out that this peak appearing well–above T_N in antiferromagnetic Ce systems should not be confused with the one that might sometimes occur at T_N due to the energy–gap discussed above. Readers are also aware that, due to the interplay between crystal–field and Kondo effect, another broad feature appears aound 100–150K and this is not a matter of discussion here). For instance, this peak appears[5] at about 15K for $CeCu_2Si_2$. While it is generally believed that the temperature (T_{max}) at which this peak appears is related to single–ion Kondo temperature (T_K) or Kondo coherence temperature (T_{coh})[6], some authors[7,8] tend to believe that T_{max} has a dominant contribution from the strength of the Ruderman–Kittel–Kasuya Yosida interaction (T_{RKKY}). From our earlier investigations[9], we arrived at the conclusion that both these views are valid and the interpretation varies from one compound to the other. We present here an experimental criterion to decide the relative importance of these interpretations for a compound of interest: (1) If T_{max} is proportional to T_K, the application of external (or positive chemical) pressure would shift T_{max} to higher values, as demanded by the well–known relationship between T_K and unit–cell volume. This is found[10] to be the case in $CeCu_2Si_2$. (2) If T_K and T_{coh} are independent of each other, and also if T_{max} is related to the latter, then, for such a compound, a decrease of the concentration of the Kondo ions by a few percent should depress the value of T_{max} significantly due to the destruction of coherence, as demonstrated[11] for the case of Y substituted UBe_{13}. The value of T_{max} may also be sensitive to slight variations of stoichiometry, as it possibly happens[12] in the case of $CeCo_xGe_2$. (3) If the gradual replacement of the Kondo ion by an iso–electronic non–mangetic ion (like La and Y) results in a decrease of T_{max}, the value of which varies proportionately to the concentration of Kondo ions, then in that compound, T_{max} is related to T_{RKKY}. One may normally encounter this situation in Ce systems with a magnetic ground state, for instance, in $CePt_2Ge_2$.[9]

MAGNETIC PRECURSOR EFFECTS IN HEAT–CAPACITY

Several Ce compounds ordering magnetically have been shown to exhibit a huge rise in the values of C/T below a certain temperature before the precipituous rise due to the onset of magnetic ordering. It is customary to attribute this rise to the mass enhancement. We feel that such a conclusion, unless supported by other experimental methods, may be erroneous. Such a rise in the values of C/T has been noticed in some rare–earth compounds

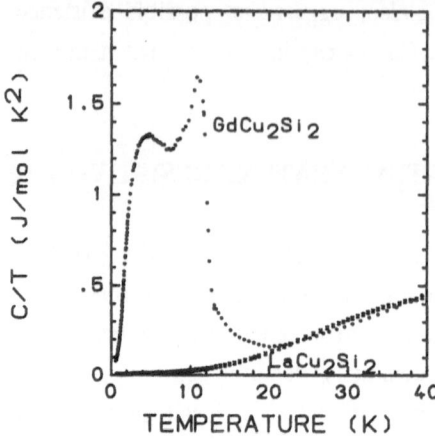

Figure 2. Heat–capacity divided by temperature as a function of temperature for the compounds, $LaCu_2Si_2$ and $GdCu_2Si_2$.

with well–localised 4f–electrons, though no serious attempt has been made to understand this feature. To illustrate this observation, we reproduce our data reported earlier[13] on $GdCu_2Si_2$ (alongwith that for $LaCu_2Si_2$ as a reference for the lattice contribution) in Fig. 2. While the data above 20K for Gd and La compounds track each other, there is a significant rise below 20K for the former, reaching a value of about 400 mJ/mol K^2 before a precipituous rise at 12K due to antiferromagnetic ordering. We believe that this anomaly arises from magnetic precursor effects, which need to be understood better. One way of verifying whether such an anomaly arises from heavy–fermion behaviour is to look for the relationship[14] between thermal expansion coefficient, heat–capacity and bulk–modulus. High pressure heat–capacity and thermal expansion studies are also helpful to clarify this point.

FINAL REMARKS

We would like to remark that we have noticed interesting transport anomalies sensitive to stoichiometry in the new Ce compounds which we have identified in the recent past[12, 15–17]. A few noteworthy features are ferromagnetic Kondo lattice behaviour and a huge increase in the electrical resistivity in the alloys of $CeNi_xGa_{4-x}$ below 20K, and a combination of logarithmic and exponential variation of resistivity above 100K in $CeRu_4Sn_6$. It is not clear at present whether energy–gap contributes to these anomalies.

ACKNOWLEDGEMENTS

The author would like to thank I. Das for his association in various experiments. He also appreciates the cooperation of M. Ishikawa and K. Hirota for heat–capacity measurements. He thanks R. Vijayaraghavan for his support.

REFERENCES

1. R.J. Elliott and F.A. Wedgewood, Proc. Phys. Soc. 81:846(1963).
2. I. Das, E.V. Sampathkumaran, and R. Vijayaraghavan, Phys. Rev. B 44:159(1991).
3. I. Das, E.V. Sampathkumaran, and R. Vijayaraghavan, J. Magn. Magn. Mater. 104–107:874(1992).
4. C. Tien, Phys. Rev. B 43:83(1991).
5. H. Spille, U. Rauchschwalbe, and F. Steglich, Helv. Phys. Acta 56:165(1983).
6. G.R. Stewart, Rev. Mod. Phys. 56:755(1984).
7. U. Larsen, J. Appl. Phys. 49:1610(1978).
8. J.S. Schilling, Phys. Rev. B 33:1667(1986).
9. E.V. Sampathkumaran, I. Das, and R. Vijayaraghavan, Z. Phys. B – Condensed Matter. 84:247(1991).
10. J. Ray, E.V. Sampathkumaran, and G. Chandra, Phys. Rev. B 35:2095(1987).
11. J.S. Kim, B. Andraka, C.S. Jee, S.B. Roy and G.R. Stewart, Phys. Rev. B 41:11073(1990).
12. I. Das and E.V. Sampathkumaran, Solid State Commun, in press.
13. E.V. Sampathkumaran, I. Das, R. Vijayaraghavan, K. Hirota and M. Ishikawa, J. Magn. Magn. Mater. 108:85(1992).
14. G. Oomi, Y. Onuki, and T. Komatsubara, J. Phys. Soc. Jap. 59:803(1990); Y. Uwatoko, G. Oomi, Y. Sakurai, and E.V. Sampathkumaran, J. Magn. Magn. Mater. 108:105(1992).
15. E.V. Sampathkumaran, and I. Das, Solid State Commun. 81:901(1992).
16. I. Das, and E.V. Sampathkumaran, Solid State Commun. 81:905(1992).
17. I. Das, and E.V. Sampathkumaran, Phys. Rev. B, in press.

SPECIFIC HEAT IN A LOW CARRIER

CONCENTRATION COMPOUND: Yb-MONOPNICTIDES

Noriaki Sato,[1] Takuo Sakon,[1] Takashi Suzuki,[1]
Takemi Komatsubara[1] and Akira Oyamada[2]

[1] Department of Physics, Tohoku University, Sendai 980, Japan
[2] College of Liberal Arts and Sciences, Kyoto University
Kyoto 606, Japan

INTRODUCTION

In rare earth compounds, 4f electrons usually tend to localize, because their wave functions are restricted within the outer closed shell in the atom. However, in some cerium and ytterbium compounds, 4f electrons are likely to fluctuate in both space and time through their mixing with conduction electrons, resulting in formation of many-body states at low temperatures. A huge value of the electronic specific heat coefficient γ is indicative of this many-body ground state. It is believed that this large γ comes from fermion-type quasi-particle excitations with the large density-of-state at the Fermi level. Because of this, these compounds are referred to as "heavy fermion" compounds. Usually, the number of the conduction electrons is comparable to that of the magnetic moments located at rare earth atoms. Recently even in a semimetallic compound with a low carrier concentration, heavy-fermion-like behavior is observed. A series of ytterbium pnictides, YbX (X=N, P, As and Sb), is one of the typical examples of these semimetallic compounds.

Yb-monopnictides crystallize in the NaCl type, and (stoichiometric) compounds of YbN and YbAs exhibit long-range antiferromagnetic ordering of type III.[1,2] The pioneer work for this series was carried out by the ETH group. The temperature dependence of the magnetic susceptibility showed that 4f electrons are in Yb^{3+} state,[3] and the inelastic neutron scattering experiment indicated crystal field splitting,[4] revealing a well-localized 4f electron state. It was also reported that these compounds exhibit a sharp peak in the specific heat corresponding to the magnetic ordering at $T_N = 0.735K(YbN)$, $0.410K(YbP)$ and $0.490K(YbAs)$.[5] A broad maximum in the specific heat appears around 4-5K, as does a sharp peak at T_N. This broad maximum was ascribed to the Kondo effect through mixing between the 4f hole state of Yb^{3+} ion and the p-states in the valence bands.[6] This assignment is supported by the fact that the cooperative transition releases only 20% of $Rln2$ (for YbP),[5] the expected molar entropy of a doublet ground state. Mòssbauer and neutron scattering measurements yielded the magnetic saturation moments significantly reduced below the value expected from the crystal field ground state Γ_6.[1,2,7] This reduction of the of the magnetic moment seems to be a common feature observed in heavy fermion compounds. Although we have several experimental results to suggest the Kondo hybridization model in Yb-monopnictides, we have no information concerning to the γ-value, which is indicative of the heavy fermion compounds. Furthermore, there

Transport and Thermal Properties of f-Electron Systems
Edited by G. Oomi *et al.*, Plenum Press, New York, 1993

exists no sign of the Kondo effect in the electrical resistivity.[8] In order to resolve these questions, experimental measurements were performed in Sendai group.[8-11] In this paper, we review these experimental investigations and also discuss the competition between the Kondo effect and the magnetic interactions.

RESULTS AND DISCUSSION

In order to obtain the γ-value, we performed the specific heat measurements of YbAs below 1K.[12] Although it was first reported by Ott et al.,[5] the electronic specific heat coefficient was not deduced. The overall features of our results[12] are consistent with the data in the literature,[5] where the increase of the specific heat below about 150mK was safely attributed to the nuclear Zeeman effect of Yb nuclei which suffer an exchange field in the antiferromagnetic ordered state. To estimate the nuclear part of the specific heat, we assumed that the temperature dependence in the lowest temperature region can be given as follows.

$$C = (\alpha/T^2) + \gamma_1 T \tag{1}$$

Here, the first term indicates the nuclear contribution to the specific heat in the high-temperature approximation, and the second one the electronic specific heat. The contribution of the antiferromagnetic spin wave excitation is small compared with the above two terms in this temperature range ($T < 150$mK). The specific heat is plotted in Fig. 1. The external magnetic field effect on α and γ_1 is small. From the fit to the experimental result, which is indicated by a broken line in the case of

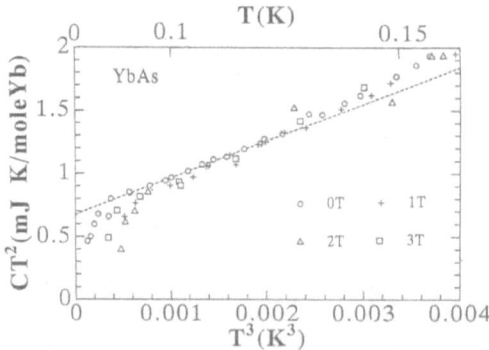

Figure 1. Plot of specific heat multiplied by T^2 as a function of T^3. Data are taken from ref. 12.

zero field, we obtain $\alpha = 0.67$mJ·K/mole and $\gamma_1 = 290$mJ/K^2 mole. The deviation from the broken line in the lowest temperature region is possibly due to the adoption of the high-temperature approximation in eq. (1). The magnitude of the internal field at a Yb (^{171}Yb and ^{173}Yb) nucleus site was obtained from the above value of α to be $H_{int} = 53$T. This value of the internal field is compatible with the hyperfine field ($H_{hf} = 87$T) derived from the measurements of the ^{170}Yb Mössbauer effect.[7]

Since the internal field is so large, α is not affected by the application of a magnetic field up to 3T, as mentioned above. By subtraction of the nuclear contribution from the measured specific heat, the electronic part of the specific heat was obtained and plotted in Fig. 2.

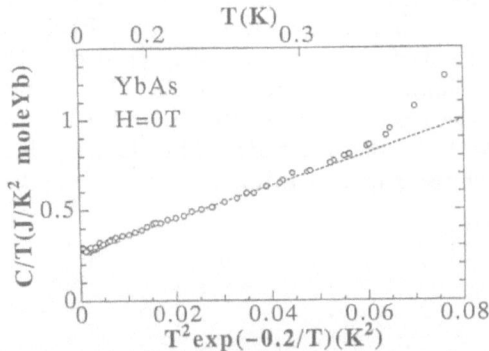

Figure 2. Electronic part of specific heat after subtraction of the contribution from the nuclear Zeeman effect. The value in the exponent (=0.2K) indicates the gap of the antiferromagnetic spin wave excitation. The values of γ and δ are obtained to be 270mJ/K^2 mole and 9.1J/K^4, respectively. Data are from ref. 12.

The temperature dependence of the specific heat was assumed to be given as follows.

$$C = \gamma T + \delta T^3 exp(-\Delta/T) \qquad (2)$$

The second term indicates the antiferromagnetic magnon contribution with an energy gap Δ in zero field. The best fit to the formula yields Δ=0.2K. Dönni et al. performed the inelastic neutron scattering experiment to obtain the spin wave dispersion,[13] and they found two magnon branches. One of them has a k-linear dispersion near the magnetic zone center Q=(1,1/2,0) with an energy gap of 0.08meV. This value of the energy gap is several times larger than the above value. This discrepancy probably originates from the difference in resolution in the experiment, and we believe that the resolution is better in the specific heat experiment than in the neutron scattering measurement. The γ value obtained from eq. (2), i.e. 270($\pm$10)mJ/K^2 mole is consistent with the above γ_1-value. This indicates the self-consistency of our analysis. The γ-value in external magnetic fields is γ=220, 210 and 200mJ/K^{2}mole for $H = 1, 2$ and 3T, respectively.

We also obtain δ=9.1 J/K^4 and further the exchange coupling constant of the same order as the Néel temperature, assuming a simple model of spin wave.[14] This result is quite contrast to the Mössbauer experiment results, which suggest that the exchange field energy (in the ordered state) is 20 times larger than the Néel temperature.[7]

It is apparent from these specific heat measurements that YbAs is a heavy fermion compound. However, the kondo behavior is not found in the electrical resistivity, as mentioned before. To resolve this contradiction, transport properties were analyzed by Oyamada,[8] as follows. Assuming both the good stoichimetry of the sample and the field independence of the mobilities and the anomalous Hall effect, one can obtain all three parameters, i.e. $n_h = n_e = n$ (carrier concentration), μ_h (hole mobility) and μ_e (electron mobility). The carrier numbers at low temperatures, i.e. $n = 0.054$/unit cell at 4.2K, is completely consistent with that derived from the dHvA effect experiment.[10] The results are illustrated in Fig. 3. This indicates the different behavior of holes and electrons at low temperatures, that is, the mobility of electrons increases strongly below 80K, whereas that of holes shows no increase.

This means that the electric current is mainly carried by the electron with larger mobility. On the other hand, it is shown from a band structure calculation that the Kondo effect is due to the mixing of the 4f state with the hole state of the conduction electrons, not due to with the electron state.[8] Therefore, the large conductivity of the electron state can mask the Kondo scattering of the hole state by the 4f moments.

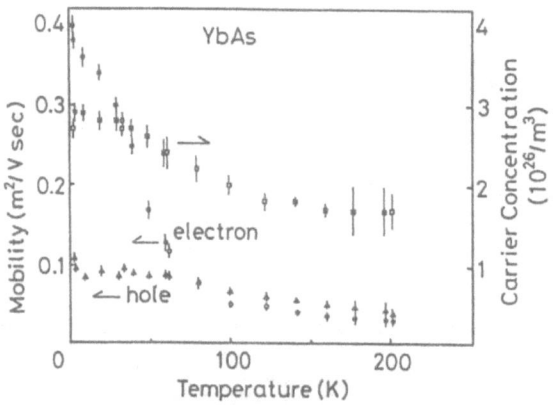

Figure 3. Temperature dependence of mobilities and the carrier number. (See text) Data are from ref. 8.

The quasi-inelastic neutron scattering experiment is useful for extracting the strength of the 4f-4f magnetic interactions. From the analysis of the line shapes of the quasi-elastic spectrum, the strength of the magnetic interaction is obtained to be about 18meV even at T=50K.[11] Surprisingly, this value is about 40 times larger than that of the Néel temperature. This implies that there exists a mechanism such as Kondo effect or some magnetic frustration which suppresses the magnetic ordering.

In this paper, the terminology of "Kondo effect" was used. However, it seems difficult to ascribe the heavy-fermion-like features in Yb-monopnictides to the Kondo effect, because the conduction-electron-density in Yb-monopnictides is two orders of magnitude smaller compared to the density of the magnetic moments. This suggests that other types of spin fluctuation is to be considered instead of the Kondo-effect-type spin fluctuation. Anyway, we conclude that Yb-monopnictides are a typical example demonstrating the competition between the "Kondo effect" and magnetic interactions.

REFERENCES

1. A. Dönni, P. Fischer, A. Furrer and F. Hulliger, Long-range f.c.c. antiferromagnetic ordering of type III in the heavy fermion compound YbAs, Solid State Commun. 71:365 (1989).
2. A. Oyamada, P. Burlet, A. Bouvet, R. Calemczuk, J. Rossat-Mignod, T. Suzuki and T. Kasuya, J. Magn. Magn. Mater. 90-91:441 (1990).
3. H. R. Ott, F. Hulliger and H. Rudigier, Absence of magnetic order in Yb monopnictides, in:"Valence Instabilities", P. Wachter and H. Boppart, eds., North-Holland (1982).
4. M. Kohgi, K. Ohoyama, A. Oyamada, T. Suzuki and M. Arai, Crystal field excitations in Yb monopnictides, Physica B 163:625 (1990).
5. H. R. Ott, R. Rudigier and F. Hulliger, Low-temperature phase transitions in Yb monopnictides, Solid State Commun., 55:113 (1985).
6. S. Takagi, A. Oyamada and T. Kasuya, ^{31}P NMR studies of the magnetically ordered heavy-electron compound YbP, J. Phys. Soc. Jpn., 57:1456 (1988).
7. P. Bonville, J. A. Hodges, F. Hulliger, P. Imbert, G. Jehanno, J. B. Marimon Da Chuha and H. R. Ott, First-order Kondo-frustrated antiferromagnetic ordering in the heavy electron compound YbAs, Hyperfine Inter. 40:381 (1988).
8. A. Oyamada, Ph.D. Thesis (Tohoku Univ.) (1991).
9. T. Sakon, Master Thesis (Tohoku Univ.) (1992).
10. N. Takeda, Ph.D. Thesis (Tohoku Univ.) (1992).
11. K. Ohoyama, Ph.D. Thesis (Tohoku Univ.) (1992).
12. T. Sakon, N. Sato, A. Oyamada, N. Takeda, T. Suzuki and T. Komatsubara, Heavy-electron behavior in a low-carrier-concentration compound YbAs, J. Phys. Soc. Jpn., 61:2209 (1992).
13. A. Dönni, A. Furrer, P. Fischer, F. Hulliger and S. M. Hayden, Spin-wave excitations and the magnetic phase transition in the ytterbium monopnictide YbAs, J. Phys. :Condens. Matter 4:1 (1992).
14. R. Kubo, Phys. Rev. 87:568 (1952).

LOW ENERGY EXCITATIONS IN LOW CARRIER CONCENTRATION SYSTEMS OF CeSb AND Yb-MONOPNICTIDES

Takashi Suzuki

Department of Physics, Tohoku University
Sendai 980, Japan

The dHvA effect has been measured in several heavy Fermion substances. There are some discrepancies between the observed cyclotron mass and the mass estimated from the linear term of low temperature specific heat. A typical example is CeB_6 in which the cyclotron mass in lower magnetic field observed by Onuki et al (1) is twice smaller than that estimated from γ. We believed that this discrepancy is due to unobservable Fermi Surfaces with a much heavier cyclotron mass at that time. $CeSn_3$ is taken as a typical example which exhibits a good agreement between the band calculated mass(2) and dHvA effect one(3). In this compound, theoretical prediction of mass enhancement factor in the specific heat is about 4, while the ratio of band mass and observed cyclotron mass is 4 only along special direction of the Fermi surface and is nearly 2-3 along other all direction. The discrepancy is clear and may be due to the error in both experimental and theoretical estimations.

There is a similar situation in CeSb. Various anomalous properties of CeSb and CeBi have been successfully analyzed on the baisis of the p-f mixing model, in which the hybridization of the hole state at the top of the valence p-band of pnictogen with the occupied f-electron plays an essential role (4). It was also shown that in the ferromagnetic phases the Fermi surfaces given by the p-f mixing model agree with the experimental ones obtained by Kitazawa(5), Aoki(6) and Goto except that for the most strongly hybridized band, which was not experimentally observed. Sakai et al predicted that the mass enhancement factor of unobserved Fermi surface is 11.6 and cyclotron mass is 5.7(9). After that, Aoki et al confirmed the existence of the largest hole Fermi surfaces, which is nearly 10% smaller than the theoretical one. However, the cyclotron masses could not be determined by their experiment.

Very recently Settai et al (11) have measured the accoustic dHvA effect down to 30mK and up to 13T in CeSb and found that the cyclotron mass is $4.2m_0$ on the [100] direction. This value is in very good agreement with $m^*=5.7m_0$ theoretically estimated by Sakai et al. The validity of the p-f mixing model has been confirmed again. Now all the Fermi surfaces were experimentally determined, and the situation becomes very clear, that is only the largest hole surface has a large mass enhancement of about $4.2m_0$, which is due to a strong p-f mixing. All the other branches at the Fermi surfaces have lighter masses ($m^*<1$). Still the discrepancy between the cyclotron mass and the mass estimated from the γ value of the specific heat remain to be very serious. The specific heat at lower temperatures in various magnetic field are shown in Fig.1(12).

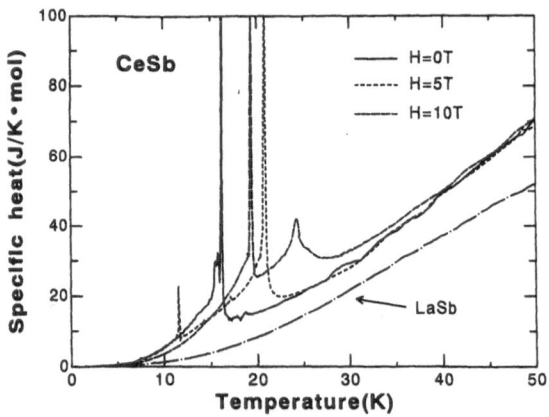

Fig.1   Specific heat of CeSb for various magnetic field

At 10T, which is nearly the same magnetic field used in the estimation of the cyclotron mass of the largest hole surface,the γ value of CeSb was obtained as 13mJ/molK^2. On the other hand, the theoretical estimation predicted a γ value of 1.9mJ/mol.K^2 with $m_0=5.7$. It is to be noted that the γ value smaller than those in usual metallic substance is due to low carrier concentration, in spite of large mass enhancement, in CeSb. The discrepancy is nearly 6 times and there is no other Fermi surface. It is sure that the all contributions to the γ value from the Fermi surfaces are about 2mJ/mol.K^2 at most if there is uniform mass enhancemen on the largest Fermi surface. We have to look for other sources of the contribution to the γ value. At first, is there any cotribution from spin fluctuation?, even though it is difficult to produce the linear term of the specific heat due to spin fluctuation. The answer is definitely no. because the magnetic excitation spectrum spectrum measured by Rossat-Mignod et al (13) and Halg et al (14) showed a large gap of about 3.6meV. This gap energy is much larger than the thermal excitation energy in the temperature range 2-5K, where the γ value was estimated from the specific heat. Next possibility is the low energy excitation corresponding to the anomalous absorption in optical conductivity of the far infrared region. This absortion peak sites at 10 meV (15) as shown in Fig. 2. This is also much larger excitation energy for the low temperature specific heat. At present, there is no experimental evidence for other sources of the contribution to low energy excitation except the contribution of the carriers on the Fermi surfaces. In order to clarify this point, the specific heat measurement at lower temperatures is necessary and may give the information about the structure of this anomalous low energy excitation. It is sure that there are some low energy excitations contributed to γ-value of specific heat without carrying the current even in present temperature region of 2-6K. This may be connected the occurrence of the heavy Fermion states in insulating materials such as Sm_3Se_4 (16) and Sm_3Te_4 (17).

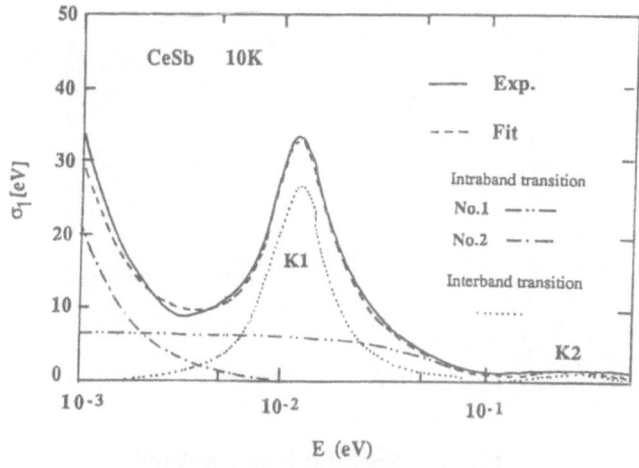

Fig.2 Optical conductivity for CeSb

Another anomalous contribution to the specific heat is found in Yb-monopnictides. Yb-monopnictides YbX (X=N,P,As,Sb) are very interesting materials which exhibit strong competition between a remarkable Kondo effect and an intersite magnetic interaction in several physical properties (18),(19). They are also low carrier concentration systems as confirmed by the result of dHvA effect for YbAs (20). A systematic study of the specific heat in the whole series of YbX has recently been carried out in the temperature range between 2K and 60K by Li et al (21). The result for YbN is shown in Fig.3. The Schottky anomaly calculated on the basis of the CEF level scheme of the neutron scattering is also shown in this figure.

For YbN no contribution from the Schottky anomaly is seen in the temperature range below 40K. The magnetic contribution to the specific heat is obtained by subtracting the specific heat of LuN. Both YbP and YbAs exhibit similar results. The magnetic entropy of the Γ_6 ground state doublet is released near 20K(22). This cannot be explained by the thermal excitation of the crystalline field level with a separation of 200-300K between the Γ_6 ground state and the Γ_8 first excited state which was determined by neutron scattering. We have to look for again the source of the contribution to this excess entropy. One possible source may come from to some kind of phonon softening effect which was observed clearly in the neutron scattering experiment of YbN(23). This softening, however, can not be seen so clearly in YbP and YbAs, where an excess entropy beyond the doublet can be seen in specific heat results.

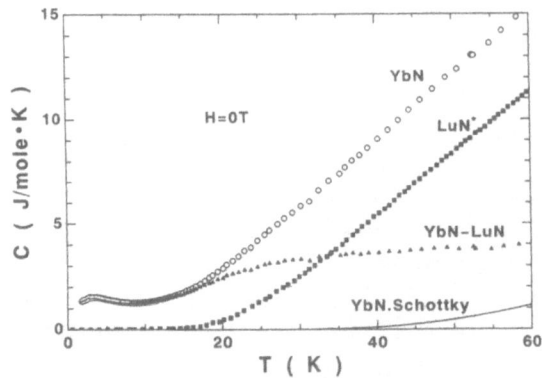

Fig.3   Specific heat for YbN

As the other possibility of the effect of a strong magnetic correlation, a phenomenological model was proposed, in which the width and separation of excited Γ_8 excited state are same as to those observed by neutron diffraction but they are assumed to have a long extended tail caused by magnetic excitation as shown in Fig.4

This model with a long tail of the first excited state could also describe well the anomalous specific heat behavior in YbP and YbAs. This long tail, however, could not be detected in simple neutron scattering experiments of YbAs. A careful temperature dependence of the scattering intensity in the neutron

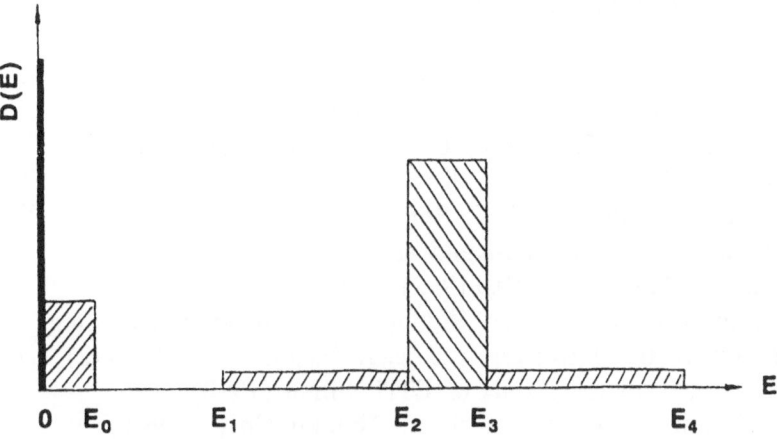

Fig. 4 Schematic energy diagram of the proposed model

and quasi-elastic scattering may give the key to solve this interesting problem. In conclusion, we would like to stress the possibility of the   existence of the low energy excitation emphasized not related to the carriers   in CeSb and Yb-monopnictides.

REFERENCES

1. Y. Onuki, T. Komatsubara, P.H.P. Reinders and M. Springford, Physica B 163 (1990) 100
2. A. Hasegawa and H. Yamagami, Prog.Theor.Pys.Suppl.108 (1992) 27
3. I . Umehara, Y. Kurosawa, N. Nagai, M.Kikuchi, K. Satoh and Y.Onuki,
   J.Phys.Soc.Jpn. 59(1990) 2848
4. H. Takahashi and T. Kasuya,
   J. Phys. C18 (1985) 2697,2709,2721,2731,2745 and 2755
5. H. Kitazawa, T. Suzuki, M. Sera, I. Oguro, A. Yanase, A. Hasegawa and T. Kasuya, J. Magn.Magn.Mat. 31-34 (1983) 421
6. H. Aoki, G.W. Crabtree,W. Joss and F. Hulliger,
   J.Magn.Magn.Mat. 52 (1985) 389
7. O. Sakai, M. Takeshige, H. Harima, K. Otaki and T.Kasuya,
   J.Magn.Magn.Mat. 52 (1985) 18
8. T. Kasuya, O. Sakai, J. Tanaka, H. Kitazawa and T. Suzuki,
   J.Magn.Magn.Mat. 63&64 (1987) 9
9. O. Sakai, Y. Kaneta and T. Kasuya ,
   Jpn.J.App.Phys. Supl. 26-3(1987) 477
10. H. Aoki, G.W. Crabtree,W. Joss and F. Hulliger,
    J.Magn.Magn.Mat. 97 (1985) 169

11. R. Settai, T. Goto, S. Sakatsume, Y.S.Kwon, T. Suzuki and T. Kasuya,
    Proc. of SCES '92 in Sendai (1992) to be published
12. T. Kasuya, Y.S.Kwon, T. Suzuki, K. Nakanishi, F. Ishiyama and K. Takegahara, J.Magn.Magn.Mat. 90&91 (1990) 389
13. J.Rossat-Mignod, J.M.Effantin, C. Vettier and O. Vogt, Physica 130B (1985) 555
14. B. Halg., A. Furrer and O. Vogt, J.Magn.Magn.Mat. 63&64 (1987) 55
15. Y.S. Kwon, M. Takeshige, T. Suzuki and T. Kasuya, Physica B 163 (1990) 328
16. K. Faas, U. Alheim, P.H.P. Reinders, C. Shank, R. Caspary, F. Steglich, A. Ochiai, T.Suzuki and T.Kasuya, J.Magn.Magn.Mat. 108(1992) 220
17. U. Alheim,K. Faas, P.H.P. Reinders, O. Nakamura , F. Steglich, T.Suzuki and T.Kasuya, J.Magn.Magn.Mat. 108(1992) 220
18. A. Donni , Doctor Thesis of ETH Zulich (1990)
19. A. Oyamada, Doctor Thesis of Tohoku Univ. (1991)
20. D.X.Li, A. Oyamada, H. Shida, T. Kasuya, A. Donni and H. Fulliger, Proc. of SCES '92 in Sendai (1992) to be published
21. H.R.Ott, H. Rudigier and F. Hulliger, Solid State Comm. 55 (1985) 113
22. M. Kohgi, K. Ohoyama, A. Oyamada, T. Suzuki and M. Arai, Physica B, 163 (1990) 625

SUMMARY, HOW FAR COULD WE UNDERSTAND

THE f-ELECTRON SYSTEMS

T. Kasuya

Department of Physics, Tohoku University
Sendai, Japan

INTRODUCTION

In this Hiroshima Workshop, mostly the transport and thermal properties of f-electron systems were discussed. We are mostly interested in some typical common anomalous properties but to establish them it is fundamentally necessary to study some typical materials in detail, and the latter becomes more important as our study extends over many materials.

It is now well known that it is possible to classify the materials by the kind of atoms. Clearly Ce and Yb compounds make a common class, the simplest system, because of the valence fluctuation between f^0 and f^1, or f_h and f_h^1, in which f_h means a f-hole state, and are most extensively studied. The other rare earth compounds make the second class, in which Sm and Tm compounds, in particular SmB_6, Tm-monochalcogenides and pressurized Sm-monochalcogenides, are known as materials to show unusual gap states, distinguished from the conventional gap states as shown later more in detail. The U-compounds may be classified as the third class, in which the f-f correlation energy U_f is much smaller than that in the rare earth compounds. Therefore many U-compounds are well described by the usual 5f-band model, similar to the 3d transition metals compounds. However, in some compounds, the 5f localized model is applicable and the compounds belonging to the boundary between them show the valence fluctuation states similar to that in the 4f systems in some sense but different in other sense because of the weaker U_f effect. The trans-uranium compounds form the forth class and are characterized so as to connect the U-compounds to the rare earth compounds.

In the following studies, however, keeping the above mentioned classification in mind, we shall classify the subjects by physical character.

Transport and Thermal Properties of f-Electron Systems
Edited by G. Oomi *et al.*, Plenum Press, New York, 1993

IMPURITY KONDO STATES

Impurity Kondo problem was the starting point for the valence fluctuating problem in the f-electron systems and still now is keeping its fundamental importance. The simplest system is a Ce or Yb impurity in a typical metal and the fundamental characters are now fairly well understood when the valence fluctuation occurs mainly between f^0 and f^1 or f_h^0 and f_h^1. When more than two f electrons occupation is involved, the situation becomes complicated. For f^2 configurations, typically observed in Tm, Pr and trivalent U compounds, a possible singlet configuration competes with the Kondo singlet constructed with the conduction electrons causing various anomalies including the multi-channel non-fermi-liquid state. This state was claimed recently to be realized in some U-compounds and a detailed debate is expected to be held in Sendai.

Two impurity Kondo problem has been studied theoretically fairly extensively because it is the simplest fundamental model to understand the Kondo lattice problem. The external field and the RKKY interation between them are introduced as fixed values but the most interesting is the induced RKKY interaction through the c-f mixing and how it competes with the impurity Kondo effect favoring the singlet state at each site. In particular the competition between the Kondo singlet at each site and the magnetic singlet state and how the RKKY interaction is renormalized by the Kondo effect are most interesting in the connection to the Kondo lattice. Some discussions were done in this workshop but a more detail is expected in Sendai.

Impurity Kondo states in U-compounds are also of controversial. Indeed, the typical impurity Kondo state seen typically in Ce and Yb impurities are not found. In the Ce and Yb compounds, the impurity Kondo character persists even in the dense Kondo systems including the Kondo lattice systems. This is not found in the U-compounds. In particular in PES-BIS, there are no difference between the heavy fermion systems and the 5f-band system, a peak at 1.5 eV above the Fermi energy with a shoulder near the Fermi energy, while in the localized regime system such as UPd_3 a PES peak is observed at about 1.5 eV below the Fermi energy. It was claimed that the diluted UPd_3 shows a Kondo like behavior, in particular at 20% U concentration a non-fermi liquid behavior as expected by the two channel Kondo system. There are some mysteries in UPd_3 systems and more detailed experimental studies are necessary to reach a final picture.

LOW CARRIER-SEMIMETAL-NARROW GAP SYSTEMS

Extensive studies for new heavy fermion systems were reported in this workshop, and in particular various low carrier exotic materials were reported. The easiest way to make a low carrier material is a doping in a insulator, typically done in the high T_c CuO_2 layered materials. In such a system, the really low carrier regime is governed by the impurity state or the magnetic impurity state as was done at first in Eu-chalcogenides doped with a trivalent rare earth atom such as Gd or La and the situation becomes complicated because of randomness.[1] To obtain the intrinsic properties, intrinsic low carrier systems such as semimetals and narrow gap semiconductors are appropriate. Actually the materials reported in this workshop belong to the latter category. However, to obtain the really intrinsic properties, the sample should be very good, that is, the impurity should be at least one order of magnitude smaller than the intrinsic carrier number. Based on this criterion, the most appropriate samples in the present stage is the rare earth monopnictides, RX, prepared by T. Suzuki group and more improvements are required for all the samples. Excepting RX and CeNiSn, all other materials are thought to be explained by the already known mechanism known first in EuB_6, a typical semimetal or narrow gap semiconductor and studied by M. Kasaya et al. experimentally and by T. Kasuya et al. theoretically.[2]

It was shown that the top of the valence band and the bottom of the conduction band are slightly overlapping, or have a narrow gap, at room temperature. The occupied 4f levels are near, about 1 eV, below the Fermi energy, while the unoccupied 4f levels are several eV above the Fermi energy. When the 4f spins on each Eu atom align ferromagnetically, the up-spin valence band is pushed up due to increasing p-f mixing interaction, while the up-spin conduction band is pulled down due to increasing intra-atomic d-f exchange interaction causing the system into a stronger semimetal with a larger carrier number. The ferromagnetic order is stabilized due to the rearranging energy of carriers from the valence band to the conduction band. The opposite thing happens for the down-spin bands. The transition is the first order due to a strong non-linearity. This phenomenon is a typical one for a weak-overlapping or a narrow-gap system and actually has been observed in many materials including the materials reported in the present workshop. A typical example is CeSb, the most extensively studied materials, or the Ce-monopnictides in general. Here, the situation is more complicated because the excited crystal field state, Γ_8, has a much stronger p-f mixing and thus the ferromagnetic order is realized by the $4f\Gamma_8$ state ordered in a layer. Ordered states are complicated bacause among the Γ_7 ground state the Γ_8 ferromagnetic layers stack in various ways.[3] More variations are, of course, possible.

A typical example in the U-compounds is U_3P_4, in which U atom is tetravalent. The reference system Th_3P_4 has a narrow gap between the valence and the conduction bands, because Th is tetravalent. Because of strong local anisotropies on three different sites, a canted ferromagnetic order is realized at low temperature. Similar to the Ce-monopnictides, U_3P_4 is a rare material in which dHvA oscillations were observed because of its high quality. Different from Ce-monopnictides, however, the Fermi surface is not yet fitted theoretically by the same model as that for Ce-monopnictides again indicating different character of 5f electrons from the 4f electrons. UNiSn reported in this workshop is a typical example belonging to the same class. Various trans-uranium compounds, in particular pnictides, also belong to the present class.

Most materials belonging to this class are pnictides, in some sense, because in the rare earth pnictides, the energy gap between the conduction band made mainly by the rare earth 5d-states and the valence band made mostly by the pnictogen p-states becomes nearly zero. In chalcogenides, the gap becomes the order of 2 eV with a strong ionic character. In some C and Sn compounds, the band overlapping is weak and thus belong to the present class. The most compounds have the form $R_xT_yX_z$, in which R means a rare earth atom or U-atom, T is a transition metal atom with nearly filled d-states and X is a pnictogen atom and C or Sn. The present situation is realized when the following equation is satisfied.

$$\alpha x = \beta y + \gamma z \quad , \tag{1}$$

This is because electrons flow from R to T and X. α is the valency of R. In Ce, when it is in a valence fluctuation regime, the usual 4f band model is applicable and thus $\alpha = 4$, while in the Kondo and localized 4f regimes, the split 4f band model, or the Hartree-Fock model, is applicable in which one 4f state is below the Fermi energy, and thus $\alpha = 3$. For the U-compounds, the valency of U-ion should be well defined. This does not exclude, however, the materials in which the 5f band model is applicable. For pnictides, $\gamma = 3$ and for C and Sn $\gamma = 4$. When z is sufficiently large, the s-band, or the OPW band, related with the s-state on T and R are pushed up above the Fermi energy due to the mixing with the closed s-shell on X. Therefore, $y = -1$ for the Cu-column, $y = 0$ for the Ni-column and $y = 1$ for the Co-column. For a small value of z, however, the OPW band is lowered below the Fermi energy and thus a careful band calculation is needed. For example in LaSb, the OPW band is above the Fermi energy but in La_2Sb it exists below the Fermi energy. In $LaPd_3$, the OPW bands appear in between the d-bands for Pd and La and thus $LaPd_3$ is a monovalent metal in the sense of the OPW bands. Because $CePd_3$ belongs to the valence fluctuation regime, it is a divalent semi metal for the OPW bands and thus belongs to the present class. Anyway, the low carrier character is explained by the usual band model and thus should be called as the ususal band gap or the ususal band semimetal, or a trivial gap. On the other hand, the gap, which is not explained by the above model, should be called as the non-trivial gap, or exotic gap. Typical examples were SmB_6, TmSe, Sm-monochalcogenides under pressure and YbB_{12}. The newly found material CeNiSn also seems to belong to this class but more studies are needed as shown later.

The main subject for the former class of materials, in particular belonging to the Kondo regime, is how far and why the Kondo-like singlet persists in a low carrier system. In a low carrier system, a more probable singlet for theorists is the magnetic singlet, in which an antiferromagnetic RKKY exchange interaction between the 4f dipole moments is dominant and the RVB-like singlet state is formed. However, this type of singlet is thought to be not realized in the systems under consideration because, first, this singlet does not cause anomalies on the transport properties as observed, second, the exchange interaction in the low carrier system is predominantly ferromagnetic, as observed, third, there are no acceptable mechanisms to cause the RVB-like singlet from the antiferromagnetic order, and forth, the Kondo-like behavior persists from the good metals to the low carrier systems continuously without any essential change. It should be noted, however, that clear impurity Kondo behavior has been not yet observed in the low carrier systems.

It is also an important issue to know what kinds of novel states, in particular magnetic and superconducting states, emerge as the competing states in the low carrier systems. It is an interesting question to ask whether any superconducting state exists in the low carrier system. A typical example may be $La_{3-x}X_4$, in which X is a chalcogen atom and $0 < x < 1/3$, where at $x = 1/3$ the system has no carrier. Clear superconductivity with a fairly large T_c, several K, is observed. Origin of this large T_c even for a low carrier density is not clear. As a typical magnetic state in a low carrier system, strong ferromagnetism due to strong non-linearity of density of state was already studied before. Another typical example is magnetic polarons, or its lattice, and this is strongly related with the Wigner crystal in the non-magnetic low carrier systems. Indeed, these were reported to be realized in the Ce- and La- monopnictides.[4] These were predicted theoretically in many years ago and investigated experimentally without success. Very high quality samples are the most important factor for the recent new observation.

Low carrier systems are also important to check the Luttinger theorem. In the metallic systems, it is difficult to obtain whole the Fermi surface by dHvA measurement, in particular for the Kondo regime compounds, because of heavy masses on a complicated fermi surface, as well as due to coexisting magnetic ordering and magnetic breakdown effect. These facts cause serious ambiguity to check agreement with the band calculation. The low carrier semimetal has a strong merit on this point because of its simple Fermi surface located at some symmetry points. Then the effects of magnetic ordering are usually expected to be much weaker except a special case that the magnetic gap boundaries cut the small Fermi surface. Furthermore, the large magnetoresistance due to semimetal can prove the Luttinger theorem very precisely. In CeSb, for example, the magnetoresistance ratio increases as high as 100, indicating that the difference of carrier numbers in the electrons and holes is less than 10^{-4}. Note that CeN in the valence fluctuation regime is monovalent metal, no holes and one conduction electron per Ce atom. Therefore, even though the effective 4f level is near the Fermi energy and the effective Kondo temperature is estimated to be about 100 K, the Luttinger theorem as the split 4f-band model is satisfied very well. Note also that for dHvA measurement both the low temperature and low Dingle temperature are required, while for the magnetoresistance only the low Dingle temperature is required. Therefore, even if whole the Fermi surface is not observed by dHvA measurement, the strong magnetoresistance is observed, as was happened really in CeSb. Indeed through Ce- and Yb- monopnictides, in which very detailed dHvA and magnetoresistivity have been measured, we have studied that the split 4f band model is applicable for the low carrier Kondo regime materials, irrespective of long range magnetic orderings, and the Luttinger theorem is applicable very accurately even if the material is near the valence fluctuation regime. The large mass enhancement is due to the electron-magnetic fluctuation interaction. The mechanism for the Kondo-like singlet state and the large effective Kondo temperature, many orderes of magnitude larger than the impurity Kondo temperature evaluated by using the usual model, are, however, not clear. More detailed studies on various different materials are necessary.

Many low carrier systems belonging to both the Kondo and valence fluctuation regimes, for example $Ce_3Au_3Sb_4$ for the former and $Ce_3Pt_3Sb_4$ for the latter, were found and reported in this Workshop. However, non of them are still not so pure enough to measure dHvA effect. Some of them show very large γ-values, such as YbPtBi, which also belongs to the low carrier system in the Kondo regime as clear from the equation shown before. Such a large γ-value is due to partly weak crystal field splitting, that is a large degeneracy, but partly related with a kind of magnetic ordering at very low temperature. In this sense, Sm_3X_4, in which X means S, Se and Te, is the extreme case, the heavy fermion without free carrier.[5] Careful measurement by using high quality sample is needed. Note that YbPtBi is essentially the same as YbBi. In the former, Pt atoms occupy the vacant space in YbBi of the Nacl type structure. Because Pt is neutral no essential change occurs. However, due to d-d mixing between the d-states on Yb and Pt, the band gap becomes wider. The crystal field splitting and the indirect f-f exchange interaction are also altered because of the f-d mixing between Yb and Pt. The same thing exists between Ce_3Sb_4 and $Ce_3Pt_3Sb_4$. Note that by introducing the transition metal atoms the crystal structure is stabilized and thus new materials are found but due to the ternary alloys it is more difficult to obtain high quality samples.

FERMI SURFACE AND MASS ENHANCEMENT

For the materials belonging to the Kondo regime, where both the occupied and unoccupied f-bands are not at the Fermi energy, the mass enhancement on the usual band states is due to mainly fluctuations of spin and orbital moments in the f-states. However, the usual second order approximation valid for the electron-phonon interaction seems to be not enough to explain the large γ-value observed in some materials. In particular, the large γ-values in low carrier systems are mystery. On the other hand in the valence fluctuation regime, the usual f-band model can explain the Fermi surface and the renormalized f-band model is claimed to explain both the Fermi surface and mass enhancement. These problems were discussed actively in this Workshop. It is clear that there are two f-peaks in BIS of Ce-compounds. One near the Fermi level is called the Kondo peak and another, several eV above the Fermi energy, is called the f^2-peak with the f^2-multiplet structure. The f-band model should be good when the f^2-peak is weak, and then the Kondo peak becomes the usual f-bands. As the f^2-peak intensity increases, the f-band becomes narrower renormalized f-band with the narrowing ratio, proportional to the intensity of Kondo peak relative to the f^2-peak. However, the intensity of the f-character at the Fermi energy does not change and thus the Fermi surface of the f-band model persists even for the renormalized band. The Kondo peak seen by BIS corresponds, however, to the high frequency Kondo state, much higher than the low frequency Kondo state, where only the lowest 4f states are mainly concerned, and thus the mass enhancement is small. To obtain the observed large mass at low temperature, the spin and charge fluctuation of f-electrons should be taken into account in the usual way, similar to the case of the 3d electrons. The renormalized f-band with the large mass enhancement corresponds to the low temperature Kondo state and thus of much lower energy. There are many ways to construct this kind of renormalized f-band depending on how to take into account various interaction effects such as the spin-orbit splitting, crystal field splitting etc. because there is no first principle description. The same intensity at the Fermi energy may be a possible guide line. The simplest way, for the present author, is to put the f-characters corresponding to the low temperature Kondo state at T_K above the fermi energy for the high frequency renormalized band. The Fermi surface is not expected to change substantially. In anyway, it is important to check the fermi surface, as well as the anisotropic mass enhancement effect carefully. For the f-band model, it is necessary to calculate the mass enhancement through the spin-charge fluctuation. It should be also noted that the fermi liquid character exists only very near the fermi energy and in the most part of the renormalized band a many body local state with strong spin and charge fluctuations is a better description.

The same thing exists in the Kondo regime. In CeSb, for an example, the f-PES splits into two peaks, named as bonding and antibonding peaks, in which the bonding peak intensity is about 40% situating at about 1 eV below the Fermi energy. Therefore, when we put one full f-state below the Fermi energy to calculate the Fermi surface of conduction band affected by the occupied f-state, the position should be accordingly chosen.[6] This corresponds to the f-band and the high frequency renormalized band in the valence fluctuation regime. Therefore, it is also possible to calculate the mass renormalization by putting the low frequency Kondo state at T_K above the fermi energy. Comparison with the usual method to use the magnetic fluctuation[7] should been done carefully.

It was reported that the fermi surface of $CeRu_2Si_2$, which has a large γ-value and thus has been thought to belong to the Kondo regime, seems to belong to the valence fluctuation regime. Both the usual f band claculation and the renormalized f-band model claimed that they could fit the experimental result fairly well. More detailed comparison is necessary.

NON TRIVIAL GAP STATES

As mentioned before, the most low carrier systems are explained by the usual band gap or weak overlap and thus called as the trivial gap formation. Of course there are many interesting physics exist in these materials but the origin of gap is clear. On the other hand, in some materials, the origin of gap remains to be controversial. Typical examples are SmB_6, pressurized Sm-monochalcogenides. TmSe and YbB_{12}. Recently found CeNiSn seems to belong to this category but still a lot of questions remain.

Among the above mentioned example, SmB_6 and YbB_{12} are most extensively studied. In TmSe, the valency of Tm depends very sensitively on the stoicheometry, which makes difficult to prepare high quality good stoicheometric samples. There is still a claim that the gap can be obtained by the usual f band model in SmB_6 and YbB_{12} . Even if it is so, the f band model is too far from the real f-state and can not be a convenient starting model to understand various anomalous properties under consideration. Actually no attempt has been done so far to understand anomalous properties based on the f-band model. The same argument is applicable for the so called c-f hybridization gap, which is essentially the same as the f-band model. The 4f states in SmB_6 and YbB_{12} are very well described by the atomic many body states with strong f-f correlation energy. In YbB_{12} , a well defined crystal field splitting is observed by the neutron scattering and the ground state is a quartet, while in SmB_6 the crystal field splitting is weak and above 50 K all the freedom of $J = 5/2$ in the trivalent Sm is free. A kind of Wigner crystal for the 4f states was proposed to be consistent with various anomalous properties, for example increasing lattice constant with decreasing temperature, opposite to the c-f hybridization model.[8] Recent neutron scattering experiment on SmB_6 also consistents with the local character of gap formation.[9] It should be noted that the study of SmB_6 started in 1970 and then the non-magnetic character was attributed to local trapping of one 5d electron at each trivalent Sm ion making non-magnetic ground state. This looks similar to the strong Kondo limit lattice but different in the point that there is no intra-atomic d-f mixing and thus the main interaction is the intra-atomic d-f Coulomb-exchange interaction. It should be noted that SmB_6 and TmSe are the mixed valence materials in which both the trivalent and divalent ions coexist in a more static sense. YbB_{12} is different in this sense because where Yb is the trivalent Kondo system. We know now that the inter-atomic d-f and p-f mixing interactions are more important and the singlet formation is due to these mixing interactions. It is clear that the transport gap, or the charge fluctuation gap, is nearly equal to the singlet bound energy, or the magnetic excitation gap, and this gap energy changes gradually to the Kondo temperature T_K as diluting Sm by trivalent ions such as La, while the gap character persists by diluting with the divalent ions such as Yb.

On the other hand various measurements on CeNiSn are causing confusion. Very clear gap formation was reported by NMR measurement. However, no other experimental results support gap formation. The γ-value is large and increases with applied field. The resistivity is rather metallic and gives a large metallic residual resistivity, which decreases with applied field. To solve this dilemma, Kasuya proposed that similar to Ce, in which both α and γ Ce coexist, both the CeNiSn belonging to the valence fluctuation regime and the Kondo regime may coexist.[10] From eq. (1), CeNiSn for valence fluctuating regime has a narrow band gap, or low carrier system, with a large energy scale. Therefore, anomalies at low temperature, in particular a narrow gap formation at low temperature should be attributed to the Kondo regime CeNiSn. The gap may collapse with applied field corresponding to $T_K \sim 10K$. Various anomalous properties are explained consistently by the above model and then the Kondo regime CeNiSn belongs to the non-trivial gap state class, the first case in the Ce-compounds. Anyway more careful experimental studies are necessary.

HEAVY FERMION SUPERCONDUCTIVITY

This is obviously one of the most interesting subject in the heavy fermion systems. However, a detailed debate is expected in Sendai and thus here we mention only briefly, mostly on the new materials. UPd_2Al_3 is the newly found heavy fermion superconductor with substantially different characteristics compared with the already known materials. First of all a clear magnetic ordering with a substantial magnetic moment appears well above Ts. Crystal field splitting is also clearly seen. Therefore, this materials looks to belong to the localized 5f regime, or the Kondo like regime in the analogy to the 4f system. As mentioned before, it is difficult to classify the U-compounds to the Kondo and valence fluctuation regimes as was done in the Ce-compounds. From FES-BIS, we are rather inclined to assume that in the most heavy fermion U-compounds the 5f band model is applicable. This is true in UPt_3 and probably in UBe_{13} and URu_2Si_2. In this sense UPd_2Al_3 is clearly different. Furthermore, we should expect that near Ts the 5f magnetic fluctuation is weak, different from other materials. The pairing scheme looks like similar to the conventional one. However, the behavior of T_1 measured by NMR is very similar to other materials indicating exotic character. UPd_2Al_3 is simpler compared with other materials and dHvA measurement seems to be possible. In this sense, this material seems to be the most convenient one to reveal the exotic heavy fermion superconductivity.

It has been a puzzling that why $CeCu_2Si_2$ is the only one heavy fermion superconductor in the 4f system. Now we have two new family members. One is $CeCu_2Ge_2$. Under applied pressure, it shows a superconducting transition, similar to $CeCu_2Si_2$. More studies should be done to check whether a mysterious phase boundary found in $CeCu_2Si_2$ also exists in $CeCu_2Ge_2$. Its pressure dependence is very interesting. It was already reported that in $CeCu_2Si_2$ Ts increases with applied pressure. A more detailed study was reported in this workshop and it was shown that the change looks like discontinuous. Therefore, the high pressure phase may be regarded as a new phase. It was proposed[11] that in the heavy fermion superconducting materials a near nesting fermi surface exists which shows strong charge and spin fluctuation near the ordered temperature preventing the superconductivity. Because of the Kondo-like behavior favoring magnetic singlet state, a charge ordering or a multipole ordering is expected with a weak magnetic ordering as a secondary effect. It is expected that with increasing pressure the nesting condition is weakened causing high T_s state. Anyway more detailed studies are expected.

Transport properties were one of the main thema in this workshop. Various properties such as thermal conductivity, thermoelectric power, magnetoresistance in general, and Hall effects were reported. In heavy fermion systems, the resistivity is usually high and then the phonon, as well as magnon, thermal conductivities have substantial contributions. Then the informations about the magnon-electron and phonon-electron interactions can be obtained. Detailed analysis to obtain these informations is, however, not so easy. Systematic studies on the low carrier systems, as well as the non-trivial gap systems, seem to be interesting.

The thermoelectric power is very sensitive to the delicate electronic structure and thus important tool to study the heavy fermion character. It is well known that in the Ce-compounds a large positive peak appears, while in the Yb-compounds a large negative peak appears due to the Kondo effect, in a general sense. In the Kondo lattice system, however, the sign is reversed for temperature below T_K and changes again at lower temperature. The magnetic scattering effect is attributed for the first sign change and the coherent effect is told for the second change but the real mechanism is not yet clear.

For the Hall effect, people are more interesting in the anomalous Hall effect than the normal Hall effect. For the anomalous Hall effect, there is a big difference depending on whether the f electrons move or not. Usually the anomalous Hall effect is treated as the scattering of conduction electrons by the localized f electrons through the intra-atomic c-f Coulomb-exchange interaction and the c-f mixing interaction and in the Kondo system the latter interaction is thought to be more important. In the U-compounds, however, the 5f-band model is thought to be applicable even in the heavy fermion system and thus the scattering mechanisms for the 5f bands should be considered in detail. For the 4f system, in particular for the Ce-compounds because the Ce-compounds are most extensively studied, however, the situation is different depending whether the system belongs to the valence fluctuation regime or to the Kondo regime. For the former, the scattering of the f-bands is the main mechanism and for the latter the scattering of conduction electron by the local f electrons is the main mechanism. However, there are no systematic studies on this stand point. The more interesting case is the non-trivial gap states, SmB_6, TmSe and YbB_{12}. From the measurement on other materials belonging to the localized 4f regime, for example CeB_6, it is clear that no anomalous Hall effect exists in these types of compounds when the 4f electrons are in the localized or Kondo regimes. Usually in the cubic crystals the anomalous Hall effect is not observed. However, in SmB_6 and TmSe, strong anomalous Hall effect is observed. In SmB_6, the anomalous Hall effect is strongly related with the gap opening, that is, the anomalous Hall effect disappears at low temperature when the gap opens.[8] This indicates clearly that the gap is formed as the 4f Wigner crystalization and as the Wigner crystal is melt, the gap disappears, the 4f electron can move contributing to the anomalous Hall effect. The same thing is expected in TmSe, where, the gap energy is much smaller. However, the situation is more complicated and more detailed experimental studies are necessary. In YbB_{12}, no anomalous Hall effect is observed. This means that SmB_6 and TmSe are mixed valent system in which both the trivalent and divalent atoms coexist and thus the 4f electrons can move but YbB_{12} is a trivalent system and thus the 4f electrons can not move even at high temperature.

CONCLUSIONS

So far some topics were picked up and considered. Many other things were also talked and discussed in this workshop and they are included in this proceedings. Anyway our main interest in the physics in the f-electron systems is how the transition from the atomic localized many body state to the itenerant one body band like state occurs, what happens at the transient region, what kind of new states with novel , exotic characteristics emerge at the boundary and whether more universal concepts eixst to explain the novel characteristics including both the limiting cases. We are now in the middle of the road to the goal and more systematic studies on more high quality samples are necessary in strong collaboration with theorists.

REFERENCES

1. For example, S. von Molnar and S. Methfessel, J. Appl. Phys. 38 (1967) 959.
2. M. Kasaya, Y. Ishikawa, K. Takegahara and T. Kasuya, J. Phys. Soc. Jpn. Supl. 49 (1980) 831.
   T. Kasuya, K. Takegahara, M. Kasaya, Y. Ishikawa and T. Fujita, J. de Physique C5 (1980) 161.
   T. Kasuya, K. Takegahara, M. Kasaya, Y. Ishikawa, H. Takahashi, T. Sakakibara and M. Date, Solid State Sciences 24 (1981) 150, Springer-Verlag, New York.
3. J. Rossat Mignod, J. M. Effantin, P. Burlet, T. Chattopadhyay, L. R. Regnault, H. Bartholin, C. Vettier, O. Vogt, D. Ravot and J. C. Achard, J. Magn. Magn. Mater. 52 (1985) 111.
   H. Takahashi and T. Kasuya, J. Phys. C, Solid State Phys. 18 (1985) 2697, 2709, 2721, 2731, 2745 and 2755.
4. T. Kasuya, J. Phys. Soc. Jpn. 61 (1992) 2206.
   T. Kasuya and T. Suzuki, J. Phys. Soc. Jpn. 61 (1992) 2628.
5. U. Ahlheimer, K. Fraas, P. H. P. Reinders, F. Steglich, O. Nakamura, T. Suzuki and T. Kasuya, J. Magn. Magn. Mater. 108 (1992) 213.
6. T. Kasuya, O. Sakai, H. Harima and M. Ikeda, J. Magn. Magn. Mater. 76&77 (1988) 46.
7. O. Sakai, Y. Kaneta and T. Kasuya, Jpn. J. Appl. Phys. 26 (1987) Supl. 26-3, 477.
8. T. Kasuya, K. Takegahara, Y. Aoki, K. Hanzawa, M. Kasaya, S. Kunii, T. Fujita, N. Sato, H. Kimura, T. Komatsubara, T. Furuno and J. Rossat-Mignod, Valence Fluctuations in Solids, ed. L. M. Falikov, W. Hanke, M. B. Maple, North Holland (1981) 215.
9. P. A. Alekseev, to be published in Proc. Intern. Conf. Strongly Correlated Electron Systems, in Sendai (1992).
10. T. Kasuya, J. Phys. Soc. Jpn. 61 (1992) 1863.
11. T. Kasuya, Prog. Theor. Phys. Supple. 108 (1992) 1.

HIROSHIMA WORKSHOP ON TRANSPORT AND THERMAL PROPERTIES OF f-ELECTRON SYSTEMS

Aug 30-Sept 2, 1992

T²P²S '92

HIROSHIMA

CONTRIBUTORS INDEX

SUBJECT INDEX